Lecture Notes in Computer Science 10810

Commenced Publication in 1973
Founding and Former Series Editors:
Gerhard Goos, Juris Hartmanis, and Jan van Leeuwen

More information about this series at http://www.springer.com/series/7151

James F. Peters · Andrzej Skowron (Eds.)

Transactions on
Rough Sets XXI

 Springer

Editors-in-Chief
James F. Peters
University of Manitoba
Winnipeg, MB, Canada

Andrzej Skowron
University of Warsaw
Warsaw, Poland

ISSN 0302-9743   ISSN 1611-3349 (electronic)
Lecture Notes in Computer Science
ISSN 1861-2059   ISSN 1861-2067 (electronic)
Transactions on Rough Sets
ISBN 978-3-662-58767-6   ISBN 978-3-662-58768-3 (eBook)
https://doi.org/10.1007/978-3-662-58768-3

Library of Congress Control Number: 2019930616

This Springer imprint is published by the registered company Springer-Verlag GmbH, DE part of Springer Nature
The registered company address is: Heidelberger Platz 3, 14197 Berlin, Germany

Preface

Volume XXI of the *Transactions on Rough Sets* (TRS) is a continuation of a number of research streams that have grown out of the seminal work of Zdzisław Pawlak[1] during the first decade of the twenty-first century.

Rough set theory in its applications to decision and knowledge engineering problems has to cope with uncertain or incomplete knowledge. Degrees of confidence as to results of decision algorithms or granulation algorithms or other procedures belong in the interval [0, 1] in spite of their finite number, hence, they fall into the realm of many-valued logics. As shown by Zdzisław Pawlak and Andrzej Skowron, many essential reasonings by means of a rough set approach can be carried out in terms of rough membership functions $\mu_X^B(x) = |[x]_B \cap X|/|[x]_B|$. This approach has been generalized by Lech Polkowski and Andrzej Skowron to a paradigm of rough mereology that treats the more general predicate $\mu(x, y, r)$, which reads: "x is a part of y to a degree at least r." The foundational role as a vehicle for reasoning modes in these approaches is played by many-valued logics proposed for the first time ever by Jan Łukasiewicz in his Farewell Lecture on March 7, 1918, at Warsaw University. We observe the centenary of this historic event with a chapter on Jan Łukasiewicz and his results by Lech Polkowski.

The paper co-authored by Jiajie Yu and Christopher J. Henry introduces an approach to simulating the task of human visual search defined in the context of descriptive topological spaces. The algorithm presented forms the basis of a descriptive visual search system (DVSS) that is based on the psychologically guided search model (GSM) for human visual search. The paper by Alexa Gopaulsingh defines and examines the consequences of double successive rough set approximations based on two generally unequal equivalence relations on a finite set. Algorithms to decompose a given defined operator into constituent parts are also presented. The paper by A. Mani presents dialectical rough logics from a semantic perspective, where a concept of dialectical predicates is formalized, connection with dialetheias and glutty negation are established, and parthood is analyzed and studied from the viewpoint of classic and dialectical figures of opposition. The paper by Lech Polkowski presents a logic for reasoning about parts and degrees of inclusion based on an abstract notion of a mass that generalizes geometric measure of area or volume and extends in the abstract manner the Łukasiewicz logical rendering of probability calculus. The paper by Łukasz Sosnowski presents mathematical foundations of a similarity-based reasoning approach for recognition of compound objects as well as applications in the fields of image recognition, text recognition, and risk management. The paper by Arkadiusz Wojna

---

[1] See, *e.g.*, Pawlak, Z., A Treatise on Rough Sets, *Transactions on Rough Sets* IV, (2006), 1–17. See, also, Pawlak, Z., Skowron, A.: Rudiments of rough sets, *Information Sciences* 177 (2007) 3–27; Pawlak, Z., Skowron, A.: Rough sets: Some extensions, *Information Sciences* 177 (2007) 28–40; Pawlak, Z., Skowron, A.: Rough sets and Boolean reasoning, *Information Sciences* 177 (2007) 41–73.

and Rafał Latkowski describes a new generation of the Rseslib library, which is available as a separate open-source library with API and with modular architecture aimed at high reusability and substitutability of its components. The new version can be used within Weka and with a dedicated graphical interface.

The editors would like to express their gratitude to the authors of all submitted papers. Special thanks are due to the following reviewers: Jan Bazan, Davide Ciucci, Ivo Düntsch, Anna Gomolińska, Ryszard Janicki, Jouni Järvinen, Piero Pagliani, Dominik Ślęzak, Sheela Ramanna, Marcin Wolski, and Wei-Zhi Wu.

The editors and authors of this volume extend their gratitude to Alfred Hofmann, Christine Reiss, and the LNCS staff at Springer for their support in making this volume of TRS possible.

The Editors-in-Chief were supported by the EU Smart Growth Operational Programme 2014–2020 under GameINN project POIR.01.02.00-00-0184/17-01 as well as the Natural Sciences and Engineering Research Council of Canada (NSERC) discovery grant 185986.

December 2018   James F. Peters
                                                                  Andrzej Skowron

LNCS Transactions on Rough Sets

The *Transactions on Rough Sets* series has as its principal aim the fostering of professional exchanges between scientists and practitioners who are interested in the foundations and applications of rough sets. Topics include foundations and applications of rough sets as well as foundations and applications of hybrid methods combining rough sets with other approaches important for the development of intelligent systems. The journal includes high-quality research articles accepted for publication on the basis of thorough peer reviews. Dissertations and monographs up to 250 pages that include new research results can also be considered as regular papers. Extended and revised versions of selected papers from conferences can also be included in regular or special issues of the journal.

Editors-in-Chief

James F. Peters, Canada
Andrzej Skowron, Poland

Managing Editor

Sheela Ramanna, Canada

Technical Editor

Marcin Szczuka, Poland

Editorial Board

Mohua Banerjee
Jan Bazan
Gianpiero Cattaneo
Mihir K. Chakraborty
Davide Ciucci
Chris Cornelis
Ivo Düntsch
Anna Gomolińska
Salvatore Greco
Jerzy W. Grzymała-Busse
Masahiro Inuiguchi
Jouni Järvinen
Richard Jensen
Bożena Kostek

Churn-Jung Liau
Pawan Lingras
Victor Marek
Mikhail Moshkov
Hung Son Nguyen
Ewa Orłowska
Sankar K. Pal
Lech Polkowski
Henri Prade
Sheela Ramanna
Roman Słowiński
Jerzy Stefanowski
Jarosław Stepaniuk
Zbigniew Suraj

Marcin Szczuka   Marcin Wolski
Dominik Ślęzak   Wei-Zhi Wu
Roman Świniarski   Yiyu Yao
Shusaku Tsumoto   Ning Zhong
Guoyin Wang   Wojciech Ziarko

Contents

Contents

Jan Łukasiewicz Life, Work, Legacy

On the Centenary of the Farewell Lecture at Warsaw University During Which Jan Łukasiewicz Introduced Multi-valued Logic and on His 140th Birth Anniversary IN THE YEAR of 100th ANNIVERSARY OF REGAINED POLISH INDEPENDENCE

Lech Polkowski[✉]

Department of Mathematics and Informatics,
University of Warmia and Mazury in Olsztyn, Słoneczna 54, 10-710 Olsztyn, Poland
polkow@matman.uwm.edu.pl

1 Introduction

Jan Łukasiewicz was one of leading logicians of the XX-th century, universally regarded as the father of many-valued logics which proved to be the language for many paradigms of Computer Science and Artificial Intelligence, inventor of the Polish notation whose dual, the Reverse Polish notation has become implemented in computers and calculators, renowned historian of logic especially of logics of Stoics school and of Aristotle, twice the Rector of Warsaw University in academic years 1922/23 and 1931/32, Minister of Religious Beliefs and Public Enlightenment in the Paderewski cabinet in 1919, earlier in Ministry of Education in provisional Jan Kanty Steczkowski cabinet in 1918, born in Lwów at the time of autonomization of Galicia, student at the Lwów University at the time of its start toward flourishing, in Warsaw between 1915 and 1944, then an exile in Germany, Belgium and finally in Ireland, far from dear Lwów and Poland. He was one of pillars of the world famous Warsaw School of Logic alongside of Warsaw School od Mathematics, Lwów School of Mathematics, Warsaw - Lwów School of Philosophy together with Kazimierz Twardowski, Alfred Tarski, Stanisław Leśniewski, Stefan Banach, Hugo Steinhaus, Juliusz Schauder, Stanisław Mazur, Stanisław Ulam, Wacław Sierpiński, Kazimierz Kuratowski, Stefan Mazurkiewicz, Adolf Lindenbaum, Mordechaj Wajsberg, Bolesław Sobociński and many others. They worked in often difficult conditions, living through two world wars, regional conflicts, many of them lost all their possessions and archives, forced to rebuild their lives anew, often overseas, but always devoted to Poland and its causes. Many of them lost their lives like Schauder, Wajsberg, Lindenbaum, many professors of Jagiellonian University arrested and sent to concentration camps in 1939 by German occupants and about 70 professors of Lwów University as well as Polytechnic, and other institutions of higher education in Lwów executed by a German commando of Krüger [1] in Lwów in 1941.

The regained independence in November 1918 gave an enormous stimulus to all areas of social activity, not leaving aside arts and sciences. Free from

© Springer-Verlag GmbH Germany, part of Springer Nature 2019
J. F. Peters and A. Skowron (Eds.): TRS XXI, LNCS 10810, pp. 1–47, 2019.
https://doi.org/10.1007/978-3-662-58768-3_1

censorship, writers, poets, artists, and scientists demonstrated great talents producing most valuable results. When the independence was endangered, they responded to call to the arms, helping the defensive efforts according to their potential. Mathematicians from Warsaw University, Wacław Sierpiński, Stanisław Leśniewski and Stefan Mazurkiewicz rendered a great service to the country by decoding Red Army's coded messages, so the General Staff of the Polish army were aware of opponent's maneuvers. Jan Łukasiewicz was also working at that time in the Royal Castle in the counterintelligence section.

Second world war was in a sense the finishing line of his life, rid of his home, living in very humble conditions, without his library and archive, he was reduced to a status of a low ranked clerk living on supplementary money sent by his friends, and, finally, leaving Poland afraid of his life when the Red Army would come, then in Germany living for three quarters of the year in bombarded Münster, in the basement of a ruined home, then in the country, but as a Pole badly tolerated by native Germans, finally able with the help of the Polish military to go to Brussels and from there to Dublin, Ireland, where he was given the post of a professor of logic at the Royal Irish Academy, in spite of adversities still able to produce work like the monograph on Aristotle's logic and some articles. He worked bravely against fate hostilities, always faithful to the maxim of René Descartes 'Cogito ergo sum', as Heinrich Scholz put it in his eulogy.

It happened so that one hundred years ago Poland rose like Phoenix from ashes of 120 years long dependence and one hundred years ago Jan Ł ukasiewicz was able to liberate us from dependence on binary thinking that lasted over 2400 years, and we celebrate that double liberation.

Many-valued logics has become the cornerstone for Approximate Reasoning, including Rough Set Theory, it is sufficient to mention Pawlak-Skowron's rough membership functions, logics for decision systems introduced by many authors, Polkowski-Skowron's rough mereology and other areas in which many-valued reasoning is an indispensable tool.

This work is dedicated to the memory of Jan Łukasiewicz and all scientists which made Poland great in science and to all who took part in the historic battles on vast territories from Harbin to Flanders and from Arkhangelsk to Odessa in order to make Poland independent and free one hundred years ago.

The special dedication is to the memory of Professor Helena Rasiowa on 25th anniversary of Her demise. Professor Helena Rasiowa was a student of Jan Łukasiewicz at Warsaw University since 1938 and prepared under His supervision Her Master Thesis which perished in days of Warsaw Uprising of August-September 1944 in burning house and had to be recapitulated and finally concluded under the guidance by Bolesław Sobociński as Jan Łukasiewicz left Poland in July 1944. Professor Helena Rasiowa initiated theoretical computer science in Poland and her students were among pioneers of Rough Set Theory when this theory was initiated in Poland.

LIFE

2 The Milieu

Jan Łukasiewicz (see Fig. 1) was born on December 21, 1878, in Lwów.

Narodowe Archiwum Cyfrowe, sygn. 1-N-358

Fig. 1. Jan Łukasiewicz (Narodowe Archiwum Cyfrowe).

Lwów, the capital of the Kingdom of Galicia and Lodomeria (see Fig. 2) with Dukedoms of Oświęcim and Zator and Archdukedom of Cracow, a province of the Hapsburg Monarchy was the centre of Ruthenia Rubra, a territory called by Polish people Ruś Czerwona and earlier Grody Czerwieńskie. We have first written information about these parts from the Kiev chronicler Nestor (ca. 1113)[2]: '[...] w roku 981 po narodzeniu Chrystusa Pana wyprawił się Włodzimierz na Lachów i zajął grody ich: Przemyśl, Czerwień i inne grody, które są aż do tego dnia pod Rusią' ('In the year 981 went Vladimir against Poles and captured their castles Przemyśl, Czerwień and others which are kept by Ruthenian princes to this day'.). This testimony certifies Polish occupancy of these parts in X-th century and Ruthenia domination over them up to XIII-th century. In mid-XIII-th century Tartar and Mongol hordes went through those parts until they were stopped in the battle at Legnica in Silesia province of Poland by prince Henry the Pious but the part of Ruthenia fell under their dominance. Ruthenian prince Danilo in search of a convenient difficult to be captured site for a new castle noticed the ridge east of Przemyśl high up to 300 meters but narrow with maximal width of a dozen kilometers and he located on it a village and a future castle called Lwów after the name Lew (Lion) of his son.

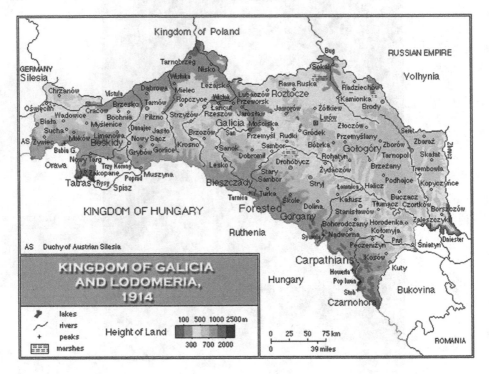

Fig. 2. This map presents the Kingdom of Galicia and Lodomeria in the year 1914. The Kingdom extended itself from Zaleszczyki on the river Dniestr in the east to Biała and Oświęcim in the west and from Tarnobrzeg and Sokal in the north to Czarnohora (Black mountain) to the south. Now, this region is divided among Poland (reaching the Curzon line on the river San) and Ukraine (basically from the river San to the east). Lwów was situated in its centre (https://en.wikipedia.org/wiki/Kingdom_of_Galicia_and_Lodomeria).

Almost from the beginnings of Lwów, due to its location on the Central European Water Division (to the west of Lwów is the basin of the Baltic sea, to the east rivers flow to Black Sea) Lwów became an important hub for traffic east-west and north-south. This dichotomy was to be the destiny of Lwów. Very soon was Lwów populated with craftsmen and merchants from Germany, Poland, Ruthenia, Armenia. Notably, a sizeable population of Armenian craftsmen and merchants populated Lwów since its beginnings as they flew from persecution by Persians in their homeland.

Lwów survived difficult times of Tartar regimes, local feuds among Ruthenian, Polish, Hungarian local rulers, until Lwów and all Ruthenia were incorporated in 1353 into the Polish Kingdom by efforts of Casimir the Great, founder of the Jagiellonian University. From that year, Lwów was in the Polish Kingdom up to the partitions of Poland when it fell into the Austrian part. King Casimir gave those parts stability and a period of peace which helped the budding city to develop and attract merchants from all parts of Europe and East. The west-eastern character of Lwów along with a local climate determined specific character of this city called 'Leopolis semper fidelis Poloniae'. Poland looked on its eastern parts through eyes of Lwów and those parts looked on Poland through Lwów as well. Three years before the Lukasiewicz was born, in 1875, in the publishing house of F. Richter in Lwów, the poet Wincenty Pol (see Fig. 3), later a professor of geography at Jagiellonian University, published his rapsod 'Mohort' in which the term 'kresy' (borderlands) designating vast territories east of the rivers Bug and San appeared for the first time:

Ktoby to myślał, ktoby się spodziewał,
Że gdzieś za światem w bodziakach Czehryńskich,
Że gdzieś na kresach niegdyś Ukraińskich
Taki świat dzielny i uroczy bywał? [3]

(Eng. Transl.: 'Who could think, who could expect
That far away at the end of world in Chehryn steppes
That far at borderlands earlier Ukrainian
Such brave and charming one place could find?')

Since that time the term 'kresy' ('borderlands') has become a catch word denoting those territories. One can say that Lwów was a 'borderland metropolis'. Through kresy, Poland contacted Turkey, far east, south of Europe: Moldova, Bukovina, Wallachia and further, Crimea, Balkans, Italy and Greece. This determined the cosmopolitan character, a sui generis 'melting pot' feature of Lwów. A very good example is Wincenty Pol, the author of 'Mohort': born as a German in the German-French family of Pohl in Reszel in Warmia of a German father and French mother, he later transferred to Lwów where his father obtained some office and changed the name to Poll von Pollenburg while young Poll studied at Lwów University, fought in November Uprising of 1830-31 alongside of Poles, became the notable patriotic poet, author of many classical songs like 'Piękna nasza Polska cała...' ('Beautiful is our whole Poland..'). He established second in the world after Berlin chair of geography at Jagiellonian University and in this character of a geography professor became a pioneer in Tatra mountains research.

Fig. 3. Wincenty Pol (Adolf Piwarski on the basis of photo by Karol Beyer - Tygodnik Ilustrowany, 1862).

As a far land metropoly Lwów became very important for the Polish Kingdom. In 1386 it became the site for the Roman Catholic archbishopric, among its XV century archbishops were the notable humanist Grzegorz of Sanok and the author of Chronicles of the Polish Kingdom Jan Długosz. But the most valuable gift of the King Jagiello was ius stapulae (prawo składu, staples). On this privilege each merchant going through Lwów was obliged to expose his goods for two weeks so the Lwów merchants could buy them, this contributed greatly to the wealth and flourishing of Lwów and of its citizens. Due to trade agreements with Moldavian and Wallachian princes, merchants from Lwów were allowed to freely trade in Moldova and Wallachia and from there to Kaffa on Crimea, Genova etc.

A blow to the prosperity came from Turks who in 1453 captured Constantinople and gradually subdued Wallachia, Moldova and Crimea, blocking trade routes and capturing people from Kresy. In 1527, a great fire devoured a large part of town but in its aftermath a law was passed that only from mortar and bricks can one build in Lwów. This was the beginning of a renessaince Lwów with its Catholic, Greek-Orthodox, Walachian and Armenian cathedrals, Boim and Kampian chapels. Lwów was the second only after Rome city in the world to boast of bishoprics and cathedrals of three and in some periods of four destinations. The true capital of Poland became Lwów in mid-XVII century when the city successfully defended itself against Khmelnitsky Tartars and Cossacks, Russian troops of Buturlin and finally Swedes. After all attackers were dispersed, in 1656 King Jan Casimir Vasa came to Lwów and bestowed on Lwów highest honors proclaiming it the King town, on par with Cracow and Vilno and granting nobility to all citizens of Lwów. In 1661 he raised the Jesuit Academy in Lwów established sixty years earlier to the rank of University (see Fig. 4).

Fig. 4. Jan Kazimierz University-formerly Galician Parliament building.

With the slow decline of power of the Polish Res Publica, fortunes of Lwów deteriorated. Heavy contributions paid over almost a century to Tartars, Khmelnitsky, Russian, Turks, Swedes impoverished the town and finally it fell along with the Kingdom, suffering partitioning in 1772 into parts belonging to Prussia, Russia and Austria. Lwów became an Austrian town in the province of Galicia and Lodomeria. This meant that the official language was German also in schools, University of Lwów became a Liceum, all institutions and offices of old Poland were abolished and Austrian institutions and laws were introduced. Yet Austria under few in line Emperors was subjected to ups and downs in politics, misfortunes in Turkish wars, the year 1848 witnessing the Spring of Peoples made peoples of the Monarchy rebellious which manifested itself in manifestations culminating in 'Sturmpetition' demanding liberalization, education and offices in national languages. This led on April 25, 1848, to the Constitution proclaimed by the Emperor. Further weakening of the Monarchy in the aftermath of the lost battle at Sadova (near Hradec Kralove) in 1866 in the Prussian war, caused the final proclamation of Constitution by Emperor Franz Josef I granting autonomy to all provinces of the Monarchy. In particular, Galicia gained a Provincial

Parliament and Government and Lwów University became a full-fledged University with Polish as the official and predominant lecturing language, save some lectures in Ukrainian and German.

As the capital of Galicia Lwów developed very quickly, especially taking care of education on which it was spending one quarter of its income. The liberal atmosphere and Polish language freely in usage and in art and literature attracted Polish artists and scientists and writers from other parts of Poland, less happy in those aspects of life. Lwów was the habitat for Maria Konopnicka, Gabriela Zapolska, Aleksander Fredro, Apollo Korzeniowski (the uncle of Joseph Conrad), Zygmunt Kaczkowski, the author of the capital gothic novel 'Murdelio' and a fictionary but very realistic portrait of old Poland in the cycle of novels 'Ostatni z Nieczujów' ('The Last of Nieczuja Family'), finally a financial agent for European monarchies, negotiating on behalf of France the problem of war of 1870 repartitions with Prussia. Also Walery Łoziński, the author of historic roman d'geste set in historic events like 'Zaklęty Dwór' ('The Enchanted Manor') showing actions of underground emissary toward the national uprising, or, 'Oko Proroka' ('The Prophet Eye') weawing the tale based on experiences of many Poles captured as slaves by Turks and his brother, a historian and archivist Władysław Łoziński author of a monumental monograph 'Prawem i Lewem' ('By Right or Wrong') based on search in judiciary archives of townships and cities and documenting the excesses of old Kresy surpassing with their bravery and often atrocities those known to us from Wild, Wild West. A great stimulus was given cultural life by the National Ossoliński Foundation established in 1817 with its collections of 300 000 books and 10000 manuscripts, collections of art, monetary collections, and with Lubomirski Art Museum, which was the place of work for many writers and historians like Łoziński brothers, Kaczkowski, Pol, later Maria Konopnicka, Gabriela Zapolska, Jan Kasprowicz, Leopold Staff. In addition to Ossolineum, libraries of Baworowski, Dzieduszycki, Pawlikowski and publishing and printing houses of Altenbergs, Gubrynowicz, Połoniecki provided a background for rapidly developing Polish art. The theatre built in 1847 and rebuilt in 1890-ties hosted many world class actors like Helena Modrzejewska, and Lwów Opera House grown from a Conservatory established in 1810 by A. Mozart, son of Wolfgang Amadeus, boasted of performers known at the greatest operatic scenes like La Scala or Metropolitan Opera; bass Adam Didur, sopranos Salomea Kruszelnicka, Jadwiga Korolewicz-Waydowa, Ewa Bandrowska - Turska.

Lwów, Leopolis semper fidelis Poloniae, was always the centre of irredenta; in 1860, on the wave of liberalization, 'Macierz Sokola' ('Falcon Companionship') (Sokół means 'falcon' in Polish) was created from which sokole units ('Falcon units'), rifle sections, and, later, scouts organizations had grown, culminating in P.O.W (Polish Military Organization) and Legions of Józef Piłsudski in the 1st World War. Since 1867, under and due to the Galicia autonomy, Lwów began its rapid growth towards European city, with many buildings, housing programs, medicare for its citizens and education. Polish language at schools and

Fig. 5. Lwów panorama.

Fig. 6. Old building of Lwów University and St. Nicolas Church.

at the University allowed for rapid growth of school system eliminating formerly dominant 1-year schools in favor of 6-year or 8-year schools like Theresianum.

Lwów University, founded as a Jesuit University in February 1661 by King Jan Casimir Vasa on the basis of the Jesuit College existing from 1608, met with vehement opposition on part of gentry opposing Jesuits which resulted in the refusal by the Pope to confirm the foundation. University existed as a College with two faculties but without academic rights of granting degrees. In 1758 King August III Saxon confirmed foundation preserving the University status of two faculties, i.e., semi-full university status. In late XVIII century, this university produced in spite of low general level, some eminent alumni like Grzegorz Piramowicz, of Armenian descent, the secretary of State Commission for Education, which underwent an ambitious program of modernizing Polish system of education, or, Kacper Niesiecki an expert in heraldry, author of Almanac of Polish gentry. In 1772, Lwów became a city in the Austrian part of partitions of Poland and University became an Academy, Theresianum. In 1784, Emperor Joseph II gave University a full status but the level was very unsatisfactory as education was restricted to very basics satisfying the demands of bureaucracy, military etc. In 1805 during Napoleonic Wars Austria, crushed at Austerlitz and Wagram, was not able to sustain university and it was closed to be opened only in 1817 with lectures in Latin and German on mediocre level.

The change came with the year 1848 of the Spring of Peoples. Powerful demonstrations, military actions on part of nations of Empire, shattered the system which however countered closing university to be opened again in 1850 with granted freedom of lectures. As Governor of province was appointed Agenor Gołuchowski which strengthened yet the Polish efforts. In effect, in 1879 and 1882 Polish language was decreed by the Emperor order as dominant and in fact obligatory with some concessions for Ruthenian students.

Fig. 7. Zakład Narodowy Ossolińskich we Lwowie. From Polish WIKIPEDIA (National Establishment Ossolineum in Lwów).

University started to develop instantly and after few years it was one of leading in the whole Monarchy. Its professors included Szymon Aszkenazy, an eminent historian, member of the Polish delegation to the Peace Conference at Versailles, Benedykt Dybowski, a veteran of Siberia, one of the first explorators of Siberia, the researcher on the flora and fauna of the lake Baikal, after whom Dybowski Ridge in Altai is named, Oskar Balzer, a historian of law, whose role was principal in the process which gave Poland the most beautiful part of Tatra mountains: Morskie Oko and Valley of Five Lakes, Ludwik Rydygier, a surgeon, the author of the first in the world surgeries on the open stomach, Marian Smoluchowski, a physicist known for his work on kinetic theory of gases.

In 1895, to Lwów from Vienna came Kazimierz ze Skrzypna h. Ogończyk Twardowski, a former student of Franz Brentano (see Fig. 8) in Vienna, already well-known in the scientific world of philosophy and psychology for his habilitationschrift 'Zum Lehre vom Inhalt und Gegenstand der Vorstellungen' ('O treści i przedmiocie przedstawienia') [4] dedicated to theory of perception. His lectures which he begun at 8 o'clock in the morning, were attended by up to 2000 listeners and for them the largest halls in the city were rented. Twardowski was raised in Vienna where his father was an Austrian official but the climate at home was Polish and patriotic. Twardowski attended the Theresianum in Vienna where teaching day lasted from 4:30 a.m. to 21:00 p.m and pupils spent at school 8 years. He studied with Franz Brentano from whom he inherited the tendency to clear and precise formulation of ideas. His role in forming Polish philosophers and logicians proved to be enormous. He co-founded the journal 'Przegląd Filozoficzny' in 1898 and the journal 'Ruch Filozoficzny' in 1911 as well as initiated Polish Philosophical Society. Over 30 professors of philosophy, psychology and logic in Poland were his former students.

Fig. 8. Left: Kazimierz Twardowski, right: Franz Brentano.

3   Years of Jan Łukasiewicz to 1939, the Outbreak of War

In such town at such time was Jan Łukasiewicz born of his father Paweł, an accounting officer in Austrian military gendarmerie and of his mother Leopoldina neé Holtzer from Austrian family. His father was of a Ruthenian origin, of Greco-orthodox denomination popularly called unionists due to the Union at Brześć by which that part of the orthodox church accepted the sovereignty of the Pope.

Paweł came to Lwów from a remote village and by his own effort reached the rank of a captain in accounting office of gendarmerie. Paweł Łukasiewicz was 56 when his only son was born. Father of Leopoldina was an Austrian official in Transilvania. It was a typical marriage for Hapsburg monarchy, a melting pot for many nationalities. Lwów at that time consisted in 50% of Polish population, 25% Jewish, 10% Ruthenic, and Armenian, German, Austrian, Greek, Italian etc. At that time Polish language school system rapidly was growing up to 5000 so called folk schools of 6 classes. Jan Łukasiewicz attended such school and then on a wish of his father willing that Jan perfected his German entered the II Gymnasium, the only German Gymnasium in Lwów. He studied there for 8 years, where students were taught Greek, Latin and modern languages like French, German. After graduation from the Gymnasium, he entered in 1897 the University with the intention of studying Law. Kazimierz Twardowski initiated on his coming to Lwów the so-called Philosophical Circle which held its meetings every Friday at 8 p.m. in the University Reading Rooms at Chorążczyzna street. Students gave there some talks and then discussed with Twardowski various problems believing as Łukasiewicz recollected that Twardowski was able to solve every problem. The highly intellectual atmosphere of meetings and personality of Twardowski decided that Łukasiewicz switched from Law to Philosophy and Mathematics.

In 1902 he defended his PhD Thesis: 'On induction as the inverse to deduction'. This theme in a sense outlined his scientific life. The promotion which was postponed a few months due to illness of Paweł[3] Łukasiewicz and his death, to November 15, 1902, was procedurally 'sub summis auspicis imperatoris' which meant the highest possible honours. This formula was possible only for doctorandi who passed every exam in high school and at the university, including doctoral exams, with the highest grades. After doctorate he was presented by the Galicia Governor professor Leon Piniński with the Emperor golden ring with 14 diamonds set in platinum donated by the Emperor Franz Josef I. After all, in Habsburg's monarchy, everyone who graduated from two faculties was entitled to the title of a baron.

The next two years were difficult as Łukasiewicz could not obtain a position at the University and was forced to accept a position of a home teacher at Bilcza Złota manor near Lwów implementing this income with some money from the task of an assistant to a librarian at the University Library. In 1904, he took part alongside of Twardowski and few fellow students of philosophy in founding Polish Philosophical Society. Twardowski selected for this ceremony February 1, the centenary of death of Immanuel Kant. Łukasiewicz recollected in his memoirs an amusing anecdote: the talk on this occasion was sent to 'Przegląd Filozoficzny' under the title 'Kant in Poland' (but the noun 'kant' means also in Polish 'a fraud') so the censor refused to allow the printing. But the editor Weryho said that in such case he was going to publish it under the title 'Kant in Warsaw district, Kant in Radom district, Kant in Kielce district,...' listing all districts in Poland and the censor gave up. In the same year 1904 Łukasiewicz obtained the fellowship from the Galician Government (Wydział Krajowy) on which he went to Berlin. In Berlin he attended courses in the Institute of Experimental Psychology and some lectures but was seemingly disappointed with lack of contacts and for the summer semester went to Louvain where he attended lectures by Mercier, later the primate of Belgium. During this stay, he prepared the monograph on causality principle 'On analysis and construction of causality principle' [5] which was accepted by Twardowski as the 'habilitationshrift' (see Fig. 9).

Fig. 9. Jan Łukasiewicz about the time of his habilitation. Polish Philosophy Page.

After habilitation, in 1906, he became a privatdozent so he could announce lectures but without payment. His first lecture was devoted to logic. An accident made him a mathematical logician: his colleague asked him for help in translating and reading an article by Bertrand Russell in 'Mind' on ordering [6]. The formal rigor and clarity of this article made deep impression on Łukasiewicz who as a library worker asked the library director to buy a copy of 'The Principles of Mathematics' [7] which he studied for many months.

This bias toward formal logic caused a rift with Twardowski as Łukasiewicz began regarding philosophy as not precise enough. However, he always acknowledged the role of Twardowski in demonstrating ability for very clear presentation of most difficult problems. In 1908 Łukasiewicz obtained a fellowship from Academy of Sciences from the Osławski Foundation. He chose Graz where Alexius Meinong, a former student of Franz Brentano, was a professor of philosophy and founder of the institute for experimental psychology. Łukasiewicz attended the Meinong seminar whose subject at that time was a monograph by J. von Kries on calculus of probability. His plans to visit Paris were cancelled by the illness of his mother and when her health improved, Łukasiewicz went again to Graz. In Graz he finished his monograph 'O zasadzie sprzeczności u Arystotelesa' ('On the contradiction principle by Aristotle') [8] printed in 1910 in Cracow. Stay at Graz resulted also in a monograph 'Die logischen Grundlagen der Warscheinlichtskeittheorie' [9] published as a monograph in Cracow in 1913. In this monograph, he proposed a logical formalism for theory of probability. It is possible that with this work the idea of fractional values came to his mind.

After the book on Aristotle's treatment of the principle of contradiction was published, Łukasiewicz applied for one of the three chairs of philosophy which were vacated at Jagiellonian University. In spite of having results already surpassing those by other candidates, he was bypassed in the procedure. It is interesting to know that one of the three chairs was given to a biologist, one to a physicist, and one to an older lawyer from Lwów.

In 1911, he obtained nomination from the Emperor for the post of a professor extraordinary at Lwów University and some lectures with a meagre salary so Twardowski joked 'You now have money for cigars'. In 1912 he met for the first time Stanisław Leśniewski who was at that time a student of Twardowski. As Łukasiewicz recollected, Leśniewski knocked on his door and when it was opened said that he came to tell that he published in 'Przegląd Filozoficzny' an article in which he opposed theses expressed by Łukasiewicz in his book on the principle of contradiction. After some discussion with Leśniewski, Łukasiewicz went to the famous later 'Szkocka Café' and told colleagues that he should probably close his scientific business as he met a man more precise than he. Łukasiewicz always very highly estimated Leśniewski stating some years later that he had learned from him what precise thinking was.

For the next two years, Łukasiewicz demanded the promotion to the rank of a full professor but to no avail and he felt that there were no chances for promotion in Galicia. In 1914, he inherited a substantial legacy after his uncle which offered possibilities for further traveling and quiet work but again fate was not favorable as in late June 1914 Gavrilo Princip shot down prince Ferdinand in Sarajevo and suddenly Central Europe found itself in the eye of a terrible tempest. Lwów witnessed the attack by Russian forces which already on 3th of September 1914 occupied Lwów. Austrian army facing a series of defeats withdrew leaving Lwów in hands of Russians. It so happened that one of generals commanding Austrian forces was feldmarschall von Koevess, a far cousin of Jan Łukasiewicz on the maternal side. Russian occupation was not very detrimental to citizens of Lwów. For scientific life, the result was that scientists born on the Russian territory and formally being Russian citizens were interned in Russia, Wacław Sierpiński found himself in Wiatka and then in Moscow, Stanisław Leśniewski and Tadeusz Kotarbiński also were interned in Moscow.

The counteroffensive of Austrian and German forces led to a few months of positional war near Gorlice in Low Beskid mountains which was concluded with the concentrated attack by them in May 1915 which led to withdrawing of Russian forces far to the East and liberation of Lwów. University which was transferred to Vienna returned to Lwów with the Rector Twardowski and

Lukasiewicz resumed his lectures in the fall of 1915. After the first lecture, he was invited to the Rectorate and Rector Twardowski informed him on behalf of the Austrian government that there is an invitation for him to become the professor at Warsaw University which offer Lukasiewicz accepted.

This was the result of the military situation in Central Poland, in Russian part of partitions. The attack of German and Austrian forces in the aftermath of a great battle by Łódź in central Poland led to liberation of Warsaw on August 5, 1915. German General governor, von Beseler, allowed for the opening of Warsaw University with Polish language in place of Russian and Citizen's Commitee led by prince Lubomirski issued invitations through Austrian government to professors in Lwów and Cracow to resettle to Warsaw. One of such invitation was passed to Jan Lukasiewicz by Twardowski. On his arrival to Warsaw the history of Warsaw-Lwów logico-philosophical school took a decisive turn.

On November 15, 1915 Warsaw University was opened. Two weeks later Lukasiewicz initiated lectures. He immediately joined the Senate of university as representative of the Department of Humane Sciences and in 1916 he became the Dean of the newly created Philosophy Department and in 1917 a pro-Rector of the University.

Variable fortunes of war, losses of Germans and Austrians weakened the grip on the Polish Kingdom and on July 3, 1917 the Council of Regents was proclaimed with prince Lubomirski, Cardinal Kakowski and Józef Ostrowski. Council started to introduce legal order independent of existing laws imposed by foreign systems. Some sequence of temporary short-lived provisional governments took place, in the second of them of Jan Kanty Steczkowski, Lukasiewicz was nominated as one of the three chairmen in the Department of Public Education in charge of high schools.

Leaving the University for this post, he gave on March 7, 1918, the Farewell Lecture, in accordance with the by then live tradition of universities, and in this lecture he announced that he successfully constructed a 3-valued logic extending classical Aristotelian 2-valued sentential calculus.

In the Ministry of Education, Lukasiewicz worked from March 1918 to December of 1919. On November 11, 1918, disarmament of Germans in Warsaw took place and Poland was proclaimed by Brigadier Józef Piłsudski in notes to foreign governments a free state. In December 1918, Ignacy Jan Paderewski, the piano virtuoso and a statesman, was nominated the next prime minister after Jędrzej Moraczewski, the first prime minister of independent Poland. In the Paderewski cabinet Lukasiewicz took the post of the Minister for Religious Beliefs and Public Enlightenment. He met with a plethora of problems as the task of sewing together three regions of Prussian, Russian and Austrian occupation which lasted 120 years was very difficult. As Lukasiewicz wrote in his recollections, in the period of eleven months, the cabinet held about 100 meetings (see Fig. 10).

Fig. 10. Marshall and Commander of the State Józef Piłsudski and Prime Minister Ignacy Jan Paderewski in 1919.

During ministerial duty of Łukasiewicz, he presided over openings of University in Poznań and Stefan Batory University in Wilno, reopened after 100 years, and of the Academy of Mining and Metallurgy in Cracow. Especially emotional was the opening in Wilno, as especially dear to Marshall Józef Piłsudski, born in Lithuania and for years an inhabitant of Wilno. A further problem was the stabilization of universities by selecting competent professors. This was done first for Warsaw University and Warsaw Polytechnical Institute as well as for new universities by the Stabilization Commission.

During his ministry the Catholic University of Lublin (KUL) was established on a large gift by Karol Jaroszyński, the owner of all banks and sugar plants in Ukraine. University in Lwów was given the patronage of the King Johannes Casimir (Jan Kazimierz) Vasa and became the Jan Kazimierz University. Particularly sensitive for Łukasiewicz was the ceremony of reopening his old school II German Gymnasium in Lwów as the Polish Gymnasium. The Paderewski government was not long living, it ceased to exist in December 1919 and from January 1, 1920 Łukasiewicz returned to the university as the professor of philosophy. After an exhausting work at the Ministry which he performed with the sense of duty but without pleasure, he took a half-year leave which he was spending in Lwów but was forced to return to Warsaw on the news of the coming battle of Warsaw with invading Russian Red Army. He enlisted himself as a volunteer and was directed to work in the Royal Castle as a help to the counterintelligence section. Due to war, courses at the University begun in the fall of 1920. Łukasiewicz lectured on logic and history of philosophy. Aside him a professor of Philosophy of Mathematics was Stanisław Leśniewski who returned from Moscow and spent first months in Warsaw as an official in the Lukasiewicz department in the Ministry of Education. Leśniewski worked already on his Ontology and studies begun Alfred Tajtelbaum later known as Alfred Tarski. In 1920, Lukasiewicz published his results on 3-valued logic giving a table of values for implication and negation in his logic. In 1922, he discovered his formulas for values of implication and negation: $v(p \longrightarrow q) = min\{1, 1 - v(p) + v(q)\}$, $v(\neg p) = 1 - v(p)$. These formulas are valid for n-valued logics as well as for infinite-valued logics.

In the academic year 1922/23 he became the Rector (at that time Rectors were elected for one year terms; this changed about 1935 with the so called Jędrzejewicz reforms and since then Rectors were elected for three year terms) (see Fig. 11).

Fig. 11. Jan Lukasiewicz as the Rector of Warsaw University (http://www. biogramy.pl/.thumbs/299x150_crop_0_260_598_330/userfiles/image/articles/scaled_ 5b0807e254ab9PIC_1-N-3319-1.jpg).

Lukasiewicz felt that lectures at University, which according to the testimony of Professor Helena Rasiowa were 'elegant with a deep historic and philosophical background' were difficult for him and he decided after his Rector's term to free himself from them and having some savings was convinced that he could spend few years solely on scientific work. Obviously, his results on 3-valued logic of 1918–1920 were compelling him to further work which was hampered by teaching and Rector's duties. With January 1, 1924, he 'became the free man', however nominated a honorary professor and that time brought forth what he regarded as one of his greatest achievements, viz., parentheses free notation later called 'the Polish notation'. With the symbol C for implication and the symbol N for negation, it became the standard notation for the Warsaw school of logic. Unfortunately again, on April 1, 1924, the monetary reform of Grabski came into effect which reduced the capital of Lukasiewicz to a meagre amount and he was forced to return to work so he found work at the Teachers College in Warsaw and in 1925 became a honorary professor working on a contract basis at the University. One can only express astonishment at the fact that a professor with already such achievements was treated in such way especially that the Dean at that time was Kotarbiński, a former student. Most probably, at the bottom were political differences as Kotarbiński was of leftist views. Since 1926, Łukasiewicz conducted a seminar at the University. The logic group consisting initially of Łukasiewicz and Leśniewski was immensely strengthened by appearance of Alfred Tajtelbaum-Tarski as well as by Adolf Lindenbaum, Mordechaj Wajsberg, Bolesław Sobociński. Among students attracted by this group were Stanisław Jaśkowski, Czesław Lejewski, Jerzy Słupecki and members of the so called Cracow Circle like Rev. Jan Salamucha, and Father J. M. Bocheński.

Łukasiewicz in the seminar felt relaxed contrary to lectures and he shared his ideas with colleagues and students. The good example is the system of Natural Deduction presented in 1932 by Stanisław Jaśkowski in his dissertation following the problem formulated at the seminar by Łukasiewicz in 1929.

In 1929 a very important event in life of Jan Łukasiewicz took place: he married Regina Barwińska. His wife was 25 while he was 50 yet he decided for the marriage. His objections that he marriage would be detrimental to his work proved fruitless. His wife was helping him with everyday problems, as Łukasiewicz wrote, she enjoyed herself in organizing their 'little house' ('domek') but according to observers she was building walls between him and his colleagues. Also in 1929, mimeographed notes of lectures by Łukasiewicz 'Elementy Logiki' (Elements of Logic) appeared in which he presented his system of multi-valued logic along with an analysis of 2-valued logic. Łukasiewicz analyzed the Frege 6-axiom system pointing to redundancy of some axioms, and, he presented his system of axioms consisting of 3 axioms. This book was reprinted by Polish Scientific Publishers in 1958 and in the English translation by Pergamon Press in 1963 and 1966 (see Fig. 12).

Meantime, Łukasiewicz was taking part in scientific life. In 1926, he travelled to the US, for a philosophical congress at Harvard. Due to late information, he came to Cambridge MA by three days too late, yet he was able to meet C.I. Lewis and Sheffer. In 1928, he took part in the mathematical congress in Bolonia where he had a talk. He records a story how once in a cafeteria were sitting at one table a professor of mathematics from Leipzig, a professor from Lille, a docent from Amsterdam, himself, Leon Chwistek and Alfred Tarski and they all conversed in Polish. One is reminded of much later saying by Hugo Steinhaus that Poland had as export goods coal and mathematicians.

The year 1930 signalled the survey 'Untersuchungen über den Aussagenkalkül' [10] of research on logic by the Warsaw School authored by Łukasiewicz and Tarski and containing main results obtained in the period 1920–1930 by both authors and Mordechaj Wajsberg, Adolf Lindenbaum, Bolesław Sobociński. This survey was to be for Łukasiewicz the source of many sad moments as many authors following a logician Lewis of Harvard University attributed many discoveries of Łukasiewicz to Tarski never energetically enough denied by Tarski (see Fig. 13).

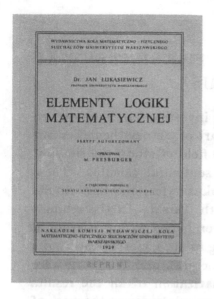

Fig. 12. Books by Jan Łukasiewicz.

In the thirties, Łukasiewicz undertook cooperation with Heinrich Scholz, the professor of logic and theology at the University of Münster in Rhein-Westfalen. This involved mutual visits, Łukasiewicz himself visited Münster four times giving lectures (see Fig. 14).

Heinrich Scholz proved a true friend which was going to be very valuable in coming years. In the result of initiated by Scholz action, University of Münster awarded Łukasiewicz the honorary doctorate (see Fig. 15).

In 1938, in December, one day before the sixtiest birthday anniversary of Łukasiewicz, at the German Embassy in Warsaw, German Ambassador von Moltke presented Łukasiewicz with the diploma of Doctor honoris causa of University at Münster in the presence of Rector of Warsaw University Włodzimierz Bołoz-Antoniewicz, Heinrich Scholz and the Dean of Philosophy Department at Münster Adolf Karsten. Present as well were some Professors from Warsaw University, Stanisław Leśniewski, Stefan Mazurkiewicz, and Tadeusz Zieliński, the eminent researcher of antiquity (see Fig. 16).

Four years earlier, Łukasiewicz attended the philosophical congress in Prague where he had a talk on history of logic of propositions published later in 'Erkenntniss' [11] and called by Scholz '20 most beautiful pages on history of mathematics'. In 1938, Łukasiewicz travelled to Zurich on invitation by Professor Gonseth to attend a congress on foundations of mathematics and he gave a lecture on 'Logics and the problem of foundations' [12] published during the war.

Łukasiewicz had many students some of them completed dissertations under his supervision: Kobrzyński, Jańkowski, Wajsberg, Słupecki, probably Sobociński which competed his habilitation shortly before war.

IV

INVESTIGATIONS INTO THE SENTENTIAL CALCULUS†

In the course of the years 1920–30 investigations were carried out in Warsaw belonging to that part of metamathematics—or better metalogic—which has as its field of study the simplest deductive discipline, namely the sentential calculus. These investigations were initiated by Łukasiewicz; the first results originated both with him and with Tarski. In the seminar for mathematical logic which was conducted by Łukasiewicz in the University of Warsaw beginning in 1926, most of the results stated below of Lindenbaum, Sobociński, and Wajsberg were found and discussed. The systematization of all the results and the clarification of the concepts concerned was the work of Tarski.

In the present communication the most important results of these investigations—for the most part not previously published—are collected together.‡

§ 1. General Concepts

It is our intention to refer our considerations to the conceptual apparatus and notation developed in article III (pp. 30–37).

† Bibliographical Note. This joint communication of J. Łukasiewicz and A. Tarski was presented (by Łukasiewicz) to the Warsaw Scientific Society on 27 March 1930; it was published under the title 'Untersuchungen über den Aussagenkalkül' in *Comptes Rendus des séances de la Société des Sciences et des Lettres de Varsovie*, vol. 23, 1930, cl. iii, pp. 30–50.

‡ To avoid misunderstandings it should be stated that the present article does not contain results discovered by both the authors jointly, but is a compilation of theorems and concepts belonging to five different persons. Each theorem and concept is ascribed to its respective originator. Theorem 3, for instance, is not a theorem of Łukasiewicz and Tarski, but a theorem of Lindenbaum. Nevertheless, some scholars mistakenly referred to both authors, Łukasiewicz and Tarski, the many-valued systems of logic ascribed in the article to Łukasiewicz alone. In spite of a correction which appeared in 1933 in the *Journal of Philosophy*, vol. 30, p. 364, this mistake persists till today. It clearly follows from § 3 and notes of this article that the idea of a logic different from the ordinary system called by Łukasiewicz the *two-valued logic*, and the construction of many-valued systems of logic described here, are entirely due to Łukasiewicz alone and should not be referred to Łukasiewicz and Tarski.

Fig. 13. A fragments from "Investigations into the Sentential Calculus" (Alfred Tarski, J.H. Woodger: Logic, Semantics, Metamathematics. Papers from 1923 to 1938. Oxford Univ. Press, Year: 1956).

Fig. 14. Announcement of the lecture by Jan Łukasiewicz on 'The old and the new logic' in Münster on February 4, 1936.

We have given excerpts from public life of Jan Łukasiewicz but we come now to the point where we have to shed some light on his private life as we enter difficult time of war.

4   Years of War and Emigration

September 1939 was fatal for the family of Łukasiewicz: in the first days of German invasion, the Professor home at Sewerynów str. 6, where they lived, was destroyed by a bomb, and they lost everything except for the truly symbolic remnant, the bound volume of reprints of research papers by Jan Łukasiewicz. They moved to another Professor house after some dozen days at Brzozowa 12, where they returned to the old apartment in which Łukasiewicz had lived earlier. Łukasiewicz through Heinrich Scholz obtained a low rank post of the translator at the City Archives with a very meagre salary. Due to Scholz who send some money transferred to Łukasiewicz by the Chief of Archives, they were able to survive.

Łukasiewicz took part in Clandestine University, one of his students was Professor Helena Rasiowa, who completed under his supervision her Master thesis

Fig. 15. Ambassador von Moltke presents Jan Łukasiewicz with Dr h.c. Diploma of University at Münster. From right: professor Heinrich Scholz, second from left: Dean of Philosophy Dept. at Münster U. Adolf Karsten.

Fig. 16. Participants at the ceremony in German Embassy at Warsaw. Jan Łukasiewicz with Dr h.c. Diploma, to his left Regina Barwińska his wife, first from left Heinrich Scholz, Dean Karsten fifth from right in the upper row, Stefan Mazurkiewicz 8th from right in the upper row, ambassador von Moltke behind Regina Barwińska, to his left rector UW professor Włodzimierz Bołoz-Antoniewicz.

which was burned during Warsaw Uprising in August-September 1944. But his failing health with the diagnosis of an early stage of angina pectoris forced him to retire from teaching and he passed his duties to Bolesław Sobociński. Lukasiewicz and wife lived in Warsaw until summer 1944, when the fates of War switched in favor of the Allies and Warsaw was witnessing a hasty retreat of German soldiers and officials. The atmosphere was dense with the news about the coming Red Army and actually from January 1944 its units entered the Polish soil and in June were on the outskirts of Warsaw.

People knew already about the fate of Polish officers murdered in Katyń Forest and other places in Soviet Union by Soviet Security Police, about the fate of the units of the Polish Home Army (AK, Armia Krajowa) at Kresy, which identified themselves helping Red Army units to take over cities like Wilno, Lwów but were disarmed and taken prisoners and in many cases simply murdered, the same fate befell many civilians in eastern Poland who were captured and went to Gulag camps, so the future seemed very uncertain and dangerous. Lukasiewicz as the minister in the government who fought soviets in 1919, whose wife's nine family members were already transported east could not feel safe. Moreover, Lukasiewicz was not sure about behavior of some of his colleagues who were of leftist views and occasionally had given him tokens of their attitude to him. In this atmosphere, Lukasiewicz decided to go west with the final destination in Switzerland where he counted on help of Professor Gonseth in Zurich. He turned to Heinrich Scholz who had given anew the helping hand which he already extended in the case of Rev. Dr Jan Salamucha. In 1939, in Cracow, Germans arrested Professors of Jagiellonian University and imprisoned them in Sachsenhausen. After few months due to protests from the civilized world, the professors who survived were liberated but two of them both priests were sent to Dachau, which was a site destined for Polish priests who were imprisoned in Priests Barracks in Dachau to the number of 1780 of which 878 were murdered there. One of the priests was father Jan Salamucha, professor of theology at Jagiellonian University, a logician in the Aristotelian spirit, a former student of Jan Lukasiewicz. Lukasiewicz asked Heinrich Scholz to intervene with the aim of freeing Jan Salamucha for whose liberation already intervened church and high officials like count Galeazzo Ciano. Heinrich Scholz wrote a letter to the Cultural Department of the German Ministry of Foreign Affairs stressing that Dr Salamucha was an eminent researcher carrying his work in the footsteps of the great German Master Gottlob Frege and in true German Spirit. As the letter met with negative answer, Scholz wrote a second letter and this time he had a success as Father Salamucha was liberated and returned to Warsaw. He was murdered in the first days of Warsaw Uprising by members of RONA (Russian National Liberation Army consisting of Russians who fought along Germans)and Wlasow units (Russian Liberation Army whose members were recruited from among Russian captives of war) who murdered almost 100 000 people before they were recalled.

Lukasiewicz did not witness these atrocities as on July 17, 1944, he and his wife left Warsaw with help of Bolesław Sobociński and on the next day found themselves in Münster. Bad luck persisted: few days later in Wolfschanze in Mazurian district, the eastern military headquarter of Hitler, an attempt on Hitler's life took place and in its aftermath all possibilities for leaving Germany were strictly forbidden so Łukasiewicz and his wife were forced to stay in Münster. Münster was heavily bombarded by Allies so they had to live for a long time in the cellar of some destroyed home, occasionally taken by Scholz to some empty flat. Only in January 1, 1945, due to winter weather, Juergen von Kempsky , a philosopher from Münster University having Polish roots, took them to his farm in Hoeltzen near Hobsten where they lived until April 1945 when they saw coming units of American soldiers.

Being free, Jan Łukasiewicz contacted general Leon Berbecki, who at that time was commanding a camp for Dipis (Displaced persons) in the former Oflag VI B. Many of these people, former prisoners of war and survivors of concentration camps were badly in need of schooling as their school time was interrupted by war and Łukasiewicz accepted the offer of teaching logic in nearby Hohenwopel.

After few months, in December 1945, he was able to obtain permission to move to Brussels where he travelled in a military train helped by Polish military authorities. In Brussels, Łukasiewicz lectured on logic at the Polish Scientific Institute. In Brussels he met an Irish officer in a Polish uniform (some sources speculate that the 'officer' was a writer Jerzy Meissner working for British government) presented as 'lieutenant Herbert' who contacted professors from Central European scientific institutions with offers to resettle to Ireland. Łukasiewicz accepted this offer and embarked on a ship at Ostend on March 1, 1946, arriving to London. From London by train and ship he came with his wife to Dublin. He was located for few weeks at the hotel 'Echo' and he was met after that lapse of time by Mr Walsh, the vice-secretary for foreign affairs. Finally in July, Łukasiewicz was invited by the prime minister (Taoiseach) Eamon de Valera and obtained the nomination for a professor of mathematical logic at the Royal Irish Academy.

In 1949, after some break, due to failing health, he resumed lectures on logic, attended by some professors from University College of Dublin and of Trinity College like Father Gwynn, Carew Meredith, who turned out to be the logician in the Polish spirit. In February 1950, he travelled to Manchester where he met Czesław Lejewski, a former student from Warsaw and had talks with Alan Turing to whom he explained his parenthesis-free notation as Turing worked on the British mainframe 'Colossus'.

In 1948, he succeeded in sending to the VI Convention of Polish Academy of Knowledge (PAU-Polska Akademia Umiejętności) a contribution dedicated to the memory of Mordechaj Wajsberg. Some time later he found himself among 18 Polish scientists who lived in the West excluded from PAU on the basis of not coming to the meetings. A bit later, in 1951, PAU was dissolved and new academy, the Polish Academy of Sciences was organized, following the Russian pattern.

In July 1955, Łukasiewicz was honored with honorary doctorate of Trinity College in Dublin.

Łukasiewicz had heart problems diagnosed as early angina pectoris when still in Warsaw during the war. For this reason, he had given up his lectures at the Underground University, passing his duties to Bolesław Sobociński. These problems happened to show in acute forms in mid-fifties, Łukasiewicz was a few times hospitalized. Finally on February 12, 1956 after the gall surgery he died of coronary thrombosis. He was buried at Grapevine Cemetery in Dublin. His wife erected a monument on which she gave the inscription: 'Prof. dr Jan Łukasiewicz, Died on 12 February 1956 Far from Dear Lwów and Poland'. She donated the archive of Jan Łukasiewicz to RIA which sent it in turn to the University of Manchester care of Czesław Lejewski. Some authors claim that part of the archive was taken by Father Gwynn. Regina Barwińska returned to Poland in 1957, she lived in Gdańsk and died in 1990.

Heinrich Scholz in the epitaph wrote

[...] he was a humanist who the principle cogito ergo sum applied in the most dramatic moments.

5 Epilogue

Jan Łukasiewicz left Lwów University in 1915. He made attempts to return but his attempt was blocked by Twardowski who opinioned that Lwów was too small for two stars. He could not witness personally the blooming albeit short (as blooming usually is) period of excellence that Lwów Jan Casimir University (UJK) reached between 1919 and 1939 never to excellence return. To give the impression of the level the Jan Kazimierz University reached in the interwar period, we briefly introduce two of its eminent scientists.

About 1903 his studies in Lwów begun Rudolf Weigl (see Fig. 17) which in 1913 as privatdozent begun lectures. In 1914, he, as a bacteriologist, was sent to military medical posts where he met with typhoid problem which determined his career. Rudolf Weigl confirmed the Nicolle discovery about Rickettsia bacteria in lice transferring typhoid and discovered a method for producing the anti-typhoid vaccine in lice themselves. It was first in biology method by which the vehicle of bacteria produced a vaccine against it. His Typhoid Institute produced vaccine freely given to those in need. An example which brought Weigl world fame was China, where Belgian Christian missionaries were dying up to eighty of them due to typhoid along thousands of Chinese but when they obtained vaccine from Weigl there was no case of death. Rudolf Weigl was to be the honorary guest at the World Fair in New York in September 1939 but refused to go in the wake of coming war. On demand of Russians and later Germans he reopened his institute which he agreed with the Polish government in exile.

The list of people saved by work in the Institute counts 457 names among them Banach, Knaster, Orlicz, Alexiewicz, Jahn, Fleck. Alltogether, about 3000 people were in touch with the Institute in some character. Weigl refused to accept German citizenship (his biological father lived in Prerov, Moravia and was of

Fig. 17. Rudolf Weigl (Narodowe Archiwum Cyfrowe).

Austrian origin) but the importance of his work provided a sufficient immunity. Threatened by Himmler's Deputy Kaufmann, he refused German citizenship saying that he preferred to be shot which would make him a hero and would free him from committing suicide. This refusal cost him the Nobel prize in 1942 as his candidacy was blocked by Germans. Same happened after war when his candidacy submitted in 1948 by the Swedish Academy was blocked by Polish communist government on grounds of false accusations about cooperation with Germans during war. Due to Weigl the medicine department at UJK was well-known in the world of science. His institute was visited by virtually all leading researchers from all over the world. After war, he was a professor at Jagiellonian University to 1948 and then a professor in Poznań. He was honored with Yad Vashem diploma for saving many Jews in Lwów and especially in Warsaw Ghetto where he clandestinely delivered the vaccine to Professor Ludwik Hirszfeld during official deliveries to ghetto German authorities which were carried out by Dr Fr. Henryk Mosing.

The second who acquired world renome was Jakub Parnas (see Fig. 18) a biochemist born near Tarnopol, who studied in Germany and worked till 1916 with leading chemists and biochemists in Germany, France and England till his return to Poland where he created the chair of medical biochemistry at Warsaw University and finally in 1920 found himself at Lwów where he created the school of biochemistry. It is very difficult to describe shortly the enormous results of Jakub Parnas, in studies on mechanism of metabolism in smooth muscles, in mechanism of ammonia production in muscles, in applications of isotopes in those studies, in phosphorene metabolism of glycogen, also in invention of some

Fig. 18. Jakub Parnas. (http://photos.geni.com/p13/c3/bd/da/4e/534448390710627f/jakub_parnas_original.jpg)

machinery for those studies. In 1939, Jakub Parnas fell probably the victim to some provocation by printed in a Lwów official daily an anti-German article with his name as an author. He, having no choice was forced to evacuate with Soviet troops in view of attacking Germans in June 1941. He landed in Ufa, the capital of Bashkir province. His son, Jan Oskar recollected a meeting which took place after the Pact Sikorski-Majski which freed Poles imprisoned in Gulag camps on the enormous territory from Workuta beyond the Arctic circle in the West to Magadan and Kolyma also far to the North on the Eastern border of Asia, and, to Kazakhstan on the south. One day, to a room in the hotel where Jakub Parnas lived with his family, came an emaciated man in tatters and said 'Good morning, Mister Professor'. Jakub Parnas responded calmly, 'Welcome, Mister President'. This man was Professor Ostrowski, the pre-war president of Lwów and later the president of the Polish Government in Exile. From Ufa, Parnas was able to move to Moscow, meantime sending his son to the Polish Army of General Anders which collected freed Polish prisoners and finally left Russia for Iran and farther west. In Moscow, Parnas was able to obtain the membership in Russian Academy of Sciences and created the Biochemical Institute of the Academy. His inability to accommodate to the Russian style of life, which required for safety open glorification of the system and avoiding any critical comments, openly shown in criticism of Trofim Lysenko at that time the leading figure in Russian biology and the pupil of Stalin, led to his imprisonment and death in 1949 in unknown circumstances. His wife was informed 10 years later that he died of heart attack when interrogated by Soviet Secret Service. Professor Parnas gave Lwów University a wordly renome in medical physiology and biology.

It is a mere formality to mention here the Lwów mathematical school led by the genial Stefan Banach, one of creators of functional analysis and his colleagues Hugon Steinhaus, Stanisław Mazur, Juliusz Schauder, Stanisław Ulam (Ulman), and others who made Lwów a mecca for mathematicians.

The Scottish Book from 'Szkocka Café', recalled by Łukasiewicz in connection with meeting with Leśniewski, in which they recorded their problems and its solutions is a legend, as a legend is Lwów itself, Warsaw-Lwów school of logic and all that world.

6 Comments

Jan Łukasiewicz was a man of a delicate psyche, sensitive, reacting inwardly acutely to symptoms of wrong. This is witnessed by breaks from work which he had taken a few times, and rather slow pace of publication of his most important discoveries. He did not make many friends in Warsaw, it seems that he was a man of Galicia, raised in totally different spirit and unable to flexible adaptation to changing winds. In effect he was left without help on the outbreak of war and had to turn to Heinrich Scholz for help. In his memoirs he mentions examples of treatment in Ireland and on trips to England bordering on unfair, seemingly he was not able to secure his interests in an adequate way. These traits of personality made his life during the war and after it a dramatic one. One cannot but agree with Heinrich Scholz who wrote in his eulogy that '... he was a humanist who the principle of cogito ergo sum applied in the most dramatic circumstances' (see Fig. 19).

Fig. 19. The tomb of Jan Łukasiewicz at St. Jerome Cemetery in Dublin.

As the letter by Rudolf Carnap testifies already in summer of 1945, when Łukasiewicz was still in Germany, some friends inquired among people working already at American universities about possibility of accommodating Łukasiewicz in the US which would certainly extend his life and work. But as Carnap writes, the serious handicap to these plans was the age of 65 of Łukasiewicz which as Carnap thought was an unsurmountable obstacle in view of administrative regulations (see Fig. 20). It seems that the obstacles were rather those facts that logic was in US rather exotic field of study at that time, that people already in the US were rather against newcomers unless they were very close to them, that authorities at US universities were ignorant as to merits of many professors from Europe, and due to just finished war, universities in US were overflown with immigrants rather in much better shape and much younger than Jan Łukasiewicz. Other motives should be left uncommented.

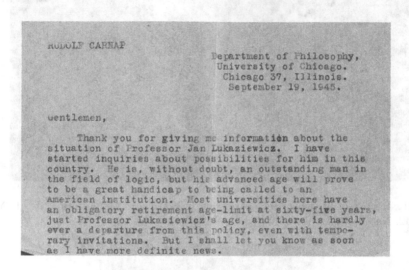

Fig. 20. Carnap letter.

His attempts to move to Oxford also ended in vain. As Rylands Archive mysteriously comments 'probably due to his age and background'. What 'background' tells against Łukasiewicz one can only guess, maybe his conservative views were against him amidst liberal views bordering on fascination and even work for communism in English academic circles. Anyway it was his fate to end his brilliant career as a cutoff from the leading world centers and abandoned on a far and unfriendly territory, an epitome of his country Poland.

WORK

Between 1902 and 1955 Jan Łukasiewicz published 92 printed works, beginning with a curtailed copy of his Dissertation in 'Vierteljahrschrift für wissenschaftliche Philosophie' (1902) [13] and ending with a letter to Polish Veterans Association (Stowarzyszenie Polskich Weteranów) published in London journal ŻYCIE (LIFE). His results are often placed in joint works, sometimes scattered in many talks at philosophical societies and on various occasions.

When in Graz, Łukasiewicz attended the seminar by Alexius Meinong and his colleagues interested in foundations of the notion of probability. In Proceedings of the Polish Academy of Lettres (PAU) in 1913 and in the monograph 'Die

Fig. 21. Rector of Warsaw University Jan Łukasiewicz. Inauguration Acad. Year 1931/32. Narodowe Archiwum Cyfrowe (NAC).

Fig. 22. President of Warsaw Stefan Starzyński is presenting Jan Łukasiewicz, Pola Gojawiczyńska and Alfons Karny with prizes of the City of Warsaw. November 1935. Narodowe Archiwum Cyfrowe (NAC).

logischen Grundlagen der Warscheinlichkeitsrechnung' published in Cracow in the same year, he gave a logical form to probability calculus. He borrowed the notion of 'truth value' from Gottlob Frege, from Peano and Russell he took the notion of an 'indefinite formula' but without distinction between free and bound variables, and, for a finite set U and an indefinite formula $q(x)$, he defined the truth value of q by letting

$$w(q) = \frac{|\{u : q(u) \text{ is true}\}|}{|U|},$$

where $|A|$ denotes the number of elements in a finite set A.

Łukasiewicz denoted the implication functor with the symbol '$<$', which would have a justification in his subsequent work and he proved for implication '$a < b$' that

$$w(b) = w(a) + w(\neg a \wedge b)$$

deriving from it all basic rules of probability. He introduced the notion of independence, defining the relative value of truth for b modulo a as

$$w_a(b) = \frac{w(a \wedge b)}{w(a)}.$$

He defined independence of b from a as the condition that

$$w_a(b) = w(b)$$

and he derived the Bayes theorem counterpart

$$(\sum_i a_i = 1)(\sum_{i \neq j} a_i \wedge a_j = 0) < w_b(a_m) = \frac{w(a_m)w_{a_m}(b)}{\sum_i w(a_i)w_{a_i}(b)}.$$

After a stint at Warsaw University at which Jan Łukasiewicz was a Dean of Philosophical Department and a pro-Rector, the new political situation, calling for independent governments, albeit short-lived, demanded from him the public activity. To the second of those governments, of Jan Kanty Steczkowski, he was called as the director in the Ministry of Education responsible for higher education. On March 7, 1918, he delivered the Farewell Lecture in the academic custom of those days (see Fig. 23). His ministerial work had been intended as a short one, the government of Steczkowski fell soon, but Łukasiewicz remained in his post. On November 11, 1918, with return from internment in Magdeburg of Brigadier Józef Piłsudski and the following disarmament of Germans, Poland proclaimed independence. The first government of Independent Poland was proclaimed under the prime minister Jędrzej Moraczewski. In December 1918, to Warsaw came via Poznań Ignacy Jan Paderewski, the legendary virtuoso of piano, and the statesman, a friend of President Woodrow Wilson, and by inter-partisan agreement, he went to form the next government. To this government, Jan Łukasiewicz was invited as the minister for Religious Beliefs and Public Enlightenment and served in this character till December 1919 when the Paderewski cabinet resigned.

In the short but emotional Farewell Lecture, did Łukasiewicz make a breaking statement. He announced that his studies on Aristotelian logic led him to discovery of some contradictions, and trying to resolve them, he was led towards necessity of accepting a new truth value between falsity and truth, i.e. the value for 'is possible'.

Łukasiewicz presented this new discovery as the fight against bi-valued confinement, the necessity of choosing 'false' or 'true' solely, and, he presented his discovery as a liberation from coercion. This lecture did not bring any formal statement, evidently the psychological difficulty of making such breakthrough had its effect. Anyway, this was the first in the history of science distinct and clear statement about existence of other than bi-valued logical calculi.

Ministerial duties kept Łukasiewicz from science. He recollected that the new government met with enormous duties of regulating all aspects of life in the new state, 'sewing together' three parts which had different laws, official languages, and mentality. Exhausted Łukasiewicz returned to University on January 1, 1920, but he took the half-year leave which he was spending in Lwów, yet, with beginning of summer, he returned to Warsaw on news that Red Army units were quickly advancing on Polish territory in particular the fearful Red Cavalry of Budionny and Stalin was advancing on Lwów. As we know now, the heroic defense of Lwów, which included the 'Polish Thermopiles' at Zadwórze, where literally all defenders up to the number of 330 were killed, saved Warsaw as Budionny on demands by Stalin kept besieging Lwów and was too late on the march to Warsaw to help his comrades who were conquered and dispersed on August 14–17, 1920.

In Warsaw Łukasiewicz met with the general call to arms and he enlisted. He was directed to counterintelligence and as he recollected, he went every day to the Royal Castle where he segregated documents. It is inconceivable that he, being a brilliant logician, did not take part in decoding tasks in which his colleagues Leśniewski, Mazurkiewicz and Sierpiński excelled. But we do not know about it. War events of course stopped lectures and only in autumn after Battle of Warsaw, after destruction of Red Cavalry in the great cavalry battle at Komarów, and after liberating Polish territory from remnants of Red Army in the aftermath of the Niemen campaign, lectures were renewed. Yet in spite of work at ministry, tense political situation and imminent danger from Soviet state, Łukasiewicz was able to prepare an article in 'Ruch Filozoficzny' 5 (1920) [14] in which he gave explicit rules for his tree-valued logical calculus (see Fig. 24). Its rules were presented as follows.

I. Identity of truth, identity of falsity, non-identity of truth and falsity: $(0 = 0) = 1, (1 = 1) = 1, (0 = 1) = 0, (1 = 0) = 0$.

II. Implication rules: $(0 < 0) = 1, (0 < 1) = 1, (1 < 1) = 1, (1 < 0) = 0$.

III. Definitions of sentence-forming functors:

Negation: $-a = a < 0$.

Disjunction: $a + b = (a < b) < b$.

Conjunction: $ab = -(-a + -b)$.

IV. Rules for 'possibility' (denoted 1/2, the first denotation was '2'):

Identity rules: $(0 = 1/2) = (1/2 = 0) = (1 = 1/2) = (1/2 = 1) = 1/2; (1/2 = 1/2) = 1$.

Implication rules: $(0 < 1/2) = (1/2 < 1) = (1/2 < 1/2) = 1; (1/2 < 0) = (1 < 1/2) = 1/2$.

FAREWELL LECTURE AT THE UNIVERSITY OF WARSAW, MARCH 7, 1918

In this farewell lecture I wish to give a synopsis of my work and for that purpose I shall use an autobiographical method, by presenting the sentimental backgroud of my theories.

I have declared a mental war against all kinds of necessity, so far as it restricts creativity. There are two kinds of necessity:

It is either a *physical* force coming from without and restricting the freedom of movement or the mental apathy which deprives a person of the will for action. One can free oneself from that necessity. One can tear the restricting fetters and by the effort of will overcome all apathy. And if everything will fail there still remains the saviour — death.

The second type of necessity is *logical*. We must accept evident principles and the theorems following from them. That necessity is much stronger than the physical one. And here there is no hope for salvation. No physical or mental strength can combat logical and mathematical principles.

That logical necessity had begun when Aristotelian logic and Euclidean geometry were created. At that time science was conceived as a system of principles and theorems, being interconnected by logical relations. That conception of science had started in Greece and prevailed then universally. The world was imagined in accordance with the pattern of a scientific system, namely that all events and phenomena are connected in a causal chain and that they follow from each other just as the theorems of a scientific discipline. All events occur according to necessary laws.

In a world so conceived there is no creative action which would not happen according to a law, but from a self-imposed desire. However desires are also ruled by necessity and they could be foreseen by a Being with universal knowledge; all actions are established before one is born.

That conception penetrated into everyday life. It became obvious that behaviour which is purposeful and ordered and occurs according to natural or common law is always causal. If a people would become mechanized and would reproduce by its behavior a scientific system, that people would be so strong as to dominate the world.

The creative mind revolts against such a conception of the world and life. The valiant individual being conscious of his value does not wish to be merely a link in the chain of causes and effects. He endeavours to influence the world independently.

Fig. 23. Jan Łukasiewicz: Farewell lecture at the University of Warsaw, March 7, 1918. The Polish Review 13(3) (Summer, 1968), 45–47.

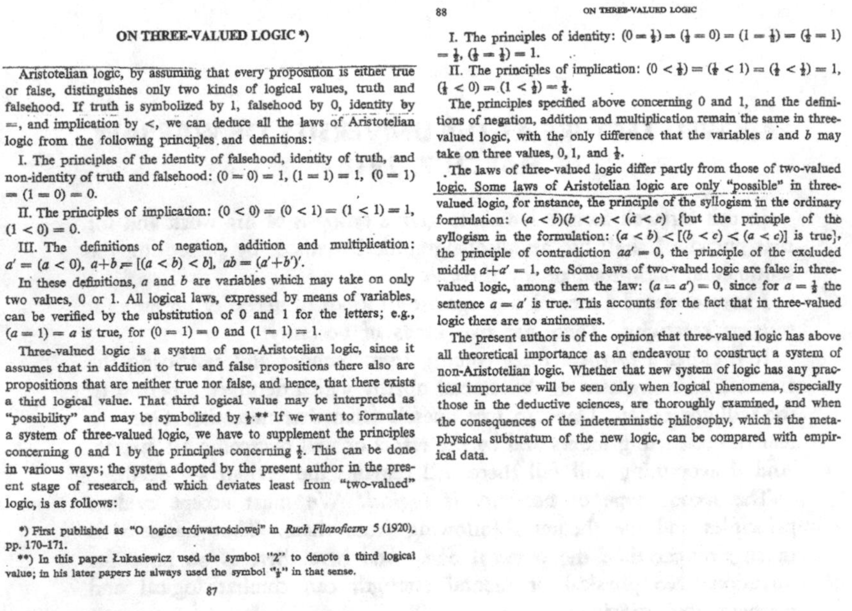

ON THREE-VALUED LOGIC *)

Aristotelian logic, by assuming that every proposition is either true or false, distinguishes only two kinds of logical values, truth and falsehood. If truth is symbolized by 1, falsehood by 0, identity by =, and implication by <, we can deduce all the laws of Aristotelian logic from the following principles and definitions:

I. The principles of the identity of falsehood, identity of truth, and non-identity of truth and falsehood: $(0 = 0) = 1$, $(1 = 1) = 1$, $(0 = 1) = (1 = 0) = 0$.

II. The principles of implication: $(0 < 0) = (0 < 1) = (1 < 1) = 1$, $(1 < 0) = 0$.

III. The definitions of negation, addition and multiplication: $a' = (a < 0)$, $a+b = [(a < b) < b]$, $ab = (a'+b')'$.

In these definitions, a and b are variables which may take on only two values, 0 or 1. All logical laws, expressed by means of variables, can be verified by the substitution of 0 and 1 for the letters; e.g., $(a = 1) = a$ is true, for $(0 = 1) = 0$ and $(1 = 1) = 1$.

Three-valued logic is a system of non-Aristotelian logic, since it assumes that in addition to true and false propositions there also are propositions that are neither true nor false, and hence, that there exists a third logical value. That third logical value may be interpreted as "possibility" and may be symbolized by $\frac{1}{2}$.** If we want to formulate a system of three-valued logic, we have to supplement the principles concerning 0 and 1 by the principles concerning $\frac{1}{2}$. This can be done in various ways; the system adopted by the present author in the present stage of research, and which deviates least from "two-valued" logic, is as follows:

*) First published as "O logice trójwartościowej" in *Ruch Filozoficzny* 5 (1920), pp. 170–171.

**) In this paper Łukasiewicz used the symbol "2" to denote a third logical value; in his later papers he always used the symbol "$\frac{1}{2}$" in that sense.

87

88 ON THREE-VALUED LOGIC

I. The principles of identity: $(0 = \frac{1}{2}) = (\frac{1}{2} = 0) = (1 = \frac{1}{2}) = (\frac{1}{2} = 1) = \frac{1}{2}$, $(\frac{1}{2} = \frac{1}{2}) = 1$.

II. The principles of implication: $(0 < \frac{1}{2}) = (\frac{1}{2} < 1) = (\frac{1}{2} < \frac{1}{2}) = 1$, $(\frac{1}{2} < 0) = (1 < \frac{1}{2}) = \frac{1}{2}$.

The principles specified above concerning 0 and 1, and the definitions of negation, addition and multiplication remain the same in three-valued logic, with the only difference that the variables a and b may take on three values, 0, 1, and $\frac{1}{2}$.

The laws of three-valued logic differ partly from those of two-valued logic. Some laws of Aristotelian logic are only "possible" in three-valued logic, for instance, the principle of the syllogism in the ordinary formulation: $(a < b)(b < c) < (a < c)$ {but the principle of the syllogism in the formulation: $(a < b) < [(b < c) < (a < c)]$ is true}, the principle of contradiction $aa' = 0$, the principle of the excluded middle $a+a' = 1$, etc. Some laws of two-valued logic are false in three-valued logic, among them the law: $(a = a') = 0$, since for $a = \frac{1}{2}$ the sentence $a = a'$ is true. This accounts for the fact that in three-valued logic there are no antinomies.

The present author is of the opinion that three-valued logic has above all theoretical importance as an endeavour to construct a system of non-Aristotelian logic. Whether that new system of logic has any practical importance will be seen only when logical phenomena, especially those in the deductive sciences, are thoroughly examined, and when the consequences of the indeterministic philosophy, which is the metaphysical substratum of the new logic, can be compared with empirical data.

Fig. 24. Jan Łukasiewicz: on three valued logic.

Let us read finishing sentences in the second page of this historic work: 'The present author is of opinion that three-valued logic has above all theoretic importance as an endeavor to construct a system of non-Aristotelian logic. Whether that new system of logic has any practical importance will be seen only when logical phenomena, especially those in deductive sciences, are thoroughly examined, and when the consequences of the indeterministic philosophy, which is the metaphysical substratum of the new logic, can be compared with empirical data. We have witnessed the prophetical truth of this statement in emergence of fuzzy and rough set theories which, indeed, apply many-valued logics in the analysis of empirical data.

In 'Przegląd Filozoficzny' 13 (1921) [15] he proposed a system for two-valued logic intended as the vehicle on which to present the three-valued extension as well. Following Brentano, he introduced rejection in addition to of assertion, after Peirce he introduced quantifiers, and he followed Leśniewski in allowing only bound ('apparent') variables. Like Peano and Russell, he was putting a dot after symbols of quantifiers, assertion and rejection. Symbol 'U' denoted 'I assert', symbol 'N' denoted 'I reject'. Hence, U:1, N:0, i.e., 'I assert truth, I reject falsehood'.

Implication rules: $U : 0, 0, 0 < 1, 1 < 1; N : 1 < 0.$

$\prod$ stood for the universal quantifier.

$U : \prod p.p < p, \prod p.0 < p, \prod p.p < 1.$

Axiom system for two-valued logic:

$T1.U : \prod p.0 < p.$

$T2.U : \prod p.p < 1.$

$Z3.N : 1 < 0.$

Definitions:

$D1\ U : \prod p. - p \equiv (p < 0).$

$D2\ U : \prod pr.p + r \equiv (p < r) < r.$

$D3\ U : \prod pr.pr \equiv -(-p + -r).$

$D4\ U : \prod pr.(p = r) \equiv (p < r)(r < p).$

He gave derivations of forty theses of two-valued calculus from this system. In 1922, he found and announced the famous truth functions of multi-valued logics:

$$w(x < y) = min1, 1 - w(x) + w(y),$$

$$w(-x) = 1 - w(x).$$

In 1924, he discovered the parentheses-free notation, which he regarded as one of his greatest discoveries. In this notation, the symbols of functors, or operators, are put before symbols for arguments. With the symbol C for implication functor and N for the negation, the implication a < b is written down as Cab, $-a$ becomes Na. This notation was to be the standard notation of the Warsaw school of logic.

Between 1920 and 1930 Łukasiewicz published on definitions, begun his historic studied on the Stoic logic, and in 1929 were issued his mimeographed notes 'Elements of Mathematical Logic. Lecture Notes' which became a best-seller reprinted by PWN (Polish Scientific Publishers) 1958, PWN and Pergamon Press, 1963,1966. In this textbook, he presented his set of three axioms for two-valued logic along with an exposition of the three-valued logic and n-valued logics for natural numbers n.

Since 1926, Łukasiewicz conducted the seminar at the University, attended by Alfred Tarski, Boleslaw Sobociński, Mordechaj Wajsberg, Aldolf Lindenbaum, and students like Stanisław Jaśkowski, who on a question by Łukasiewicz came with the system of Natural Deduction. Their results were summed up in the publication by Łukasiewicz and Tarski of the famous article 'Untersuchungen über den Aussagenkalkul' w C.R. Soc. Sci. Lettr. Varsovie, cl.iii, 23 (1930). On twenty pages, authors gave the survey of results by Lukasiewicz, Tarski, Sobociński, Wajsberg, Lindenbaum. In spite of clear separation of results by distinct authors, due to misreadings by authors like Lewis, Moore and others, the discovery of multi-valued logics was by many authors attributed to Tarski. Authors mentioned the analysis by Łukasiewicz of Frege axiom system for two-valued logic:

1. $CpCqp.$
2. $CCpCqrCCpqCpr.$
3. $CCpCqrCqCpr.$
4. $CCpqCNqNp.$
5. $CNNpp.$
6. $CpNNp.$

Łukasiewicz demonstrated that axiom 3 is implied by axioms 1 and 2; axioms 4,5,6 can be replaced by one: $CCNpNqCpq.$ They recalled the Łukasiewicz axiom system, given earlier in 'Elements':

1. $CCpqCCqrCpr$.
2. $CCNppp$.
3. $CpCpNpq$.

They mentioned the Tarski theorem that each system of two-valued calculus in which formulas $CpCqp$, $CpCqCCpCqrr$ are theses, can be given axiom system consisting of one formula only. Tarski proposed later a formula on 53 symbols (shown in: Sobociński, "Z badań nad teorią dedukcji" in 'Przegląd Filozoficzny' (1932)). Łukasiewicz bettered that result by showing a formula of 33 symbols:

$$CCCpCqpCCCNrCsNtCCrCsuCCtsCtuvCwv.$$

To make things more difficult, they observed that those formulas were not organic, i.e., they contained sub-formulas which were theses of the system. Sobociński produced an organic formula of 47 symbols. In 1936, Łukasiewicz discovered an organic formula of 23 symbols which was bettered in 1952 by Carew A. Meredith, the Irish logician imbued with the Polish spirit, who discovered the formula of 21 symbols. Łukasiewicz was also the discoverer of the technique of logical matrices (independently applied by Paul Bernays) and authors applied them in general characteristics of multi-valued logics discovered by Łukasiewicz:

Definition 7: n-valued logical calculus L_n, where n is either a natural number or n is aleph zero, is the set of sentences satisfied by the logical matrix

$$M = [A, B, f, g],$$

where:
For $n = 1$: $A = \emptyset, B = \{1\}$.
For $1 < n < \aleph_0$ (where $\aleph_0$ denotes aleph zero): $A = \{k : 0 <= k < n - 1\}, B = \{1\}$.
For $n = \aleph_0$: A is an infinite countable subset of the unit interval $[0,1]$ closed on functions $f(x,y) = min1, 1 - x + y$, $g(x) = 1 - x$, $B = 1$. Systems L_n were shown to be axiomatizable for $n < \aleph_0$ as shown by Wajsberg and Lindenbaum. The Wajsberg axiomatics for L_3 was given as:

1. $CpCqp$.
2. $CCpqCCqrCpr$.
3. $CCNpNqCpq$.
4. $CCCpNppp$.
5. $CCNpNqCqp$.

Mordechaj Wajsberg ("Beitrage zum Metaausagenkalkul I". Monatshefte für Mathematic und Physik 42 (1935)) announced without proof that the hypothesis was true. In 1974 in Notre Dame J. Logic, Goldberg, Leblanc and Weaver proved completeness of L_3 in the Wajsberg axiom system. The editor of Notre Dame J. Formal Logic was then Bolesław Sobociński. For infinite valued logics the proof gave Rose and Rosser in TAMS 1958. In 'Untersuchungen ...', authors considered

Fig. 25. Chairmen of III Polish Philosophical Convention with Jan Łukasiewicz (third from right). Narodowe Archiwum Cyfrowe (NAC).

inter alia the narrow (implicational) logical calculus L_+. Łukasiewicz, later on, gave the only axiom

$$CCCpqrCCrpCsp$$

and proved that it was the shortest possible axiom (Proc. RIA 52 sect A no. 3 (1948)). Łukasiewicz published a sequel 'Philosophische Bemerkungen zu mehrwertigen Systemen des Aussagenkalkuls' in C.R. Soc. Sci. Lettr. Varsovie 23 (1930) [16] and then turned himself to his historic studies on calculus of sentences studying Stoics and their commentators. The results of his studies were published in 'Z historii logiki zdań' ('From history of logic of sentences') in 'Przegląd Filozoficzny' 37 (1934) [17] (see Fig. 26).

Łukasiewicz proved on the basis of his studies that the school of Stoa created the calculus of sentences in distinction to Aristotle, the creator of the logic of names. Studies on history of logic earned Lukasiewicz the title of the first sensu stricto historian of mathematics and his works are regarded as exemplary as based on original sources and rooted in formal logic and as a pattern for clarity of style.

Lukasiewicz publications between 1935 and 1939 encompass among others 'Zur vollen Aussagenlogik' in 'Erkenntnis' 1935-36 [18], 'Bedeutung der logischen Analyse für die Erkenntnis' in Actes VIII congr. Int. Phil at Prague (1936), 'O sylogistyce Arystotelesa' in 'Sprawozdania PAU' (1939), [19] 'Der Aequivalenz calcul' in 'Collectanea Logica' 1 (1939) [20]. The monograph on Aristotle's Syllogistics was finished by Łukasiewicz in 1939, and, in June 1939, he presented it in the meeting of the Polish Academy of Lettres (PAU). Ready for print, it perished in the first days of war in the bombarded and burnt down house at Sewerynów 6. In Ireland, Lukasiewicz undertook the task of reproducing the monograph and

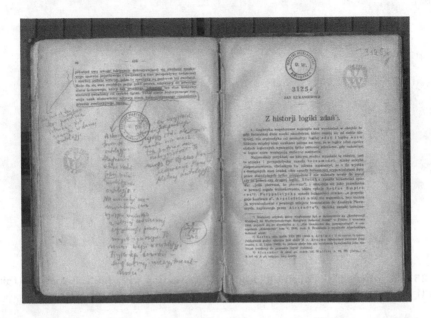

Fig. 26. An extended version of the contribution to the conference at Prague on 1.09.1934, published as 'Zur Geschichte der Aussagenlogik' in 'Erkenntnis' V(2), 1935-36 and in Author's translation in 'Przegląd Filozoficzny' 37, 1934, pp 417–437.

it was published in 1951 by Clarendon Press as 'Aristotle's Syllogistic from the Standpoint of Modern Logic' (see Fig. 27 and 28).

Second edition was posthumously edited in 1957 by Czesław Lejewski. Syllogistic of Aristotle operates with four predicates: Aab (each a is b), Iab (some a is b), Oab (some a is not any b), Eab (no a is b), and a syllogism figure is a formula of the form P and Q imply R. Łukasiewicz established this contrary to earlier historians who presented syllogisms as reasoning schemes. Axioms for syllogistics presented by Łukasiewicz are the following (K stands for the conjunction operator):

1. Aaa.
2. Iaa.
3. $CKAmbAamAab$ (Barbara).
4. $CKAmbImaIab$ (Datisi).
5. Deduction rules: substitution, modus ponens, definition induced substitution.
6. $D1Eab=NIab$.
7. $D2Oab=Naab$.
Rejection rules:
8. $\dashv CKAbmAamIab$.
9. $\dashv CKEbmEamIab$.

Fig. 27. Jan Łukasiewicz on Aristotle from Clarendon 1951.

Fig. 28. Second edition.

If in effect of a substitution a rejected rule is obtained then we reject the rule. With these axioms and rules Łukasiewicz deduced all 24 true figures and 232 rejected ones. It turned out that there existed formulas neither deductible nor rejected like

$$CKOabObaEab.$$

In this respect, Słupecki (1937) added a new rejection rule:
(S) If α and β denote Eab or Oab and γ is Aab, Iab, Eab or Oab, or, Cuv, where U is Aab, Iab, Eab or Oab, and, v is a certain conjunction of the former four, then if $C\alpha\gamma$ and $C\beta\gamma$ can be rejected then $CK\alpha\beta\gamma$ can be rejected.
    With (S) the system is decidable.

In 1938, Łukasiewicz took part in the conference 'Les entretiens de Zurich sur les fondements et la metode des sciences mathematiques' materials of which were published only in 1941 (Editeurs S.A. Leeman Freres & Cie., pp 82–100) [21] when Łukasiewicz lived in occupied Warsaw. His chapter entitled 'Die Logik und das Grundlagen problem' discusses basic problems of logic (Polish translation in 'Filozofia Nauki' 5(3), 1997, pp 147–162) in particular the problematics of logical matrices invented by Łukasiewicz (and independently used by Bernays) and their applications in establishing independence.

The period of war and subsequent emigration was too hard on him to think of scientific work, the matter was to survive. Yet, when in Ireland, he resumed activity by publishing 'The shortest axiom of the implicational calculus of propositions' in Proceedings RIA (A 52 (1948)) [22]. In the same year 1948, he sent 'On the system of axioms of the implicational propositional calculus' to the VI Congress of Polish Mathematicians, dedicated to the memory of Mordechaj Wajsberg who perished in the II World War, published as 'W sprawie aksjomatyki implikacyjnego rachunku zdań' in Ann. Soc. Polon. Math. 22 (1950) [23]. In 1950 he also published 'On the principle of the least number' in Polish as 'O zasadzie najmniejszej liczby' in 'Rocznik Polskiego Towarzystwa Matematycznego' 21, 1948-49, presented at the Vth Congress of Polish Mathematicians in 1947 [24].

In 1948, Łukasiewicz found himself among seventeen colleagues, all of them Polish scientists who left Poland for the West, which became excluded from the Polish Academy of Lettres (PAU) as ones who did not attend its meetings. PAU itself was dissolved in 1951 and few years later the Polish Academy of Sciences was called to life. This act of exclusion put the end to contacts of Łukasiewicz with the Country enveloped in darkness behind the Iron Curtain. He published in foreign journals: 'On variable functors of propositional arguments' in 'Proc. RIA' A 54 (1951) [25], 'On the Intuitionistic Theory of Deduction' in 'Indagationes Mathematicae. Proc. Ned. Koninklijke Academie van Wetenschappen' A no. 3 (1952) [26], 'Arithmetic and Modal Logic' in 'The Journal of Computing Systems 1 (4) (1954) [27]. The main work of the after-war period was the monograph 'Aristotle's Syllogistic from the Standpoint of Modern Formal Logic' (1951) [28] which was the summary of lasting for many years studies on Aristotle and his commentators.

LEGACY

Jan Łukasiewicz was throughout his life faithful to the line of research initiated in his Dissertation on induction as reverse to deduction and theory of deduction was to be his specialty. On this way, he obtained brilliant results of which the first place is certainly taken by many-valued logics, their axiomatizations, and applications in Computer Science and Artificial Intelligence as the basis for Reasoning under Uncertainty, in Fuzzy Set Theory, in Rough Set Theory and in Rough (Fuzzified) Mereology. Many-valued logics entered the intersection of mathematical logic and computer oriented logics as fuzzy logics of which the Łukasiewicz logic is based on the Łukasiewicz t-norm $T(x, y) = max\{0, x+y-1\}$

and the Łukasiewicz implication $x \longrightarrow y = min\{1, 1 - x + y\}$, and it is one of the three main fuzzy logics aside of logics of Gödel and Goguen. The Polish notation, invented by Łukasiewicz in 1924, along with its dual, the Reversed Polish notation, was applied in calculators and in theoretical computer science as computations in this format are in analogy to computations using stacks. Still important and will stay as such are his studies on Stoic logic and logic of Aristotle. He established the proper form of their rules and deduction methods and with his beautiful style and precision, he established an example for later authors.

As twice Rector of Warsaw University, Minister for education in 1918 and Minister for Religious Beliefs and Public Enlightenment in the Paderewski government through whole year 1919, he put enormous work towards creation of the University system and whole school system in independent Poland (see Fig. 29).

His scientific work created many problems whose solution took years and stimulated monographs. It is sufficient to mention Mc Naughton's 'A theorem about infinite-valued sentential logic' in 'J. Symbolic Logic' 16 (1951) [29], Goguen's 'The logic of inexact concepts' in 'Synthese' 19 (1968) [30], Meredith's 'The dependence of an axiom of Łukasiewicz' in TAMS 87 (1954) [31], Rose and Rosser's 'Fragments of many-valued statement calculi' in TAMS 87 (1958) [32] - the proof of completeness of infinite-valued logics with Łukasiewicz's implication and negation, Goldberg, Leblanc and Weaver's 'A strong completeness theorem for 3-valued logic' in 'Notre Dame J. Formal Logic' 15 (1974) [33], Mundici's 'Satisfiability in many-valued sentential logic is NP-complete' in Theoretical Computer Science 52 (1987) [34], Hájek's 'Metamathematics of Fuzzy Logic' published at Kluwer (1998) [35].

Fig. 29. Paderewski cabinet in 1919. Jan Łukasiewicz is second from left in the first row. Paderewski is standing (http://i2.wp.com/upload.wikimedia.org/wikipedia/commons/0/0c/Rzad_RP_Ignacego_Paderewskiego_1919.jpg).

Fig. 30. Statues in the Library of Warsaw University. From left to right: Kazimierz Twardowski, Jan Łukasiewicz, Alfred Tarski, Stanisław Leśniewski (http://portal. uw.edu.pl/image/image_gallery?uuid=c05f6dc1-6d2d-4775-8691-5fe5898d7bc6& groupId=11186509&t=1393538657785).

Fig. 31. Jan Łukasiewicz. Sculpture by Alfons Karny. Till 1939 in the Hall of Warsaw University.

Fuzzy set theory got its final form in 1965 in the article 'Fuzzy sets' [36] by Lotfi Zadeh. A fuzzy set was defined as a mapping m_A from a domain U into the interval $[0, 1]$ and the value $m_A(u)$ for $u \in U$ was interpreted as the degree to which the element u belonged in the set (concept) A. In this way, infinitely valued logics entered reasoning under uncertainty.

The Łukasiewicz implication along with its dual t-norm of Łukasiewicz $L(x, y) = max\{0, x + y - 1\}$ and formulas of Łukasiewicz for sentence-forming functors delineate the Łukasiewicz fuzzy logic. As proved by Menu and Pavelka, if a fuzzy (residual) implication is continuous then the t-norm which does induce it is isomorphic to the Łukasiewicz t-norm.

In 1982, Zdzisław Pawlak introduced the idea of a rough set in the article 'Rough sets' [37]. Rough set based logics employ as a tool many-valued logics.

Ideas of Łukasiewicz and Leśniewski were blended by introduction of rough mereology in which mereology created by Leśniewski in his 'Podstawy Ogólnej Teoryi Mnogości' [38] was in a sense fuzzified and entered the realm of many-valued logics in Polkowski and Skowron 'Rough mereology' [39,40].

The planetoid Łukasiewicz and statues in the Warsaw University Library express recognition of him and of his colleagues (see Fig. 30).

One more legacy which Jan Łukasiewicz passed us is to be firm in view of misfortunes which befell us, to keep working, to be friendly, understanding and magnanimous.

Acknowledgments. This work is based on the lecture delivered by the author at the 1st Warmian-Masurian Mathematics and Computer Science Colloquium at the Department of Mathematics and Informatics of the University of Warmia and Mazury in Olsztyn in May 2018 which commemorated centenaries of the announcement of many-valued logics by Łukasiewicz and of Poland's independence. Author thanks Professors Andrzej Skowron and Jan Bazan who took part in the Colloquium and presented lectures on rough set theory.

Thanks go to Professor Andrzej Skowron for his invitation to publish this text in Transactions on Rough Sets and to Professor Andrzej Skowron and Dr Soma Dutta for help with technical preparation of this text for publication.

Reproduced in this text photos come from the Narodowe Archiwum Cyfrowe as well as from commonly accessible internet sources. Author was taking care to attribute sources of photos. Reproduced in this text facsimiles of selected pages of Jan Łukasiewicz works come from the collection 'Jan Łukasiewicz. Selected Works', edited by Ludwik Borkowski and published by PWN (Polish Scientific Publishers) and North Holland Publishing Company, Warsaw-Amsterdam, 1970.

Additional notes

Valuable reviews and information on Jan Łukasiewicz's life and legacy are given in:

Peter Simmons. Jan Łukasiewicz. In: Stanford Encyclopedia of Philosophy (SEP): https://plato.stanford.edu/entries/lukasiewicz/

Jacek Jadacki, Piotr Surma. Jan Łukasiewicz: Pamiętnik (Jan Łukasiewicz. Memoirs). Wydawnictwo Naukowe Semper, Warszawa, 2013.

Piotr Surma. Jan Łukasiewicz. Pamiętnik (fragmenty)(Memoirs (fragments)). Rocznik Historii Filozofii Polskiej T. 2/3 , 2009/2010, 313-380.

Selected works of Jan Łukasiewicz were edited by Ludwik Borkowski for PWN and North Holland, 1970.

A review of ideas of propositional, modal and many-valued logics along with discussion of fuzzy and rough reasoning methods and accounts of mereology and rough mereology with applications to analysis of data, granular computing and behavioral robotics are given in

Lech Polkowski. Approximate Reasoning by Parts. An Introduction to Rough Mereology. Springer Verlag, Berlin-Heidelberg, 2011.

Recollections of the lost world of Lwów and its University can be found in:

Recollections of Professor Tomasz Cieszyński on Rudolf Weigl: http://www.lwow.home.pl/weigl/cieszynski.html

Recollections of Professor Tomasz Cieszyński on Jakub Karol Parnas: http://www.lwow.home.pl/parnas.html

Recollection of Tadeusz Riedl on Stefan Banach: http://kielich.amu.edu.pl/Stefan_Banach/riedl.html

Alfred Jahn on Banach, Schauder, Institute of Rudolf Weigl: Z Kleparowa w Świat Szeroki (From Kleparów into the Wide World). Zakład Narodowy Im. Ossolińskich. Wrocław, 1991.

Roman Ingarden on Schauder: Juliusz Schauder - personal reminiscences. Topol. Methods Nonlinear Anal. 2 (1993), no. 1, 1–14. https://projecteuclid.org/euclid.tmna/1479287168

Józef Wittlin on Lwów: 'Mój Lwów' ('My Lwów'). Czytelnik. Warszawa, 1991. See also: http://www.lwow.home.pl/przekroj/wittlin.html

https:\\www.youtube.com/watch?v=uiekjno91WA

https://www.youtube.com?watch?v=neDnpgZPPvY

Notes on authors of non-scientific quoted here texts:

Karolina Lanckorońska was a professor of University of Lwów in 1939. During the war she acted as a liason to Home Army and an officer to the Polish Help Committee (Rada Główna Opiekuńcza, RGO). In this character she travelled through Kresy and in Tarnopol was arrested by Krüger, responsible for murdering in Lwów of 40 professors of University and Polytechnic who confirmed to her his participation in that act. She was sent to the Ravensbrück concentration camp from which she was liberated by demands of the Italian Royal Family. She was the witness at the Krüger process after the war. She spent the rest of her life in Rome where she edited the monumental series of documents on Poland and the relations with the Papacy 'Antemurale' ('Bulwarks').

Tomasz Cieszyński was the son of the renowned professor of medicine in Lwów University, murdered by the Krüger commando. He was the lice feeder at the Rudolf Weigl Institute and after war in 1945 in Kraków met with professor Jakub Parnas who was allowed to travel to Poland, of course accompanied by the Security officer. From him we know some details of Parnas life in Russia as told by Parnas himself.

Tadeusz Riedl met Banach at the Rudolf Weigl Institute and became more familiar with him when Banach came to live in his parents house in Lwów in 1945.

Alfred Jahn was an assistant at the University in 1939 and along many others became a worker in Typhoid Institute of Rudolf Weigl. He personally knew Schauder who was his mathematics teacher at Gymnasium. He was most probably the last who had seen Schauder alive, walking along the street in Lwów. After war he became a professor at Wrocław University and its Rector in the critical time of student protests in 1968.

Roman Ingarden as Jahn, had Schauder as his professor of mathematics in Gymnasium and at the University. He met Schauder during the war and Schauder's life in hiding and was able by his contacts with the Home Army to provide Schauder with documents allowing travel. This was the primary cause of Schauder death as Schauder was walking when seen by Jahn to a bath in a factory in which Ingarden worked because he wanted bath before travel to Warsaw. On this walk Schauder was apprehended by a patrol and shot when trying to escape.

Józef Wittlin, a poet (collection 'Hymny' ('Hymns') written during the defense of Lwów in November 1918) and writer ('Sól Ziemi' ('The Salt of the Earth') published in 1935 which brought him a nomination to Nobel prize in literature in 1937), wrote 'My Lwów" when on immigration in the US during the 2nd World War and it was first published in 1946.

References

1. Lanckorońska, K.: Wspomnienia Wojenne (War Memoirs). Społeczny Instytut Wydawniczy Znak, Kraków (2001)
2. Nestor: The Tale of Bygone years. Cross, S.H., Sherbowitz-Wetzor, O.P. (translators & eds.) The Russian Primary Chronicle, Laurentian Text. The Mediaeval Academy of America, Cambridge (1930 & 1953)

3. Pol, W.: Mohort: rapsod rycerski z podania (A Knight's Romance Retold from a Legend). Wyd, Akademickie WSSP (2017)
4. (Kazimierz) Twardowski, C.: Zur Lehre vom Inhalt und Gegenstand der Vorstellungen: Eine Psychologische Untersuchung (Classic Reprint). Fb&c Limited (2018)
5. Łukasiewicz, J.: Analiza i konstrukcja pojęcia przyczyny. W: Łukasiewicz, J. (ed.) Z zagadnień logiki i filozofii: pisma wybrane. PWN, Warszawa, pp. 9–62
6. Russell, B.: On the notion of order. Mind 10(37), 30–51 (1901)
7. Russell, B.: The Principles of Mathematics. London (1903)
8. Łukasiewicz, J.: O zasadzie sprzeczności u Arystotelesa (On the contradiction principle in Aristotle). PWN, Warszawa (1987)
9. Łukasiewicz, J.: Die logischen Grundlagen der Wahrscheinlichkeitsrechnung. Kraków. Również w: Z zagadnień logiki i filozofii: pisma wybrane, str. 76–113 (1913)
10. Łukasiewicz, J., Tarski, A.: Untersuchungen über den Aussagenkalkül. C.R. Soc. Sci. Lett. Varsovie 23, 1–21 (1930)
11. Łukasiewicz, J.: Zur Geschichte der Aussagenlogik. Erkenntnis V(2) (1935)
12. Łukasiewicz, J.: Logika i problem podstaw matematyki (Logic and the problem of foundations of mathematics). W: Logika i Metafizyka: miscellanea. J. Jadacki, edytor, WFiS UW, str. 70–85 (1998)
13. Łukasiewicz, J.: O indukcji jako inwersji dedukcji (Induction as the reverse to deduction). Przegląd Filozoficzny 6, 9–24; 138–152 (1903)
14. Łukasiewicz, J.: O logice trójwartościowej (On the 3-valued logic). Ruch Filozoficzny 5, str. 170–171 (1920)
15. Łukasiewicz, J.: Logika dwuwartościowa (On 2-valued logic). Przegląd Filozoficzny 13 (1921)
16. Łukasiewicz, J.: Philosophische bemerkungen zu mehrwertige Systemen des Aussagenkalküls. C. R. Soc. Sci. Lett. Varsovie 23, 51–77 (1930)
17. Łukasiewicz, J.: Z historii logiki zdań (On the history of the logic of sentences). Przegląd Filozoficzny 37, 417–437 (1934)
18. Łukasiewicz, J.: Bedeutung der logischen Analyse für die Erkenntnis. W: Actes du VIII Congres International de Philosophie, Prague, pp. 75–84 (1936)
19. Łukasiewicz, J.: O sylogistyce Arystotelesa (On the syllogistic of Aristotle). Sprawozdania PAU 44, 220–227 (1939)
20. Łukasiewicz, J.: Der Äquivalenzkalkül. Collectanea logica 1, 145–69 (1939). (This journal was established by B. Sobociński but the outbreak of war prevented publication save one offprint). Published also in L. Borkowski (ed.): J. Łukasiewicz. Selected Works, pp. 250–277 (1970)
21. Łukasiewicz, J.: Die Logik und das Grundlagenproblem. Les entretiens de Zurich sur les fondements et la méthode des sciences mathématiques, December 1938, pp. 82–100. Leemann Freres, Zurich (1941)
22. Łukasiewicz, J.: The shortest axiom of the implicational calculus of propositions. In: Proceedings of the Royal Irish Academy, vol. 52 (A), pp. 25–33 (1948)
23. Łukasiewicz, J.: W sprawie aksjomatyki implikacyjnego rachunku zdań (On the axiomatics of the implicational calculus of sentences). Ann. Soc. Polon. Math. 22, 87–92 (1950)
24. Łukasiewicz, J.: O zasadzie najmniejszej liczby (On the principle of the least number). Rocznik Polskiego Towarzystwa Matematycznego 1948-9. Praca nadesłana na V Kongres Matematyków Polskich (1947)
25. Łukasiewicz, J.: On variable functors of propositional arguments. In: Proceedings of the Royal Irish Academy, vol. 54 (A), pp. 25–35 (1951)

26. Łukasiewicz, J.: On the intuitionistic theory of deduction. In: Indagationes Mathematicae. Koninklijke Nederlandse Academie van Wetenschappen, Proceedings Series A, vol. 14, pp. 201–212 (1952)
27. Łukasiewicz, J.: Arithmetic and modal logic. J. Comput. Syst. **1**, 213–219 (1954)
28. Łukasiewicz, J.: Aristotle's Syllogistic from the Standpoint of Modern Formal Logic. Clarendon Press (1951). 2nd edn. Oxford University Press (1957) (Ed. by, C. Lejewski)
29. McNaughton, R.: A theorem about infinite-valued sentential calculus. J. Symb. Log. **16**, 1–13 (1951)
30. Goguen, J.A.: The logic of inexact concepts. Synthése **19**, 325–373 (1968)
31. Meredith, C.: The dependence of an axiom of Łukasiewicz. TAMS **87**, 54 (1958)
32. Rose, A., Rosser, J.B.: Fragments of many-valued statement calculi. TAMS **87**, 1–53 (1958)
33. Goldberg, H., Leblanc, H., Weaver, G.: A strong completeness theorem for 3-valued logic. Notre Dame J. Form. Log. **15**, 325–332 (1974)
34. Mundici, D.: Satisfiability in many-valued sentential logic is NP-complete. Theor. Comput. Sci. **52**, 143–153 (1987)
35. Hájek, P.: Methamathematics of Fuzzy Logic. Kluwer, Rotterdam (1998)
36. Zadeh, L.A.: Fuzzy sets. Inf. Control **8**, 338–353 (1965)
37. Pawlak, Z.: Rough sets. Int. J. Comp. Inf. Sci. **11**, 341–356 (1982)
38. Leśniewski, S.: O podstawach Ogólnej Teoryi Mnogości. Moskwa (1916)
39. Polkowski, L., Skowron, A.: Rough mereology. In: Raś, Z.W., Zemankova, M. (eds.) ISMIS 1994. LNCS, vol. 869, pp. 85–94. Springer, Heidelberg (1994). https://doi.org/10.1007/3-540-58495-1_9
40. Polkowski, L., Skowron, A.: A new paradigm for approximate reasoning. Int. J. Approx. Reason. **15**(4), 333–365 (1996)

Descriptive Topological Spaces for Performing Visual Search

Jiajie Yu and Christopher J. Henry[✉]

University of Winnipeg, 515 Portage Avenue, Winnipeg, MB, Canada
yu-j83@webmail.uwinnipeg.ca, ch.henry@uwinnipeg.ca
http://www.acs.uwinnipeg.ca

Abstract. This article presents an approach to performing the task of visual search in the context of descriptive topological spaces. The presented algorithm forms the basis of a descriptive visual search system (DVSS) that is based on the guided search model (GSM) that is motivated by human visual search. This model, in turn, consists of the bottom-up and top-down attention models and is implemented within the DVSS in three distinct stages. First, the bottom-up activation process is used to generate saliency maps and to identify salient objects. Second, perceptual objects, defined in the context of descriptive topological spaces, are identified and associated with feature vectors obtained from a VGG deep learning convolutional neural network. Lastly, the top-down activation process makes decisions on whether the object of interest is present in a given image through the use of descriptive patterns within the context of a descriptive topological space. The presented approach is tested with images from the ImageNet ILSVRC2012 and SIMPLIcity datasets. The contribution of this article is a descriptive pattern-based visual search algorithm.

Keywords: Human visual search · Guided search model ·
Bottom-up attention · Top-down attention · Salient objects ·
Visual field · Descriptive topological space · Descriptive proximity ·
Descriptive set intersection · Convolutional neural network

1 Introduction

The problem considered in this article is the automation of visual search motivated by behaviour performed by the human visual system. The problem of visual search is the process of identifying an object in our field-of-view (FOV) amongst many distractor objects, *i.e.* objects that are not the object of interest. This visual search and it is dependent on our ability to direct visual attention.

Jiajie Yu—This research has been supported by the Natural Sciences and Engineering Research Council of Canada (NSERC) Discovery Grant 418413, and the Faculty of Graduate Studies at the University of Winnipeg. Also, special thanks to Keith Massey for developing the code that produced the VGG object descriptions.

© Springer-Verlag GmbH Germany, part of Springer Nature 2019
J. F. Peters and A. Skowron (Eds.): TRS XXI, LNCS 10810, pp. 48–67, 2019.
https://doi.org/10.1007/978-3-662-58768-3_2

This is a complex task that humans perform seamlessly. The aim of visual search systems are to automatically mimic human behaviour when processing the output of optical sensors, whether images or frames in a video sequence. The task of visual search performed by the human visual system is an active area of research in psychology [1]. The solution presented here is based on the definition of visual search as a type of perceptual task that directs attention [2]. In this context, perceiving particular objects is an act of selective attention, where the selective attention mechanism serves to link the processes of perception, action and learning [1].

The attention model used in this work is the guided search model (GSM) [3], which consists of modelling two types of visual search-based selective attention mechanisms: namely, bottom-up and top-down models. With respect to the GSM, the bottom-up approach focuses on modelling salient regions in the FOV. On the other hand, the top-down attention approach models the selection of a desired object from among salient regions identified by the bottom-up method, where salient regions are matched to some representation of the desired object in human memory. From a systems point of view, the top-down approach is a user-guided attention model.

Practical application of the GSM requires a theoretical framework to relate information captured in digital images to perceptual objects in the FOV. Inspired by [4–7], this work implements the GSM within the context of descriptive topological spaces, where the perceptual objects are pixels obtained from digital images. Here, the bottom-up model is implemented by the graph-based visual saliency (GBVS) method [8], and the top-down model is achieved with descriptive topological spaces and a descriptive proximity relation. Results are generated by using a digital image to represent an object of interest, called a query image, which is compared with other images from an image dataset. The solution consists of generating saliency maps (using the approach in [8]) from two images under consideration and using a descriptive proximity relation – defined within a descriptive topological space – in making decisions on the presence of query objects.

This article is based on the work reported in [9], and the contribution of this work is the Descriptive Visual Search System (DVSS) defined in the context of descriptive topological spaces. The DVSS represents the first attempt of using a descriptive pattern-based visual search algorithm, as well as the first use of convolutional neural networks to generate perceptual object descriptions within a perceptual system. Similarly, this article introduces a novel tolerance-based extension of descriptive intersection for use in the DVSS. Finally, the article is organized as follows. Section 2 presents background material on the visual search program and near set theory. Section 3 gives the theoretical framework implemented in the DVSS. Section 4 describes the implementation details of the DVSS, and Sect. 5 presents results and discussions. Finally, the article is concluded with Sect. 6.

2   Background

This section provides context for visual search models and descriptive near set theory, both of which are used to produce the DVSS.

2.1   Visual Search Psychological Model

Visual search is the act of finding a visual object[1] of interest in an FOV containing many distractor objects. This problem is considered a perceptual task, where the focus, within the human visual system, is to direct attention. In our case, we aim to mimic the human ability to quickly and effortless find objects of interest in our FOV [2]. The model of human visual attention used in this article has two aspects, namely bottom-up [10] and top-down attention [1]. The bottom-up approach is a type of instinctual, non-guided attention mechanism and is scene-dependent. In contrast, the top-down approach is a user-guided attention mechanism, or task-dependent.

There are four main psychological models that could be a basis for a computational model for visual attention, namely feature integration theory (FIT) [11], biased competition (BC) hypothesis [12], integrated competition hypothesis (IC) [13], and guided search model (GSM) [3]. Based on the application, there are two approaches to modelling the human approach to searching for objects in our FOV. One is object-based [14]. It involves the analysis of parts of an object and it may be used to recognize an object. The other is space-based [14], and it is concerned with the location or position of an object. Feature integration theory [11] is usually used for space-based searching tasks and is typically associated with bottom-up attention models. It assumes that features come first in the perceptual process and are later combined to form objects. Here, the visual scene is coded along multiple feature dimensions, including colour, orientation, texture, and intensity. These features are then fused together when attention is directed at a specific location (hence the *space-based* moniker) in the FOV to form an object. In contrast, the biased competition hypothesis [12] maintains that, regardless of space-based or object-based, the selection according to attention is a biased, competitive process. The competition is among different objects or local area to determine which of them is a reasonable selection according to the relevant task. Further, [12] asserts that the competition is biased toward bottom-up attention in order to benefit local inhomogeneity, *i.e.* locations most distinct from their surroundings are likely to be the winner of the competition for attention. On the other hand, the top-down attention model in the BC hypothesis shifts the bias based on items relative to the current task, such as visual search. Finally, the integrated competition hypothesis [13] is an extension of the BC hypothesis, which posits that any property of the object could be a basis for guiding attention. In the other words, object properties are also treated as task-relevant features.

---

[1] The term *perceptual object* has specific meaning in descriptive set theory and perceptual systems. Hence, we will use *visual object* to represent any salient object in an FOV.

2.2 Near Sets

In this work, visual objects inherent to digital images are denoted by sets within a descriptive topological space, and the aim is to assess the nearness or apartness between these disjoint sets. In other words, quantifying the nearness between sets is the basis for determining the similarity of visual objects. Inherent to the study of perceptual similarity is the idea of nearness and tolerance, both of which have a rich and rigorous mathematical history [4,15]. The idea of sets of similar sensations was first introduced by Poincaré in which he reflects on experiments performed by Weber in 1834, and Fechner's insight in 1850 [16–19]. Poincaré's work was inspired by Fechner, but the key difference is Poincaré's work marked a shift from stimuli and sensations to an abstraction in terms of sets together with an implicit idea of tolerance. Next, the idea of tolerance is formally introduced by Zeeman [20] with respect to the brain and visual perception. This idea of tolerance is important in mathematical applications, where systems deal with approximate input and results are accepted with a tolerable level of error, an observation made by Sossinsky [16], who also connected Zeeman's work with that of Poincaré's. In addition to these ideas on tolerance, Riesz first published a paper in 1908 on the nearness of sets [21,22], initiating the mathematical study of proximity spaces and the eventual discovery of descriptively near sets. Specifically, in 2002, Pawlak and Peters considered an informal approach to the perception of the nearness of physical objects such as snowflakes that was not limited to spatial nearness [23]. In 2006, a formal approach to the descriptive nearness of objects was considered by Peters, Skowron and Stepaniuk in the context of proximity spaces [22,24–26]. In 2007, descriptively near sets were introduced by Peters [27,28], followed by the introduction of tolerance near sets [29,30]. Recently, the study of descriptively near sets has led to algebraic [31,32], topological and proximity space [5–7] foundations of such sets.

Originally, the notion of nearness between sets, introduced by Riesz, was based on a spatial relationships between sets, called proximity. As has been mentioned, this idea of proximity between sets was recently expanded to include both spatial quantitative interpretation and a non-spatial qualitative interpretation, called descriptive proximity [4,5,27,28,33]. In this article, the qualitative interpretation of the notion of proximity between sets (*i.e.* description-based) is used. This idea of descriptive proximity between sets is also know as descriptive near set theory.

3 Preliminaries

This section presents descriptive near set theory, which is primarily used in the last stage of the proposed system. However, the ideas presented in this section form a narrative that underlies the entire approach. In particular, visual search is a type of perceptual task that is complementary to the basic inspiration of descriptive near set theory, namely that humans make decisions on nearness of disjoint sets of objects based on perceived features associated with the objects. In fact, objects in descriptive set theory are labelled *perceptual objects* and the fundamental structure that introduces descriptive near set theory is a perceptual system, which is where this section begins.

3.1 Perceptual System

Sets of perceptual objects and their descriptions form a perceptual system. To begin, a perceptual object [33] is something perceivable that has its origin in the physical world. Thus, perceptual objects are objects which can be perceived in the physical world, using the senses of sight, touch, taste, smell and hearing. However, the focus of this work is visual search and the sense of sight. Thus, in this work the descriptions of the objects are all extracted from digital images. In general, a description is a real-valued tuple representing features of a perceptual object. Each description is a vector of real-valued features associated with each respective object. Continuing on, a perceptual system consists of both perceptual objects and probe functions [33]. Typically, these values are extracted by a series of functions, called a *probe function* [27,34]. A probe function is a real-valued function representing a feature of a perceptual object [4]. A set of probe functions are used to generate the feature vector that provide descriptions. In this work, probe functions are defined in the context of deep convolutional neural networks [35] and are used to produce feature vectors for each object.

Definition 1 Perceptual System [33]. *A perceptual system $\langle O, \mathbb{F} \rangle$ consist of a non-empty set O of sample perceptual objects and a non-empty tuple $\mathbb{F}$ of real-valued functions $\phi \in \mathbb{F}$ such that $\phi : O \to \mathbb{R}$.*

Next, there is a need within a perceptual system to characterize perceptual objects in O. As a result, an object description is given as follows.

Definition 2 Object Description [36,37]. *Let $\langle O, \mathbb{F} \rangle$ be a perceptual system, then the description of a perceptual object $x \in O$ is a feature vector given by*

$$\Phi_{\mathbb{F}}(x) = (\phi_1(x), \phi_2(x), ..., \phi_i(x)...\phi_l(x)),$$

where l is the length of the vector $\Phi_{\mathbb{F}}$, and each $\phi_i(x)$ in $\Phi_{\mathbb{F}}(x)$ is a probe function value that is part of the description of the object $x \in O$.

Typically, object descriptions are also referred to as feature vectors in other disciplines. Finally, a descriptive neighbourhood of an object is given by the following.

Definition 3 Descriptive Neighbourhood [5]. *Let $x, y \in O$ be perceptual objects with object descriptions given by $\Phi(x), \Phi(y)$, and let $\varepsilon \in \mathbb{R}$. Then, a description-based neighbourhood is defined as*

$$N_{\Phi(x)} = \{y \in O : |\Phi(x) - \Phi(y)| < \varepsilon\}.$$

A point y is a member of $N_{\Phi(x)}$, if and only if, $|\Phi(x) - \Phi(y)| < \varepsilon$.

This definition will be used to produce neighbours of a certain point.

3.2   Descriptive Topologies

Descriptive topological spaces are a significant portion of the proposed visual search system, and, since the human FOV is simulated with digital images, this section introduces a descriptive topological framework defined in the context of digital images [5]. Recently, much work has been reported regarding descriptive topological spaces that are defined with respect to the descriptive intersection and union of open sets [4–6,38,39]. Keeping this in mind, a topology is defined as follows.

Definition 4 Topology [39]. *For a given set, X, a topology, τ, on X is a family of subsets of X (called open sets) such that:*

1. *X and $\emptyset$ are in τ,*
2. *unions of members of τ are in τ, and*
3. *finite intersections of members of τ are in τ.*

The pair (X, τ) is called topological space, or, in the other words, a nonempty set X with a topology τ on it is a topological space [39]. Correspondingly, a descriptive topological space is obtained when considering set descriptions and descriptive-based set operators [5], which are defined below.

Definition 5 Set Description [5,36,38]. *Let $A \subseteq O$ be a set within a perceptual system $\langle O, \mathbb{F} \rangle$, then the set description of A is defined as*

$$\mathcal{D}(A) = \{\Phi(a) : a \in A\}.$$

A new form of topology – called *descriptive topology* – requires new operators analogous to union and intersection. These are given next, both of which use Definition 5. Note, the descriptive union and the union operators are equivalent (see, *e.g.*, [38]). Thus, a new definition is not given for set union. On the other hand, the descriptive set intersection operator is defined as follows.

Definition 6 Descriptive Set Intersection [4,5]. *Let A and B be any two sets. The* descriptive (set) intersection *of A and B is defined as*

$$A \underset{\Phi}{\cap} B = \{x \in A \cup B : \Phi(x) \in \mathcal{D}(A) \text{ and } \Phi(x) \in \mathcal{D}(B)\}.$$

Moreover, the formal properties of descriptive intersection depend upon the perceptual system. Based on the definitions given above, a descriptive topology [39] is defined next.

Definition 7 Descriptive Topology [39]. *For a given set X, a descriptive topology, τ_Φ, on X is a family of subsets of X such that:*

1. *X and $\emptyset$ are in τ_Φ,*
2. *descriptive unions of members of τ_Φ are in τ_Φ, and*
3. *finite descriptive intersections of members of τ_Φ are in τ_Φ.*

Here, it is interesting to note that a descriptive topology depends on the underlying perceptual system and it may, in fact, not be a topology. As a result, Definition 7 can be considered a straightforward translation of the standard definition of topology into the presented descriptive framework and nothing more.

3.3 Descriptive Proximities

A descriptive topology defines a structure that is a collection of sets containing objects with comparable descriptions. The aim in developing the presented visual search system is to make decisions based on the degree of shared descriptions within the topology. As a result, our system relies on the ability to quantify the nearness of members of a descriptive topology. The first step in achieving this goal is the definition of a proximity relation. Proximities are nearness relations among the subsets of X in a topological space (X, τ). In other words, a proximity is a closeness or apartness relation on pairs of subsets of X. In [39], two basic types of proximities are defined, namely traditional spatial proximity and descriptive proximity. In [39], the traditional spatial proximity is considered when nonempty sets that have spatial proximity are close to each other, either asymptotically or with common points. In contrast, [39] defines descriptive proximity as nonempty sets are close provided the sets contain one or more elements that have matching descriptions.

There are a number of well-known proximities [5–7] such as the Čech [40], Efremovič [41], Lodato [42], and Wallman [43] proximities. An example of the simplest proximity, a Čech proximity, δ_c, satisfies the following.

1. $\emptyset \mathbin{\not\delta_c} A, \forall A \subset X$,
2. $A \, \delta_c \, B \Leftrightarrow B \, \delta_c \, A$,
3. $A \cap B \neq \emptyset \Rightarrow A \, \delta_c \, B$,
4. $A \, \delta_c \, (B \cup C) \Leftrightarrow A \, \delta_c \, B$ or $A \, \delta_c \, C$.

Finally, a specific descriptive proximity relation is defined below.

Definition 8 Descriptive Proximity Relation [44]. *Given a perceptual system $\langle X, \mathbb{F} \rangle$, with $A, B \in \mathcal{P}(X)$, the descriptive proximity relation is defined by*

$$\delta_\Phi = \{(A, B) \in \mathcal{P}(X) \times \mathcal{P}(X) : A \mathbin{\underset{\Phi}{\cap}} B \neq \emptyset\}, \tag{1}$$

where the notation $A \, \delta_\Phi \, B$ reads A is descriptively close to B.

3.4 Descriptive Patterns

Patterns play a pivotal role in the presented approach to measuring the nearness or apartness of visual objects. The descriptive intersection of member sets from two respective patterns is the basis for decisions regarding the presence of visual query objects. These patterns, in turn, are created via pattern generators. Beginning with patterns, the definitions for spatial and descriptive set patterns are given as follows.

Definition 9 Spatial Set Pattern [6]. *A spatial set pattern $\mathcal{P}$ contains sets that are spatially near each other.*

Definition 10 Descriptive Set Pattern [6]. *A descriptive set pattern, $\mathcal{P}_\Phi$, contains sets that are descriptively near a given set and possibly near each other.*

Relying on the definition of a set pattern, a pattern generator is defined as follows.

Definition 11 Pattern Generator [6]. *A pattern generator is a distinguished set that is close to each set in the collection of sets in a set pattern.*

3.5 Tolerance-Based Descriptive Intersection Operator

This section presents a new descriptive set operator based on tolerance spaces and relations [16,20,45]. As is discussed below, sets formed in this work are extracted from digital images, where decisions on the similarity of visual objects contained in these images are based on features values extracted from image pixels. The result is that comparison between patterns – generated from sets representing visual objects – is the pivotal step. However, the output of probe functions for two objects perceived to be *the same* is rarely an exact match [46]. As a result, the following operator was defined out of necessity for producing results in presented real-world application.

Definition 12 Tolerance Descriptive Set Intersection [9]. *Let A and B be any two sets. The* tolerance descriptive (set) intersection *of A and B is defined as*

$$A \underset{\Phi,\varepsilon}{\cap} B = \{a \in A, b \in B : \parallel \Phi(a) - \Phi(b) \parallel_2 \leq \varepsilon\},$$

where $\parallel \cdot \parallel_2$ *is the L^2 norm.*

This new definition of descriptive set intersection provides for the introduction of a nuanced version of the descriptive proximity relation. Recall Definition 3 produced a set of points that are neighbours of a certain point. However, the Definition 12 gives a similarity measurement between two sets of points.

Definition 13 Descriptive Tolerance Proximity Relation. *Given a perceptual system* $\langle X, \mathbb{F} \rangle$, *with* $A, B \in \mathcal{P}(X)$, *the descriptive tolerance proximity relation is defined by*

$$\delta_{\Phi,\varepsilon} = \{(A,B) \in \mathcal{P}(X) \times \mathcal{P}(X) : A \underset{\Phi,\varepsilon}{\cap} B \neq \emptyset\}. \tag{2}$$

4 Descriptive Visual Search System

The proposed descriptive visual search system (DVSS) consists of both the GSM bottom-up and top-down attention models and is implemented in three distinct stages. First, the bottom-up activation process is used to generate saliency maps and to identify salient objects. Second, perceptual objects, defined in the context of descriptive topological spaces, are identified and associated with feature vectors (*i.e.* object descriptions) obtained from a VGG [47] deep learning convolutional neural network. Lastly, the top-down activation process makes decisions on whether the object of interest is present in a given image through the use of descriptive patterns within the context of a descriptive topological space.

4.1 Bottom-Up Attention

The aim of bottom-activation is to guide attention to salient regions of the FOV. These regions can be mapped with respect to the FOV producing *saliency maps*. In other words, a saliency map indicates the degree in which a particular region is unusual or different from its surrounding regions. Within the DVSS, saliency maps are implemented using digital images, which means a region's saliency value is quantified by grey levels. More specifically, each pixel in the saliency map represents a saliency value. The DVSS uses GVBS [8] to translate the RGB pixel values into saliency values using graph theory and ultimately produce these maps. Further, the salient regions of these maps are then used to identify perceptual objects, *i.e.* pixels from the original image, that will subsequently be used to create a pattern generator. It is these generated patterns that are finally used to make decisions on whether or not the (visual) object is contained in a given FOV, *i.e.* a digital image. Finally, a predefined threshold is used to determine which pixels in a saliency map constitute visual objects. Examples of this process are given in Fig. 1.

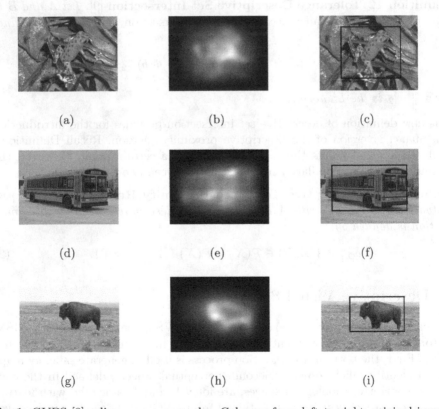

(a)   (b)   (c)

(d)   (e)   (f)

(g)   (h)   (i)

Fig. 1. GVBS [8] saliency map examples. Columns from left to right: original image, saliency map, detected salient object with bounding box.

4.2   Convolutional Neural Network-Based Probe Functions

A pre-trained VGG network [47] was used to extract features in this work. In general, convolutional neural networks (ConvNets) consist of one or more layers, which are labelled as convolutional, fully connected, and pooling layers [48]. Convolutional layers are so named since each neuron develops (*i.e.* learns) a filter during the training process which identifies different types of features within the image. They are named *convolution* since the operation performed by each neuron is analogous to the convolution operation between the filter and the input to the neuron. Pooling layers down-sample the results from the former layer, and neurons in fully connected layers connect all neurons in the previous layer in order to infer classes from the output of the penultimate layer. Figure 2 present a visual example of a multilayer ConvNet. The VGG ConvNet was developed by the Visual Geometry Group [47], and the VGG network used in this article was pre-trained using the ILSVRC-2012 dataset [49]. This data set contains 1000 categories of images, split into training (1.3 million images), validation (50 thousand images), and testing (100 thousand images with missing class labels) sets.

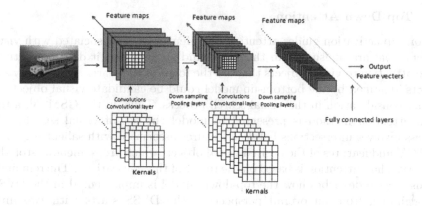

Fig. 2. Example of a standard multilayer convolutional neural network.

In this work, a trained ConvNet was used to extract features (*i.e.* object descriptions) for perceptual objects, where output from the layers were used to produce probe function values (see, *e.g.*, [50]). Any layer in the network could be used as a feature extractor since ConvNets can generate features of visual objects ranging from low-level to high-level. Low-level features are associated with lower levels of a ConvNet and usually describe some low-level digital image characteristic (*e.g.* colour, texture, or edge orientation). They are termed *low-level* features as they are closely related to pixel information and are far removed from global representations of the visual objects in the image. On the other hand, high-level features are associated with higher levels of the network and are related to the perceptual knowledge of the visual objects within the image. This knowledge is represented by the classes identified by the neural network (*e.g.* people, dog, bus). Further, current deep convolutional neural networks are able to process large numbers of classes with very small differences between categories. For example, the classes of people, dog, and bus can be further expanded to include categories such as seniors, kids, Husky, Chihuahua, school bus, charter bus, *etc.* In all cases, features are produced by presenting an image to the ConvNet input and forwarding outputs through the network until the desired layer is reached. At this point, the features are the outputs of the neurons at that level. In this paper, a 19 layer VGG network consisting of 16 convolutional layers and 3 fully connected layers was used, and the features were extracted from the 13th convolutional layer of the VGG network [47]. Further, layer 13 produces object descriptions of length 512 for each perceptual object. Layer 13 was selected for the DVSS since this level corresponds to higher perceptual representation of visual objects. This is important since the act of visual search does not only rely on low-level features, such as colour or shape, but it also relies on category knowledge and information. In other words, DVSS searches for visual objects of the same category, but which may have different colours or shapes. In this case, high-level features produce better performance.

4.3   Top-Down Attention

Bottom-up activation guides attention to salient regions, associated with visual objects, that are unusual from their surrounding area, but it does not actively guide attention to visual objects that are the focus of the search. However, visual objects identified by the bottom-up model could be candidate visual objects for further consideration in the visual search process. Thus, in the GSM [3], a top-down attention method is presented to model the act of visual search. This process involves intersections between features associated with salient regions in the FOV and features of the desired visual object, which are, somehow, stored in memory. Thus, attention is based on the result of this comparison. The remainder of this section describes how the top-down model is implemented in the DVSS.

Beginning from an overall perspective, the DVSS starts with two input images, where one is the query image, denoted Q, and the other is a candidate image C. The query image is a representation of the visual object that is the focus of the search process, and it may contain several visual objects. However, to start a searching task, only one query object is selected by the user. Typically, query objects are identified by some domain expert, and this process is analogous to the long term memory discussed in [1]. The candidate image represents the FOV, containing one or more visual objects, all of which are considered as candidate objects for the searching task.

Secondly, GBVS [8] is applied to both query and candidate images, and the output is two saliency maps [10]. The saliency maps are denoted as QS and CS for the query image and candidate image, respectively. As was mentioned, saliency maps are represented by greylevel images, and the pixels associated with salient visual objects are determined by defining a greylevel threshold on the saliency maps. Any saliency map pixels with values above this threshold are considered part of a salient visual object. Keeping this in mind, only one visual object is stored in memory for the query image, while all the candidate visual objects are stored into memory. In terms of storing the visual objects, the coordinates of pixels associated with each object are stored in the memory.

Continuing on, the top-down activation model is simulated and implemented as follows. Let $X = Q \cup S$, and let $K \subset X$ be a set of perceptual objects (*i.e.* pixels) representing a salient visual object identified using QS or CS. Then, a spatial pattern generator, G, is formed by selecting every l^{th} member of K, where l is some application dependent quantity. Next, descriptive neighbourhoods are found for each member of G to produce a descriptive pattern $\mathcal{P}_\Phi = \{N_{\Phi(x)} : x \in G\}$. In other words, all the descriptive neighbourhoods are generated by members of G to form a descriptive pattern, $\mathcal{P}_\Phi$. In this light, G_q and G_c represent pattern generators formed from query and candidate visual objects, respectively, and they generate descriptive set patterns for query and candidate objects that are denoted by $\mathcal{P}_\Phi^q$ and $\mathcal{P}_\Phi^c$, respectively.

Recall, the GSM relates bottom-up and top-down attention models through the intersection of features of salient regions identified from bottom-up mechanisms with knowledge or memory of the visual object that is the focus of the search. Thus, the DVSS operates in the same manner by searching for non-empty intersections between members of the query and candidate patterns. In particular, each member of $\mathcal{P}_\Phi^q$ is compared to each member of $\mathcal{P}_\Phi^c$ using Definition 12. The result of these intersections (*i.e.* the cardinalities) are then accumulated using a nearness measure inspired by [38]. This process is formalized in Algorithm 1.

Here, it is important to make several observations regarding Algorithm 1. First, as stated in line 14, the nearness of members of the respective patterns is determined by the fraction of objects present in the tolerance descriptive set intersection versus the number of objects in the union of the two sets. Moreover, any value of $s > 0$ implies that the two sets satisfy the descriptive tolerance proximity relation given in Eq. 2. Next, each member of the query object pattern, $\mathcal{P}_\Phi^q$, is compared with all the members of the candidate pattern, $\mathcal{P}_\Phi^c$ in order to find the most descriptively close (*near*) part of the candidate object, and only the maximum value of s will be used to represent this relationship. Therefore, the final value S is the sum of all the maximum values which are normalized by the total number of members in $\mathcal{P}_\Phi^q$. This value S is the final basis for determining whether the query visual object is present in the FOV represented by the candidate image.

5   Results and Discussion

The DVSS was tested by performing image retrieval. In this setup, the query image acts as the object that is the focus of a visual search task and the other images in the dataset represent different visual scenes presented to the FOV. Particularly, images from categories other than the query image represent distractor objects and images from the same category as the query represent the objective of the search. Image retrieval of this nature, *i.e.* based on the content

---

Algorithm 1. Pattern Based Visual Search

Input   : A query image QI, a threshold T
Output: A candidate image CI with objects in bounding box

1 Bottom up;
2 $QS \leftarrow GBVS(Q)$ (Section 4.1);
3 $CS \leftarrow GBVS(C)$;
4 $Q_k \leftarrow$ Selected query object in QS (*i.e.* $Q_k \subseteq QS$);
5 $\mathcal{C} \leftarrow$ All visual objects in CS;
6 Top down; $G_q \leftarrow Generator(Q_k)$ (Def 11);
7 **for** $C \in \mathcal{C}$ **do**
8 | $G_c \leftarrow Generator(C)$;
9 | $\mathcal{P}_{\Phi}^q \leftarrow PatternGenerator(G_q)$ (Section 4.3);
10 | $\mathcal{P}_{\Phi}^c \leftarrow PatternGenerator(G_c)$;
11 | **for** $PQ \in \mathcal{P}_{\Phi}^q$ **do**
12 | | $max_s \leftarrow 0$;
13 | | **for** $PC \in \mathcal{P}_{\Phi}^c$ **do**
14 | | | $s = \dfrac{\left| PQ \underset{\Phi,\varepsilon}{\bigcap} PC \right|}{|PQ \bigcup PC|}$;
15 | | | **if** $s > max_s$ **then**
16 | | | | $max_s \leftarrow s$;
17 | | $S \leftarrow S + max_s$;
18 | $S \leftarrow \frac{S}{|\mathcal{P}_{\Phi}^q|}$;
19 | **if** $S > T$ **then**
20 | | $CI \leftarrow Boundingbox(C)$

---

contained in the images, is called content-based image retrieval (CBIR) [51]. Further, CBIR systems are typically evaluated using precision vs. recall plots [52], which is also the case here. Finally, two data sets were used to generate the reported results, namely the ImageNet ILSVRC2012 [53,54] and SIMPLIcity [55] datasets.

5.1   Experimental Setup

The ImageNet dataset experiment consisted of 10 categories from the ILSVRC20-12 training set, where each category contains 1300 images. Moreover, 10 images from each category were randomly selected as query images. These query images were compared to the remaining 12,900 images, where each comparison produced an S value from line 18 of Algorithm 1. The SIMPLIcity dataset experiment also consisted of 10 categories where each category contains 100 images. In this experiment, each image, in turn, is considered a query image and compared to all the other images in the dataset for a total of 500,500 comparisons. The SIMPLIcity dataset was also used since the resolution is lower than the ImageNet dataset, which allowed for the larger number of comparisons in a realistic timespan. Again, the images are ranked based on line 18 of Algorithm 1. These ranked S values are sorted in descending order, where the largest value represents the results of the first query, the second value the results of the second query, *etc.* Precision/recall plots are then created based on these values. In the ideal case, all images from the same category as the query are retrieved before any images from other categories. In this case, precision is 100% until recall reaches 100%.

5.2   ImageNet Results

Figures 3 and 4 contain the average precision vs. recall plots for each category of the ImageNet dataset. Notice, the results in Figs. 3(c), 4(a), and 4(e) indicate that categories *bison, school bus,* and *pepper* performed quite well, while the remaining categories did not. Specifically, most curves experience a step drop before recall reaches 20%. These plots could be interpreted as poor performance, however 20% recall corresponds to at least 260 images retrieved out of 12,900 images and is more than a typically user would be interested in an image retrieval system. As a result, the average precision of the top 20 queries for each category is presented in Table 1. Observe, the results are much better, and the top categories (*i.e. bison, school bus,* and *strawberry*) correspond to the best plots mentioned above.

Table 1. ImageNet precision values for top 20 retrieved images from each category averaged over 10 query images.

Category	Precision	Category	Precision
Frogs	0.655	Socks	0.625
Turtles	0.635	Teapot	0.655
Bison	0.985	Umbrella	0.465
Cellphones	0.610	Bell pepper	0.935
School bus	0.920	Strawberry	0.760
Average 0.7245			

5.3   SIMPLIcity Results

The SIMPLIcity dataset was used to further demonstrate the utility of the proposed approach since only 10 query images per category were used in the ImageNet experiment, whereas all images in the SIMPLIcity dataset were used as query images. Additionally, the SIMPLIcity dataset allowed for comparison with four other CBIR methods [56–59]. These methods are briefly summarized in [9], and they all use low features to represent global content. In contrast, the DVSS focuses on salient visual objects and is more localized. Moreover, the features are

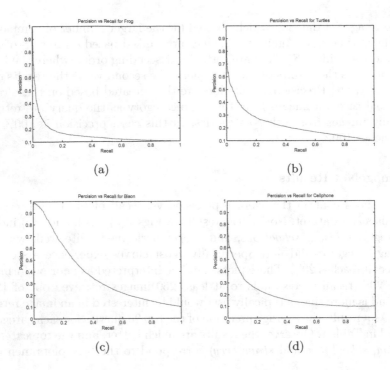

(a)   (b)

(c)   (d)

Fig. 3. ImageNet precision vs recall plots for Categories: (a) Frog, (b) turtle, (c) bison, and (d) cellphone.

extracted from higher network levels of the VGG ConvNet that typically produce features associated with visual objects in the images rather than low-level features such as texture or edges, and they are processed within the context of a descriptive topological space. These differences mean the DVSS makes judgements more in line with human perception and understanding of the content within the images. The results of SIMPLIcity experiment are given Table 2.

Notice the proposed approach performs well in the categories *Africans* and *buses*. Markedly, in these categories the DVSS is better than almost all of the other methods, and the DVSS produces the best results with the buses category. On the other hand, the other methods performed very well on the *dinosaurs* category, but the DVSS does not. Upon further investigation, the *Africans* and *buses* categories have analogous categories contained in the training set of the VGG ConvNet [47]. For instance, the ILSVRC-2012 dataset contains images in categories *people*, *school buses*, and *minibuses*. What is more, based on the first experiment, the *school bus* category had very good performance on both precision versus recall plots and top 20 retrieval. Therefore, the DVSS performs very well on the SIMPLIcity *buses* category. On the other hand, the VGG network was not trained with, for example, dinosaur images. Furthermore, the other methods perform quite well on this category because all these images are very similar to each other and are very different from the other categories, thus making for a

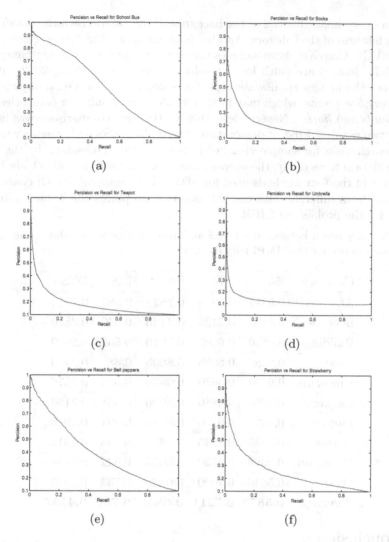

Fig. 4. ImageNet precision vs recall plots for Categories: (a) School bus, (b) socks, (c) teapot, (d) umbrella, (e) bell pepper, and (f) strawberry.

very clear separation in low-level feature space. Similarly, the other SIMPLIcity categories do not exist in the VGG neural network either [47].

There are other fundamental reasons explaining why the DVSS did not perform as well as the other methods on the SIMPLIcity dataset. For instance, the *beaches* category is characterized by images primarily consisting of background information representing the beaches, sea, and sky. However, the DVSS was designed to search for salient regions and visual objects in the FOV. Examples in the *beaches* category include people, or umbrellas. As a result, the DVSS uses these objects for quantifying the nearness or apartness of a candidate image with that of a query image instead of the background information which contains the defining features of the category. Another issue compounding the problem is that 19-layer VGG ConvNet down-samples the original input image many times, and SIMPLIcity images are much lower resolution than the ILSVRC-2012 dataset. Thus, once the images are down-sampled, the corresponding visual objects have only a very few points, which may further weaken the results for categories such as *elephants* and *horses*. Nevertheless, this SIMPLIcity comparisons was important for two reasons. First, it demonstrated that the DVSS performs very well on visual search tasks for sample visual objects that were represented by the VGG training data set. Secondly, the average precision values from Table 1 are better than three of the four methods used for SIMPLIcity comparison. Of course, the results are from different datasets, this observation places the Table 1 values in context for the problem of CBIR.

Table 2. Comparison between the DVSS and four existing approaches using the top 20 precision values for the SIMPLIcity dataset

Category	[56]	[57]	[58]	[59]	DVSS
Africans	0.5315	0.6975	0.7825	0.683	0.7115
Beaches	0.4385	0.5425	0.4425	0.540	0.3870
Buildings	0.4870	0.6395	0.5910	0.562	0.2950
Buses	0.8280	0.8965	0.8605	0.888	0.9420
Dinosaurs	0.9500	0.9870	0.9870	0.993	0.2570
Elephants	0.3485	0.4880	0.5900	0.658	0.3180
Flowers	0.8835	0.9230	0.8535	0.891	0.4915
Horses	0.5935	0.8945	0.7495	0.803	0.1310
Mountains	0.3080	0.4730	0.3655	0.522	0.3730
Food	0.5040	0.7090	0.6440	0.733	0.2305
Average	**0.5873**	**0.7211**	**0.6866**	**0.730**	**0.4137**

6   Conclusion

This article presented the DVSS that automates the task of visual search using descriptive topological spaces defined in a perceptual system based on probe functions from a convolutional neural network. The results indicate that the proposed solution works very well when employed on data that was also part of the ConvNet training set. As a result, the proposed approach has potential for widespread use and application as machine learning methods based on convolutional neural networks are becoming extremely popular and prevalent. Future work will consist of improvements in execution runtime as the current approach demonstrates inherent parallelism, but was implemented serially on a CPU rather than using highly parallel co-processors (such as GPUs). Additionally, more investigation will be performed on improving accuracy through using larger convolutional neural networks and larger training datasets.

References

1. Yu, Y., Mann, G.K.I., Gosine, R.G.: A goal-directed visual perception system using object-based top-down attention. IEEE Trans. Auton. Ment. Dev. **4**(1), 87–103 (2012)
2. Duncan, J., Humphreys, G.W.: Visual search and stimulus similarity. Psychol. Rev. **96**(3), 433 (1989)
3. Wolfe, J.M.: Guided search 2.0 a revised model of visual search. Psychon. Bull. Rev. **1**, 202–238 (1994)
4. Peters, J.F., Naimpally, S.A.: Applications of near sets. Not. Am. Math. Soc. **59**(4), 536–542 (2012)
5. Naimpally, S.A., Peters, J.F.: Topology with Applications Topological Spaces via Near and Far. World Scientific, Singapore (2013)
6. Peters, J.F.: Topology of Digital Images. Visual Pattern Discovery in Proximity Spaces. Intelligent Systems Reference Library, vol. 63. Springer, Berlin (2014). https://doi.org/10.1007/978-3-642-53845-2
7. Peters, J.: Computational Proximity: Excursions in the Topology of Digital Images. Intelligent Systems Reference Library. Springer, Berlin (2016). https://doi.org/10.1007/978-3-319-30262-1
8. Harel, J., Koch, C., Perona, P.: Graph-based visual saliency. In: Advances in Neural Information Processing Systems, pp. 545–552 (2006)
9. Yu, J.: A descriptive topological framework for performing visual search, Masters thesis, University of Winnipeg (2017)
10. Itti, L., Koch, C., Niebur, E.: A model of saliency-based visual attention for rapid scene analysis. IEEE Trans. Pattern Anal. Mach. Intell. **20**(11), 1254–1259 (1998)
11. Treisman, A.M., Gelade, G.: A feature-integration theory of attention. Cogn. Psychol. **12**(1), 97–136 (1980)
12. Dismone, R., Duncan, J.: Neural mechanisms of selective visual attention. Annu. Rev. Neurosci. **18**, 193–222 (1995)
13. Duncan, J., Humphreys, G., Ward, R.: Competitive brain activity in visual attention. Curr. Opin. Neurobiol. **7**(2), 255–261 (1997)
14. Fink, G.R., Dolan, R.J., Halligan, P.W., Marshall, J.C., Frith, C.D.: Space-base and object-based visual attention: shared and specific neural domains. Brain **120**(11), 2013–2028 (1997)
15. Peters, J.F.: Near sets. In: Henry, C.J. (ed.) Wikipedia, The Free Encyclopaedia (2015)
16. Sossinsky, A.B.: Tolerance space theory and some applications. Acta Applicandae Mathematicae: Int. Surv. J. Appl. Math. Math. Appl. **5**(2), 137–167 (1986)
17. Poincaré, H.: Science and Hypothesis. The Mead Project, Brock University (1905). L. G. Ward's translation
18. Benjamin Jr., L.T.: A Brief History of Modern Psychology. Blackwell Publishing, Malden (2007)
19. Hergenhahn, B.R.: An Introduction to the History of Psychology. Wadsworth Publishing, Belmont (2009)
20. Zeeman, E.C.: The topology of the brain and the visual perception. In: Fort, K.M. (ed.) Topoloy of 3-Manifolds and Selected Topics, pp. 240–256. Prentice Hall, New Jersey (1965)
21. Naimpally, S.A.: Near and far. A centennial tribute to Frigyes Riesz. Siberian Electron. Math. Rep. **6**, A.1–A.10 (2009)

22. Naimpally, S.A., Warrack, B.D.: Proximity spaces. In: Cambridge Tract in Mathematics No. 59. Cambridge University Press, Cambridge (1970)
23. Pawlak, Z., Peters, J.F.: Jak Blisko (how near). Systemy Wspomagania Decyzji **I**, 57–109 (2002)
24. Mozzochi, C.J., Naimpally, S.A.: Uniformity and proximity. In: Allahabad Mathematical Society Lecture Note Series, vol. 2, p. 153 pp. The Allahabad Mathematical Society, Allahabad (2009)
25. Naimpally, S.A.: Proximity Approach to Problems in Topology and Analysis. Oldenburg Verlag, München (2009). ISBN 978-3-486-58917-7
26. Hocking, J.G., Naimpally, S.A.: Nearness-a better approach to continuity and limits. In: Allahabad Mathematical Society Lecture Note Series, vol. 3, p. 153 pp. The Allahabad Mathematical Society (2009)
27. Peters, J.F.: Near sets. General theory about nearness of objects. Appl. Math. Sci. **1**(53), 2609–2629 (2007)
28. Peters, J.F.: Near sets. Special theory about nearness of objects. Fundamenta Informaticae **75**(1–4), 407–433 (2007)
29. Peters, J.F.: Tolerance near sets and image correspondence. Int. J. Bio-Inspired Comput. **1**(4), 239–245 (2009)
30. Peters, J.F.: Corrigenda and addenda: tolerance near sets and image correspondence. Int. J. Bio-Inspired Comput. **2**(5), 310–318 (2010)
31. İnan, E., Öztürk, M.A.: Near groups on nearness approximation spaces. Hacettepe J. Math. Stat. **41**(4), 545–558 (2012)
32. Peters, J.F., İnan, E., Öztürk, M.A.: Spatial and descriptive isometries in proximity spaces. Gen. Math. Notes **21**(2), 1–10 (2014)
33. Peters, J.F., Wasilewski, P.: Foundations of near sets. Inf. Sci. **179**(18), 3091–3109 (2009)
34. Peters, J.F.: Classification of perceptual objects by means of features. Int. J. Inf. Technol. Intell. Comput. **3**(2), 1–35 (2008)
35. Li, F., Karpathy, A.: CS231n: Convolutional Neural Networks for Visual Recognition, Course Lecture Notes, Standford University (2015)
36. Henry, C.J., Smith, G.: Proximity system: a description-based system for quantifying the nearness or apartness of visual rough sets. In: Peters, J.F., Skowron, A. (eds.) Transactions on Rough Sets XVII. LNCS, vol. 8375, pp. 48–73. Springer, Heidelberg (2014). https://doi.org/10.1007/978-3-642-54756-0_3
37. Henry, C.J.: Near sets: theory and applications, Ph.D. thesis, University of Manitoba (2010)
38. Henry, C.J.: Metric free nearness measure using description-based neighbourhoods. Math. Comput. Sci. **7**(1), 51–69 (2013)
39. Peters, J.F.: Computational proximity. In: Computational Proximity. ISRL, vol. 102, pp. 1–62. Springer, Cham (2016). https://doi.org/10.1007/978-3-319-30262-1_1
40. Čech, E.: Topological Spaces. Wiley, London (2014). fr seminar, Brno, 1936–1939; rev. ed. Z. Frolik, M. Katětov
41. Efremovič, V.A.: The geometry of proximity I (in Russian). Mat. Sb. (N.S.) **31**(73)(1), 189–200 (1952)
42. Lodato, M.: On topologically induced generalized proximity relations, Ph.D. dissertation, Rutgers University (1962). supervisor: S. Leader
43. Wallman, H.: Lattices and topological spaces. Ann. Math. **39**(1), 112–126 (1938)
44. Peters, J.F.: Local near sets. Pattern discovery in proximity spaces. Math. Comput. Sci. **7**(1), 87–106 (2013)

45. Peters, J.F., Wasilewski, P.: Tolerance spaces: origins, theoretical aspects and applications. Inf. Sci. **195**, 211–225 (2012)
46. Henry, C.J.: Perceptual indiscernibility, rough sets, descriptively near sets, and image analysis. In: Peters, J.F., Skowron, A. (eds.) Transactions on Rough Sets XV. LNCS, vol. 7255, pp. 41–121. Springer, Heidelberg (2012). https://doi.org/10.1007/978-3-642-31903-7_3
47. Simonyan, K., Zisserman, A.: Very deep convolutional networks for large-scale image recognition. In: ICLR, pp. 1–14 (2015). http://arxiv.org/abs/1409.1556
48. Karpathy, A.: CS231n: Convolutional Neural Networks for Visual Recognition. Stanford University, Stanford (2015)
49. Krizhevsky, A., Sutskever, I., Hinton, G.E.: ImageNet classification with deep convolutional neural networks. In: Advances in Neural Information Processing Systems, pp. 1–9 (2012)
50. Perone, C.: Deep learning - convolutional neural networks and feature extraction with python (2015). http://blog.christianperone.com/2015/08/convolutional-neural-networks-and-feature-extraction-with-python//
51. Smeulders, A.W.M., Worring, M., Santini, S., Gupta, A., Jain, R.: Content-based image retrieval at the end of the early years. IEEE Trans. Pattern Anal. Mach. Intell. **22**(12), 1349–1380 (2000)
52. Yates-Baeza, R., Ribeiro-Neto, B.: Modern Information Retrieval. ACM Press/Pearson Addison Wesley, New York (1999)
53. Deng, J.D.J., et al.: ImageNet: a large-scale hierarchical image database. In: 2009 IEEE Conference on Computer Vision and Pattern Recognition, pp. 2–9 (2009)
54. Russakovsky, O., et al.: ImageNet large scale visual recognition challenge. Int. J. Comput. Vision (IJCV) **115**(3), 211–252 (2015)
55. Wang, J.Z., Li, J., Wiederholdy, G.: SIMPLIcity: semantics-sensitive integrated matching for picture libraries. In: Laurini, R. (ed.) VISUAL 2000. LNCS, vol. 1929, pp. 360–371. Springer, Heidelberg (2000). https://doi.org/10.1007/3-540-40053-2_32
56. Jhanwar, N., Chaudhuri, S., Seetharaman, G., Zavidovique, B.: Content based image retrieval using motif cooccurrence matrix. Image Vis. Comput. **22**(14), 1211–1220 (2004)
57. Subrahmanyam, M., Jonathan Wu, Q.M., Maheshwari, R.P., Balasubramanian, R.: Modified color motif co-occurrence matrix for image indexing and retrieval. Comput. Electr. Eng. **39**(3), 762–774 (2013)
58. Vadivel, A., Sural, S., Majumdar, A.K.: An integrated color and intensity co-occurrence matrix. Pattern Recogn. Lett. **28**(8), 974–983 (2007)
59. Lin, C.-H., Chen, R.-T., Chan, Y.-K.: A smart content-based image retrieval system based on color and texture feature. Image Vis. Comput. **27**(6), 658–665 (2009). https://doi.org/10.1016/j.imavis.2008.07.004

Double Successive Rough Set Approximations

Alexa Gopaulsingh$^{(\boxtimes)}$

Central European University, Budapest, Hungary
alexa3e@hotmail.com

Abstract. We examine double successive approximations on a set, which we denote by L_2L_1, U_2U_1, U_2L_1, L_2U_1 where L_1, U_1 and L_2, U_2 are based on generally non-equivalent equivalence relations E_1 and E_2 respectively, on a finite non-empty set V. We consider the case of these operators being given fully defined on its powerset $\mathcal{P}(V)$. Then, we investigate if we can reconstruct the equivalence relations which they may be based on. Directly related to this, is the question of whether there are unique solutions for a given defined operator and the existence of conditions which may characterise this. We find and prove these characterising conditions that equivalence relation pairs should satisfy in order to generate unique such operators.

Keywords: Double approximations · Successive approximations ·
Double, successive rough set approximations

1 Successive Approximations

Double successive rough set approximations here, are considered using two, generally different equivalence relations. These are interesting because one can imagine a situation or model where sets/information to be approximated is input through two different approximations before returning the output. It is possible that for example heuristics in the brain can be modelled using such layered approximations. Decomposing successive approximations into constituent parts is somewhat analogous to decomposing a wave into sine and cosine parts using Fourier analysis.

In our case, we have two equivalence relations E_1 and E_2 on a set V with lower and upper approximations operators acting on its powerset $\mathcal{P}(V)$, denoted by L_1, U_1 and L_2, U_2 respectively. What if we knew the results of passing all the elements in $\mathcal{P}(V)$ through L_1 and then L_2, which we denote by L_2L_1. Could we then reconstruct E_1 and E_2 from this information? In this paper, we will investigate this question and consider the four cases of being given a defined L_2L_1, U_2U_1, U_2L_1, L_2U_1 operators. We will find that two equivalence relations do not always produce unique such operators but that some pairs do. We find and characterise conditions which the pairs of equivalence relations must satisfy for them to produce a unique operator. Cattaneo and Ciucci found that

© Springer-Verlag GmbH Germany, part of Springer Nature 2019
J. F. Peters and A. Skowron (Eds.): TRS XXI, LNCS 10810, pp. 68–95, 2019.
https://doi.org/10.1007/978-3-662-58768-3_3

preclusive relations are especially useful for rough approximations in information systems in [1]. For the L_2L_1 case we will show that these conditions form a preclusive relation between pairs of equivalence relations on a set and so we can define a related notion of independence from it. After this, we will find a more conceptual but equivalent version of the conditions of the uniqueness theorem. These conditions are more illuminating in that we can easier see why these conditions work while the conditions in the first version of the theorem are easier to use in practice. Lastly, we will consider the cases of the remaining operators, U_2U_1, U_2L_1 and L_2U_1. We note that the L_2L_1 and U_2U_1 cases are dual to each other and similarly for the U_2L_1 and L_2U_1 cases.

Rough set theory has quite a large number of practical applications. This is due in part to the computation of reducts and decision rules for databases. Predictions can be made after the data is mined to extract decision rules of manageable size (i.e. attribute reduction). In this way, rough set theory can be used to make decisions using data in the absence of major prior assumptions as argued in more detail in [2]. Hence in retrospect, it is perhaps not so surprising that this leads to tremendous applications. Therefore, rough set analysis adds to the tools of Bayes' Theorem and regression analysis for feature selection and pattern recognition in data mining [3–5]. The resulting applications include in artificial intelligence and machine learning [6–8], medical databases [9–11], and cognitive science [12–14]. Indeed in [15], Yao noted that there is currently an imbalance in the literature between the conceptual unfolding of rough set theory and its practical computational progress. He observed that the amount of computational literature currently far exceeds the amount of conceptual, theoretical literature. Moreover, he made the case that the field would prosper from a correction of this imbalance. To illustrate this, he began his recommendation in [15] by formulating a conceptual example of reducts that unifies three reduct definitions used in the literature which on the surface look different. We strongly agree that more efforts to find conceptual formulations of notions and results would increase the discovery of unifying notions. This would greatly aid the aim of making a cohesive and coherent map of the present mass of existing literature. In this direction, we have developed Subsects. 4.2 and 4.3 in Sect. 4.

2   Basic Concepts of Rough Set Theory

We go over some basic notions and definitions which can be found in [16]. Let V be a set and E be an equivalence relation on V. Also, let the set of equivalence classes of E be denoted by V/E. If a set $X \subseteq V$, is equal to a union of some of the equivalence classes of E then it is called E-exact. Otherwise, X is called E-inexact or E-rough or simply *rough* when the equivalence relation under consideration is clear from the context. Inexact sets may be approximated by two exact sets, the lower and upper approximations as is respectively defined below:

$$l_E(X) = \{x \in V \mid [x]_E \subseteq X\},$$

$$u_E(X) = \{x \in V \mid [x]_E \cap X \neq \emptyset\}. \tag{1}$$

Equivalently, we may use a granule based definition instead of a pointwise based definition:

$$l_E(X) = \bigcup \{Y \subseteq V/E \mid Y \subseteq X\},$$

$$u_E(X) = \bigcup \{Y \subseteq V/E \mid Y \cap X \neq \emptyset\}. \tag{2}$$

The pair (V, E) is called an *approximation space*. It may be the case that several equivalence relations are considered over a set. Let $\mathcal{E}$ being a family of equivalence relations over a finite non-empty set V. The pair, $K = (V, \mathcal{E})$ is called *knowledge base*. If $\mathcal{P} \subseteq \mathcal{E}$, we recall that $\bigcap \mathcal{P}$ is also an equivalence relation. The intersection of all equivalence relations belonging to $\mathcal{P}$ is denoted by $IND(\mathcal{P}) = \bigcap \mathcal{P}$. This is called the *indiscernibility relation* over $\mathcal{P}$.

For two equivalence relations E_1 and E_2 on a set X, we say that $E_1 \leq E_2$ iff $aE_1b \Rightarrow aE_2b$ for any $a, b \in X$. In this case we say that E_1 is *finer* than E_2 or that E_2 is *coarser* than E_1.

We draw your attention to the fact that there are two commonly used ways to represent equivalence relations. Set-theoretically an equivalence relation is identified as the set of ordered pairs of its relation. Another way to identify an equivalence relation is to represent it as the set of its equivalence classes. For example consider two equivalence relations, X and Y which set-theoretically are $E_1 = \{(a, a), (b, b), (c, c), (d, d), (a, b), (b, a), (a, c), (c, a), (b, c), (c, b)\}$ and $E_2 = \{(a, a), (b, b), (c, c), (d, d), (a, b), (b, a), (a, c), (c, a), (a, d), (d, a), (b, c), (c, b), (b, d), (d, b), (c, d), (d, c)\}$ respectively. As sets of ordered pairs they are:

$E_1^* = \{\{a, b, c\}, \{d\}\}$ and $E_2^* = \{\{a, b, c, d\}\}$ respectively. As we can see, the second representation is much shorter and helps us to picture the equivalence relations much quicker than the first representation. For these reasons, when we use specific examples of equivalence relations in this paper, we will use the second representation. For this representation, $E_1 \leq E_2$ iff for any $a, b \in X$, $aE_1b \Rightarrow \exists Z \in E_2 : \{a, b\} \subseteq Z$.

We recall from [17] some definitions about different types of roughly definable and undefinable sets. Let V be a set then for $X \subseteq V$:

(i) If $l_E(X) \neq \emptyset$ and $u_E(X) \neq V$, then X is called *roughly E-definable*.

(ii) If $l_E(X) = \emptyset$ and $u_E(X) \neq V$, then X is called *internally roughly E-undefinable*.

(iii) If $l_E(X) \neq \emptyset$ and $u_E(X) = V$, then X is called *externally roughly E-undefinable*.

(iv) If $l_E(X) = \emptyset$ and $u_E(X) = V$, then X is called *totally roughly E-undefinable*.

2.1   Properties Satisfied by Rough Sets

In [16], Pawlak enlists the following properties of lower and upper approximations. Let V be a non-empty finite set and $X, Y \subseteq V$. Then, the following holds:

(1) $l_E(X) \subseteq X \subseteq u_E(X),$
(2) $l_E(\emptyset) = u_E(\emptyset) = \emptyset; \quad l_E(V) = u_E(V) = V,$
(3) $u_E(X \cup Y) = u_E(X) \cup u_E(Y),$
(4) $l_E(X \cap Y) = l_E(X) \cap l_E(Y),$
(5) $X \subseteq Y \Rightarrow l_E(X) \subseteq l_E(Y),$
(6) $X \subseteq Y \Rightarrow u_E(X) \subseteq u_E(Y),$
(7) $l_E(X \cup Y) \supseteq l_E(X) \cup l_E(Y),$
(8) $u_E(X \cap Y) \supseteq u_E(X) \cap u_E(Y),$
(9) $l_E(-X) = -u_E(X),$
(10) $u_E(-X) = -l_E(X),$
(11) $l_E(l_E(X)) = u_E(l_E(X)) = l_E(X),$
(12) $u_E(u_E(X)) = l_E(u_E(X)) = u_E(X).$

2.2   Dependencies in Knowledge Bases

A database can also be represented in the form of a matrix of *Objects* versus *Attributes* with the entry corresponding to an object attribute pair being assigned the value of that attribute which the object satisfies. From the following definition, we can form equivalence relations on the objects for each given attribute. The set of these equivalence relations can then be used as our knowledge base.

Definition 1. *Let V be the set of objects and P be the set of attributes. Let $Q \subseteq P$, then V/Q is an equivalence relation on U induced by Q as follows: $x \sim_Q y$ iff $q(x) = q(y)$ for every $q \in Q$.*

To construct decision rules, we may fix two sets of attributes called *condition attributes* and *decision attributes* denoted by C and D respectively. We then use these to make predictions about the decision attributes based on the condition attributes. *Decision rules* are made by recording which values of decision attributes correlate with which values of condition attributes. As this information can be of considerable size, one of the primary goals of rough set theory is to reduce the number of condition attributes without losing predictive power. A minimal set of attributes which contains the same predictive power as the full set of decision attributes is called a *reduct* with respect to D.

   Next we give the definition of the positive region of one equivalence relation with respect to another.

Definition 2. *Let C and D be equivalence relations on a finite non-empty set V. The* positive region *of the partition D with respect to C is given by,*

$$POS_C(D) = \bigcup_{X \in D} l_C(X), \tag{3}$$

Definition 3. *It is said that* D *depends on* C *in a degree k, where $0 \leq k \leq 1$, denoted by $C \Rightarrow_k D$, if*

$$k = \gamma(C, D) = \frac{|POS_C(D)|}{|V|}. \tag{4}$$

If $k = 1$, then we say that C depends totally on D i.e. $C \Rightarrow D$.

Let $K_1 = (V, \mathcal{P})$ and $K_2 = (V, \mathcal{Q})$. We now give the definitions dependency of knowledge and then partial dependency. We say that $\mathcal{Q}$ *depends on* $\mathcal{P}$ i.e. $\mathcal{P} \Rightarrow \mathcal{Q}$ iff $IND(\mathcal{P}) \subseteq IND(\mathcal{Q})$.

Proposition 1. $I_{IND(\mathcal{P})} \leq I_{IND(\mathcal{Q})}$ *iff* $\mathcal{P} \Rightarrow \mathcal{Q}$.

Proposition 2. $POS_{IND(\mathcal{P})}IND((\mathcal{Q})) = U$ *iff* $\mathcal{P} \Rightarrow \mathcal{Q}$.

Otherwise, in the above case, $\gamma(IND(\mathcal{P}), IND(\mathcal{Q})) = k < 1$ and then we say that $\mathcal{P} \Rightarrow_k \mathcal{Q}$.

3   Properties of Successive Approximations

Next, we see that in general, approximating with respect to E_1 and then approximating the result with respect to E_2 gives a different result than if we had done it in the reverse order. That is, successive approximations do not commute. We consider some properties of successive approximations below.

Proposition 3. *Let V be a set and E_1 and E_2 be equivalence relations on V. Then for $Y \in \mathcal{P}(V)$, the following holds,*

1. $l_{E_1}(l_{E_2}(Y)) = Z \quad \not\Rightarrow \quad l_{E_2}(l_{E_1}(Y)) = Z$,
2. $u_{E_1}(u_{E_2}(Y)) = Z \quad \not\Rightarrow \quad u_{E_2}(u_{E_1}(Y)) = Z$,
3. $u_{E_1}(l_{E_2}(Y)) = Z \quad \not\Rightarrow \quad l_{E_2}(u_{E_1}(Y)) = Z$,

Proof. We give a counterexample to illustrate the proposition. Let $V = \{a, b, c, d\}$ and let $E_1 = \{\{a, b, c\}, \{d\}\}$ and $E_2 = \{\{a, b\}, \{c, d\}\}$.
To illustrate 1., let $Y = \{a, b, c\}$. Then $l_{E_1}(l_{E_2}(Y)) = \emptyset$ while $l_{E_2}(l_{E_1}(Y)) = \{a, b\}$.
For 2., let $Y = \{a\}$. Then $u_{E_1}(u_{E_2}(Y)) = \{a, b, c\}$ while $u_{E_2}(u_{E_1}(Y)) = \{a, b, c, d\}$.
For 3., let $Y = \{a, b\}$. Then $u_{E_1}(l_{E_2}(Y)) = \{a, b, c\}$ while $l_{E_2}(u_{E_1}(Y)) = \{a, b\}$.

From Properties (1), (5) and (6) of lower and upper approximations in Sect. 2.1, we immediately get that,
(i) $l_{E_1}(l_{E_2}(Y)) \subseteq l_{E_2}(Y)$, $u_{E_1}(u_{E_2}(Y)) \supseteq u_{E_2}(Y)$,
     $u_{E_1}(l_{E_2}(Y)) \supseteq l_{E_2}(Y)$ and $l_{E_1}(u_{E_2}(Y)) \subseteq u_{E_2}(Y)$.
If we do not know anything more about the relationship between E_1 and E_2 then nothing further may be deduced. However, if for example we know that $E_1 \leq E_2$ then the successive approximations are constrained as follows:

Proposition 4. *If $E_1 \leq E_2$ then the following properties hold;*

(ii) $l_{E_1}(l_{E_2}(Y)) = l_{E_2}(Y)$
(iii) $l_{E_2}(l_{E_1}(Y)) \subseteq l_{E_2}(Y)$
(iv) $u_{E_1}(u_{E_2}(Y)) \supseteq u_{E_1}(Y)$
(v) $u_{E_2}(u_{E_1}(Y)) = u_{E_2}(Y).$

Proof. Straightforward.

Proposition 5. *Let V be a finite non-empty set and let E_1 and E_2 be equivalence relations on V. Let $x \in V$. Then $l_{E_1}(u_{E_2}(\{x\})) \subseteq POS_{E_1}(E_2)$.*

Corollary 1. *Let V be a finite non-empty set and let E_1 and E_2 be equivalence relations on V. Let $X \subseteq V$. Then $POS_{E_1}(E_2) \cap X \subseteq \bigcup_{x \in X} l_{E_1}(u_{E_2}(\{x\}))$.*

Corollary 2. *Let V be a finite non-empty set and let E_1 and E_2 be equivalence relations on V. Then $POS_{E_1}(E_2) = \bigcup_{x \in V} l_{E_1}(u_{E_2}(\{x\}))$.*

Next, we give a graph which illustrates what happens when we take successive approximations on a set.

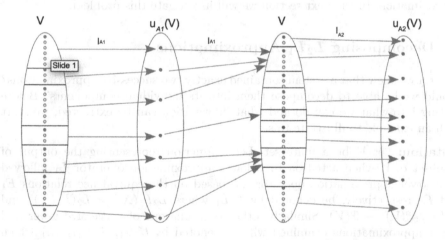

Fig. 1. Illustrates that successive approximations get more coarse when iterated.

Proposition 6. *Let G be a graph with vertex set V and E an equivalence relation on V. Let S_E be the set containing equivalence classes of E and taking the closure under union. Let $F : \mathcal{P}(V) \rightarrow S_E$ be such that $F(X) = \bigcup_{x \in X} [x]_E$ and let $Id : S_E \rightarrow \mathcal{P}(V)$ be such that $Id(Y) = Y$ Then F and Id form a Galois connection.*

Proof. It is clear from definitions that both F and Id are monotone. We need that for $X \in \mathcal{P}(V)$ and $Y \in S_E$, $F(X) \subseteq Y$ iff $X \subseteq Id(Y)$. This is also the case because from the definition of F, we have the $X \subseteq F(X)$.

Remark 3.1. Successive approximations break the Galois structure of single approximations. We can imagine that single approximations are a kind of sorting on the domain of a structure. We partition objects in the domain into boxes and in each box there is a special member (the lower or upper approximation) which identifies/represents any member in its respective box. We may say that objects are approximated by their representative.

For successive approximations, we have two different sortings of the same domain. Objects are sorted by the first approximation and only their representative members are then sorted by the second approximation. An object is then placed in the box that its representative member is assigned to in the second approximation, even though the object itself may be placed differently if the second approximation alone was used. Hence the errors 'add' in some sense. In Fig. 1, the final grouping as seen by following successive arrows, may be coarser than both the first and second approximations used singly. An interesting problem is how to correct/minimise these errors. It is also interesting how much of the individual approximations can be reconstructed from knowledge of the combined approximation. In the next section we will investigate this problem.

4   Decomposing $L_2 L_1$ Approximations

What if we knew that a system contained exactly two successive approximations? Would we be able to decompose them into its individual components? Before getting into what we can do and what information can be extracted, we start with an example to illustrate this.

Notation: Let V be a finite set. Let a function representing the output of a subset of V when acted on by a lower approximation operator L_1 followed by a lower approximation operator L_2, based on the equivalence relations E_1 and E_2 respectively, be denoted by $L_2 L_1$ where $L_2 L_1(X) = L_2(L_1(X))$ and $L_2 L_1 : \mathcal{P}(V) \to \mathcal{P}(V)$. Similarly, other combinations of successive lower and upper approximations examined will be denoted by $U_2 U_1$, $L_2 U_1, U_2 L_1$ which denotes successive upper approximations, an upper approximation followed by a lower approximation and a lower approximation followed by an upper approximation respectively.

Sometimes when we know that the approximations are based on equivalence relations P and Q we may use the subscripts to indicate this for example; $L_Q L_P$.

Lastly, if for a defined $L_2 L_1$ operator there exists a pair of equivalence relation solutions E_1 and E_2 which are such that the lower approximation operators L_1 and L_2 are based on them respectively, then we may denote this solution by the pair (E_1, E_2). Also, (E_1, E_2) can be said to produce or generate the operators based on them.

Example 4.1. Let $V = \{a, b, c, d, e\}$. Let a function representing the output of a subset of V when acted on by a lower approximation operator L_1 followed by a lower approximation operator L_2, which are induced by equivalence relations E_1 and E_2 respectively and let $L_2L_1 : \mathcal{P}(V) \to \mathcal{P}(V)$ be as follows:

$L_2L_1(\{\emptyset\}) = \emptyset$

$L_2L_1(\{a\}) = \emptyset$

$L_2L_1(\{b\}) = \emptyset$

$L_2L_1(\{c\}) = \emptyset$

$L_2L_1(\{d\}) = \emptyset$

$L_2L_1(\{e\}) = \emptyset$

$L_2L_1(\{a, b\}) = \emptyset$

$L_2L_1(\{a, c\}) = \emptyset$

$L_2L_1(\{a, d\}) = \emptyset$

$L_2L_1(\{a, e\}) = \emptyset$

$L_2L_1(\{b, c\}) = \emptyset$

$L_2L_1(\{b, d\}) = \emptyset$

$L_2L_1(\{b, e\}) = \emptyset$

$L_2L_1(\{c, d\}) = \emptyset$

$L_2L_1(\{c, e\}) = \emptyset$

$L_2L_1(\{d, e\}) = \{e\}$

$L_2L_1(\{a, b, c, d, e\}) = \{a, b, c, d, e\}$

$L_2L_1(\{b, c, d, e\}) = \{e\}$

$L_2L_1(\{a, c, d, e\}) = \{c, d, e\}$

$L_2L_1(\{a, b, d, e\}) = \{e\}$

$L_2L_1(\{a, b, c, e\}) = \{a, b\}$

$L_2L_1(\{a, b, c, d\}) = \{a, b\}$

$L_2L_1(\{c, d, e\}) = \{e\}$

$L_2L_1(\{b, d, e\}) = \{e\}$

$L_2L_1(\{b, c, e\}) = \emptyset$

$L_2L_1(\{b, c, d\}) = \emptyset$

$L_2L_1(\{b, c, d\}) = \emptyset$

$L_2L_1(\{a, d, e\}) = \{e\}$

$L_2L_1(\{a, c, d\}) = \emptyset$

$L_2L_1(\{a, b, e\}) = \emptyset$

$L_2L_1(\{a, b, d\}) = \emptyset$

$L_2L_1(\{a, b, c\}) = \{a, b\}$

We will now try to reconstruct E_1 and E_2. The minimal sets in the output are $\{e\}$ and $\{a, b\}$. Clearly, these are either equivalence classes of E_2 or a union of two or more equivalence classes of E_2. Since $\{e\}$ is a singleton it must be an equivalence class of E_2. So far we have partially reconstructed E_2 and it is equal to or finer than $\{\{a, b\}, \{c, d\}, \{e\}\}$.

Let us consider the pre-images of these sets in L_2L_1 to try to reconstruct E_1. Now, $L_2L_1^{-1}(\{e\}) = \{\{d, e\}, \{a, d, e\}, \{b, d, e\}, \{c, d, e\}, \{a, b, d, e\}, \{b, c, d, e\}\}$. We see that this set has a minimum with respect to containment and it is $\{d, e\}$. Hence either $\{d, e\}$ is an equivalence class of E_1 or both of $\{d\}$ and $\{e\}$ are equivalent classes of E_1.

Similarly, $L_2L_1^{-1}(\{a, b\}) = \{\{a, b, c\}, \{a, b, c, e\}, \{a, b, c, d\}\}$. We, see that this set has a minimum which is $\{a, b, c\}$ hence either this set is an equivalence class or is a union of equivalence classes in E_1. Now, $L_2L_1^{-1}(\{c, d, e\}) = \{\{a, c, d, e\}\}$. Hence, $\{a, c, d, e\}$ also consists of a union of equivalence classes of E_1. Since we know from above that $\{d, e\}$ consists of the union of one or more equivalence classes of E_1, this means that $\{a, c\}$ consists of the union of one or more equivalence classes of E_1 and $\{b\}$ is an equivalence class of E_1. So far we have that E_1 is equal to or finer than $\{\{a, c\}, \{b\}, \{d, e\}\}$.

Now we consider if $\{a, c\} \in E_1$ or both of $\{a\}$ and $\{c\}$ are in E_1. We can rule out the latter for suppose it was the case. Then $L_2L_1(\{a, b\})$ would be equal to $\{a, b\}$ since we already have that $\{b\} \in E_1$ and $\{a, b\}$ is the union of equivalence classes in E_2. Since this is not the case we get that $\{a, c\} \in E_1$. By a similar analysis of $L_2L_1(\{a,c,d\}) \neq \{c,d\}$ but only $\emptyset$ we get that $\{d,e\} \in E_1$. Hence, we have fully constructed E_1 and $E_1 = \{\{a,c\},\{b\},\{d,e\}\}$.

With E_1 constructed we can complete the construction of E_2. Recall, that we have that $\{a,b\}$ is a union of equivalence classes in E_2. Suppose that $\{a\} \in E_2$. Then $L_2L_1(\{a,c\})$ would be equal to $\{a\}$ since $\{a,c\} \in E_1$ but from the given list we see that it is not. Hence, $\{a,b\} \in E_2$. Similarly, we recall that $\{c,d\}$ is a union of equivalence classes in E_2. Suppose that $\{d\} \in E_2$. Then $L_2L_1(\{d,e\})$ would be equal to $\{d,e\}$ since $\{d,e\} \in E_1$ but it is only equal to $\{e\}$. Hence, $\{c,d\} \in E_2$. We have now fully reconstructed E_2 and $E_2 = \{\{a,b\},\{c,d\},\{e\}\}$.

The next example shows that we cannot always uniquely decompose successive approximations.

Example 4.2. Let $V = \{a,b,c,d\}$ and let $E_1 = \{\{a,b\},\{c,d\}\}$, $E_2 = \{\{a,c\},\{b,d\}\}$ and $E_3 = \{\{a,d\},\{b,c\}\}$. We see that $L_1L_2(X) = L_1L_3(X) = \emptyset$ for all $X \in (\mathcal{P}(V) - V)$ and $L_1L_2(X) = L_1L_3(X) = V$ when $X = V$. Then for all $X \subseteq U$, $L_1L_2(X) = L_1L_3(X)$ even though $E_2 \neq E_3$. Hence, if we are given a double, lower successive approximation on $\mathcal{P}(V)$ which outputs $\emptyset$ for all $X \in (\mathcal{P}(V) - V)$ and V for $X = V$ then we would be unable to say that it was uniquely produced by L_1L_2 or L_1L_3.

In the following we start to build a picture of what conditions are needed for the existence of unique solutions for double, successive approximations.

Proposition 7. *Let V be a set with equivalence relations E_1 and E_2 on V. If for each $[x]_{E_1} \in E_1$, $[x]_{E_1}$ is such that $L_2([x]_{E_1}) = \emptyset$ i.e. $[x]_{E_1}$ is either internally E_2–undefinable or totally E_2–undefinable, then the corresponding approximation operator, L_2L_1 on $\mathcal{P}(V)$ will be such that $L_2L_1([x]_{E_1}) = \emptyset$.*

Proof. Here, $L_1([x]) = \emptyset$. Hence $L_2L_1([x]) = L_2(\emptyset) = \emptyset$.

Remark 4.1. We note that the union of E-undefinable sets is not necessarily E-undefinable. Consider Example 4.2. Here, $\{a,b\}$ and $\{c,d\}$ are both totally E_2–undefinable but their union, $\{a,b,c,d\}$ is E_2–definable.

Algorithm 4.1: For Partial Decomposition of Double Successive Lower Approximations

Let V be a finite set. Given an input of a fully defined operator $L_2L_1 : \mathcal{P}(V) \to \mathcal{P}(V)$, if a solution exists, we can produce a solution (S,R), i.e. where L_1 and L_2 are the lower approximation operators of equivalence relations S and R respectively, by performing the following steps:

1. Let J be the set of output sets of the given L_2L_1 operator. We form the relation R to be such that for $a,b \in V$, $a \sim_R b \iff (a \in X \iff b \in X)$ for any $X \in J$. It is clear that R is an equivalence relation.
2. For each $Y \neq \emptyset$ output set, find the minimum pre-image set with respect to $\subseteq$, Y_m, such that $L_2L_1(Y_m) = Y$. Collect all these minimum sets in a set K. If there is any non-empty output set Y, such that the minimum Y_m does not exist, then there is no solution to the given operator and we return 0 signifying that no solution exists.

3. Using K, we form the relation S to be such that for $a, b \in V$, $a \sim_S b \iff$ $(a \in X \iff b \in X)$ for any $X \in K$. It is clear that S is an equivalence relation.

4. Form the operator $L_R L_S : \mathcal{P}(V) \to \mathcal{P}(V)$ generated by (S, R). If for all $X \in \mathcal{P}(V)$, the given $L_2 L_1$ operator is such that $L_2 L_1(X) = L_R L_S(X)$, then (S, R) is a solution proving that a solution exists (note that it is not necessarily unique). Return (S, R). Otherwise, discard S and R and return 0 signifying that no solution exists.

We will prove the claims in step 2 and step 4 in this section. Next, we prove step 2.

Proposition 8. *Let V be a set and $L_2 L_1 : \mathcal{P}(V) \to \mathcal{P}(V)$ be a given fully defined operator on $\mathcal{P}(V)$. If for $Y \neq \emptyset$ in the range of $L_2 L_1$, there does not exist a minimum set Y_m, with respect to $\subseteq$ such that $L_2 L_1(Y_m) = Y$, then there is no equivalence relation pair solution to the given operator.*

Proof. Suppose to get a contradiction that a solution (E_1, E_2) exists and there is no minimum set Y_m such that $L_2 L_1(Y_m) = Y$. Since V is finite, then there exists at least two minimal sets Y_k and Y_l say, such that $L_2 L_1(Y_s) = Y$ and $L_2 L_1(Y_t) = Y$. Since Y_s and Y_t are minimal sets with the same output after two successive lower approximations, then Y_s and Y_t must each be unions of equivalence classes in E_1 which contain Y. Since they are unequal, then without loss of generality, there exists $[a]_{E_1} \in E_1$ which is such that $[a]_{E_1} \in Y_s$ but $[a]_{E_1} \notin Y_t$. Since Y_s is minimal, then $[a]_{E_1} \cap Y \neq \emptyset$ (or else $L_2 L_1(Y_s) = L_2 L_1(Y_s - [a]_{E_1}) = Y$). So let $x \in [a]_{E_1} \cap Y$. Then $Y_t \not\supseteq x$ which contradicts $Y_t \supseteq Y$.

We now prove three lemmas on the way to proving the claim in step 4.

Lemma 1. *Let V be a set and $L_2 L_1 : \mathcal{P}(V) \to \mathcal{P}(V)$ be a given fully defined operator on $\mathcal{P}(V)$. Let R and S be equivalence relations defined on V as constructed in the previous algorithm. If (E_1, E_2) is a solution of $L_2 L_1$ then $E_2 \leq R$ and $E_1 \leq S$.*

Proof. We first prove $E_2 \leq R$. Now the output set of a non-empty set in $\mathcal{P}(V)$ is obtained by first applying the lower approximation L_1 to it and after applying the lower approximation, L_2 to it. Hence by definition of L_2, the non-empty output sets are unions of equivalence classes of the equivalence relation which corresponds to L_2. If a is in an output set but b is not then they cannot belong to the same equivalence class of E_2 i.e. $a \not\sim_R b$ implies that $a \not\sim_{E_2} b$. Hence $E_2 \leq R$.

Similarly, the minimal pre-image, X say, of a non-empty output set which is a union of equivalence classes in E_2, has to be a union of equivalence classes in E_1. For suppose it was not. Let $Y = \{y \in X \mid [y]_{E_1} \not\subseteq X\}$. By assumption, $Y \neq \emptyset$. Then $L_1(X) = L_1(X - Y)$. Hence $L_2 L_1(X) = L_2 L_1(X - Y)$ but $|X - Y| < |X|$ contradicting minimality of X. Therefore, if a belongs to the minimal pre-image of a non-empty output set but b does not belong to it, then a and b cannot belong to the same equivalence class in E_1 i.e. $a \not\sim_S b$ which implies that $a \not\sim_{E_1} b$. Hence $E_1 \leq S$.

Remark 4.2. The above lemma implies that for a given L_2L_1 operator on $\mathcal{P}(V)$ for a set V, that the pair of solutions given by the algorithm S and R for corresponding to L_1 and L_2 are the coarsest solutions for E_1 and E_2 which are compatible with the given, fully defined L_2L_1 operator. That is, for any other possible solutions, E_1 and E_2 to the given L_2L_1 operator, $E_1 \leq S$ and $E_2 \leq R$.

Lemma 2. *Let V be a finite set and $L_2L_1 : \mathcal{P}(V) \to \mathcal{P}(V)$ be a fully defined operator. If there exists equivalence pair solutions to the operator (E_1, E_2) which is such that there exists $[x]_{E_2}, [y]_{E_2} \in E_2$, such that $[x]_{E_2} \neq [y]_{E_2}$ and $\boldsymbol{u}_{E_1}([x]_{E_2}) = \boldsymbol{u}_{E_1}([y]_{E_2})$, then there exists another solution, (E_1, H_2), where H_2 is an equivalence relation formed from E_2 by combining $[x]_{E_2}$ and $[y]_{E_2}$ and all other elements are as in E_2. That is, $[x]_{E_2} \cup [y]_{E_2} = [z] \in H_2$ and if $[w] \in E_2$ such that $[w] \neq [x]_{E_2}$ and $[w]_{E_2} \neq [y]_{E_2}$, then $[w] \in H_2$.*

Proof. Suppose that (E_1, E_2) is a solution of a given L_2L_1 operator and H_2 is as defined above. Now, $L_2L_1(X) = Y$ iff the union of E_1-equivalence classes in X contains the union of E_2-equivalence classes which is equal to Y. So, in the (E_1, H_2) solution, the only way that $L_{H_2}L_{E_1}(X)$ could be different from $L_{E_2}L_{E_1}(X)$(which is $= L_2L_1(X)$) is if (i) $[x]_{E_2}$ is contained in $L_{E_2}L_{E_1}(X)$ while $[y]_{E_2}$ is not contained in $L_{E_2}L_{E_1}(X)$ or if (ii) $[y]_{E_2}$ is contained in $L_{E_2}L_{E_1}(X)$ while $[x]_{E_2}$ is not contained in $L_{E_2}L_{E_1}(X)$. This is because in H_2, $[x]_{E_2}$ and $[y]_{E_2}$ always occur together in an output set if they are in it at all (recall that output sets are unions of equivalence classes) in the equivalence class of $[z] = [x]_{E_2} \cup [y]_{E_2}$ and all the other equivalence classes of H_2 are the same as in E_2. However, neither (i) nor (ii) is the case since $\boldsymbol{u}_{E_1}([x]_{E_2}) = \boldsymbol{u}_{E_1}([y]_{E_2})$. That is, the equivalence classes of $[x]_{E_2}$ are contained by exactly the same union of equivalences in E_1 which contains $[y]_{E_2}$. Thus, any set X which contains a union of E_1-equivalences which contains $[x]_{E_2}$ also must contain $[y]_{E_2}$ and therefore $[z]_H$. Hence, if (E_1, E_2) is a solution for the given vector, then so is (E_1, H_2).

Lemma 3. *Let V be a finite set and $L_2L_1 : \mathcal{P}(V) \to \mathcal{P}(V)$ be a fully defined operator. If there exists equivalence pair solutions to the operator (E_1, E_2) which is such that there exists $[x]_{E_1}, [y]_{E_1} \in E_2$, such that $[x]_{E_1} \neq [y]_{E_1}$ and $\boldsymbol{u}_{E_2}([x]_{E_1}) = \boldsymbol{u}_{E_2}([y]_{E_1})$, then there exists another solution, (H_1, E_2), where H_1 is an equivalence relation formed from E_1 by combining $[x]_{E_2}$ and $[y]_{E_2}$ and all other elements are as in E_1. That is, $[x]_{E_1} \cup [y]_{E_1} = [z] \in H_1$ and if $[w] \in E_2$ such that $[w] \neq [x]_{E_1}$ and $[w]_{E_1} \neq [y]_{E_1}$, then $[w] \in H_1$.*

Proof. Suppose that (E_1, E_2) is a solution of a given L_2L_1 operator and H_1 is as defined above. Now, $L_2L_1(X) = Y$ iff the union of E_1-equivalence classes in X contains the union of E_2-equivalence classes which is equal to Y. So, in the (H_1, E_2) solution, the only way that $L_{E_2}L_{H_1}(X)$ could be different from $L_{E_2}L_{E_1}(X)$(which is $= L_2L_1(X)$) is if the union of equivalence classes in X which is needed to contain Y, (i) contains $[x]_{E_2}$ but not $[y]_{E_2}$ or (ii) contains $[y]_{E_2}$ but not $[z]_{E_2}$. However, this is not the case since $\boldsymbol{u}_{E_2}([x]_{E_1}) = \boldsymbol{u}_{E_2}([y]_{E_1})$. That is, $[x]_{E_1}$ intersects exactly the same equivalence classes in E_2 as $[y]_{E_1}$. So if $[x]_{E_1}$ is needed to contain an equivalence class in E_2, then $[y]_{E_1}$ is also needed. In other words, if $L_2L_1(X) = Y$, then for any minimal set such $Y_m \subseteq X$ such that $L_2L_1(Y_m) = Y$, $[x]_{E_1}$ is contained in Y_m iff $[y]_{E_1}$ is contained in Y_m iff $[z] \in H_1$ is contained in Y_m. Hence, if (E_1, E_2) is a solution for the given vector, then so is (H_1, E_2).

We now have enough to be able to prove the claim in step 4 of Algorithm 4.1 (actually we prove something stronger because we also show conditions which the solutions of the algorithm must satisfy).

Theorem 1. *Let V be a finite set and $L_2L_1 : \mathcal{P}(V) \to \mathcal{P}(V)$ be a fully defined operator. If there exists equivalence pair solutions to the operator, then there exists solutions (E_1, E_2) which satisfy,*

(i) for each $[x]_{E_2}, [y]_{E_2} \in E_2$, if $[x]_{E_2} \neq [y]_{E_2}$ then $\boldsymbol{u}_{E_1}([x]_{E_2}) \neq \boldsymbol{u}_{E_1}([y]_{E_2})$,
(ii) for each $[x]_{E_1}, [y]_{E_1} \in E_1$, if $[x]_{E_1} \neq [y]_{E_1}$ then $\boldsymbol{u}_{E_2}([x]_{E_1}) \neq \boldsymbol{u}_{E_2}([y]_{E_1})$.

Furthermore, $E_1 = S$ and $E_2 = R$ where (S, R) are the solutions obtained by applying Algorithm 4.1 to the given L_2L_1 operator.

Proof. Suppose that there exists a solution (C, D). Then, either (C, D) already satisfies condition (i) and condition (ii) or it does not. If it does, take $(E_1, E_2) = (C, D)$. If it does not satisfy condition (i) then use repeated applications of Lemma 2 until we arrive at an (C, E_2) solution which does. Similarly, if (C, E_2) does not satisfy condition (ii), use repeated applications of Lemma 3 until it does. Since $\mathcal{P}(V)$ is finite this will take at most finite applications of the lemmas until we obtain a solution, (E_1, E_2) which satisfies the conditions of the theorem. Since there is a solution, using Proposition 8 we will at least be able to reach step 4 of Algorithm 4.1. So let S and R be the relations formed by the algorithm after step 3. Next, we will show that $E_1 = S$ and $E_2 = R$. Now, by Lemma 1, we have that $E_1 \leq S$ and $E_2 \leq R$.

Consider the output sets of the given L_2L_1 operator. It is clear that these sets are unions of one or more equivalence classes of E_2. Let $[y]_{E_2} \in E_2$ then $L_2L_1(\boldsymbol{u}_{E_1}([y]_{E_2})) \supseteq [y]_{E_2}$.

Claim 1. $L_2L_1(\boldsymbol{u}_{E_1}([y]_{E_2}))$ *is the minimum output set of L_2L_1 such that it contains $[y]_{E_2}$ and $\boldsymbol{u}_{E_1}([y]_{E_2})$ is the minimum set X such that $L_2L_1(X) \supseteq [y]_{E_2}$.*

To see this we first note that L_2L_1 is a monotone function on $\mathcal{P}(V)$ since L_1 and L_2 are monotone operators and L_2L_1 is the composition of them. Then, if we can show that $\boldsymbol{u}_{E_1}([y]_{E_2})$ is the minimum set $X \in \mathcal{P}(V)$, such that $L_2L_1(X) \supseteq [y]_{E_2}$, then $L_2L_1(\boldsymbol{u}_{E_1}([y]_{E_2}))$ will be the minimum set output set which contains $[y]_{E_2}$. This is true because for $L_2L_1(X) \supseteq [y]_{E_2}$, then $L_1(X)$ must contain each member of $[y]_{E_2}$. We note that the range of L_1 contains only unions of equivalence classes of E_1 (counting the emptyset as a union of zero sets). Hence for $L_1(X)$ to contain each element of $[y]_{E_2}$, it must contain each equivalence class in E_1 which contains any of these elements. In other words, it must contain $\boldsymbol{u}_{E_1}([y]_{E_2})$. Suppose that X is such that $X \not\supseteq \boldsymbol{u}_{E_1}([y]_{E_2})$ and $L_2L_1(X) \supseteq [y]_{E_2}$. Then for some $v \in [y]_{E_2}$, v is not in X and so $\boldsymbol{u}_{E_1}([v]_{E_2}) \not\subseteq L_1(X)$. Hence $L_2L_1(X) \not\supseteq v$ and so does not contain $[y]_{E_2}$ which is a contradiction.

Claim 2. $L_2L_1(\boldsymbol{u}_{E_1}([y]_{E_2}))$ is not the minimum output set with respect to containing any other $[z]_{E_2} \neq [y]_{E_2}$.

Suppose that for some $[z]_{E_2} \neq [y]_{E_2} \in E_2$, that $L_2L_1(\boldsymbol{u}_{E_1}([y]_{E_2}))$ is the minimum output set containing $[z]_{E_2}$. Then by the previous Claim, we get that $L_2L_1(\boldsymbol{u}_{E_1}([y]_{E_2})) = L_2L_1(\boldsymbol{u}_{E_1}([z]_{E_2}))$ and that $\boldsymbol{u}_{E_1}([y]_{E_2}) \supseteq \boldsymbol{u}_{E_1}([z]_{E_2})$. But since $\boldsymbol{u}_{E_1}([y]_{E_2})$ is the minimum set such that $L_2L_1(X) \supseteq [y]_{E_2}$, then the stated equality also gives us that $\boldsymbol{u}_{E_1}([y]_{E_2}) \subseteq \boldsymbol{u}_{E_1}([z]_{E_2})$. Hence we have $\boldsymbol{u}_{E_1}([y]_{E_2}) = \boldsymbol{u}_{E_1}([z]_{E_2})$ which is a contradiction to the assumption of condition (i) of the theorem.

Now we can reconstruct E_2 by relating elements which always occur together in the output sets. That is, $a \sim_R b \iff (a \in X \iff b \in X)$ for each X in the range of L_2L_1. From the previous proposition we have that $E_2 \leq R$. We claim that $R \leq E_2$, hence $R = E_2$. To show this, suppose that it is not the case. Then there exists $a, b \in V$ such that $a \sim_R b$ but $a \nsim_{E_2} b$. By Claim 1, $L_2L_1(\boldsymbol{u}_{E_1}([a]_{E_2}))$ is the minimum set which contains $[a]_{E_2}$ and since $a \sim_R b$ then it must contain b, and consequently $[b]_{E_2}$ as well. Similarly by Claim 1, $L_2L_1(\boldsymbol{u}_{E_1}([b]_{E_2}))$ is the minimum set which contains $[b]_{E_2}$ and since $a \sim_R b$ then it must contain a, and consequently $[a]_{E_2}$ as well. By minimality we therefore have both $L_2L_1(\boldsymbol{u}_{E_1}([a]_{E_2})) \subseteq L_2L_1(\boldsymbol{u}_{E_1}([b]_{E_2}))$ and $L_2L_1(\boldsymbol{u}_{E_1}([a]_{E_2})) \supseteq L_2L_1(\boldsymbol{u}_{E_1}([b]_{E_2}))$ which implies that $L_2L_1(\boldsymbol{u}_{E_1}([a]_{E_2})) = L_2L_1(\boldsymbol{u}_{E_1}([b]_{E_2}))$. This contradicts Claim 2 since $[a]_{E_2} \neq [b]_{E_2} \in E_2$. Hence, $E = R$ and we can reconstruct E_2 by forming the equivalence relation R which was defined by using the output sets.

It remains to reconstruct E_1. Next, we list the pre-images of the minimal output sets which contain $[y]_{E_2}$ for each $[y]_{E_2}$ in E_2 and by Claim 1 this exists and is equal to $\boldsymbol{u}_{E_1}([y]_{E_2})$. This implies that each such set is the union of some of the equivalence classes of E_1. Now using this pre-image list we relate elements of V in the following way: $a \sim_S b \iff (a \in X \iff b \in X)$ for each X in the pre-image list. From the previous proposition we have that $E_1 \leq S$. We claim that $S \leq E_1$ and hence $S = E_1$. Suppose that it was not the case. That is, there exists $a, b \in V$ such that $a \sim_S b$ but $a \nsim_{E_1} b$. Hence $[a]_{E_1} \neq [b]_{E_1}$. By condition (ii) of the theorem, we know that $\boldsymbol{u}_{E_2}([a]_{E_1}) \neq \boldsymbol{u}_{E_2}([b]_{E_1})$. So without loss of generality, suppose that $d \in \boldsymbol{u}_{E_2}([a]_{E_1})$ but $d \notin \boldsymbol{u}_{E_2}([b]_{E_1})$. Since these sets are unions of equivalence classes in E_2 this implies that (1), $[d]_{E_2} \subseteq \boldsymbol{u}_{E_2}([a]_{E_1})$ and (2) $[d]_{E_2} \cap \boldsymbol{u}_{E_2}([b]_{E_1}) = \emptyset$. Now by Claim 1, $\boldsymbol{u}_{E_1}([d]_{E_2})$ is the minimum set, X such that $L_2L_1(X)$ contains $[d]_{E_2}$ and so is on the output list from which the Relation S was formed. However, (1) implies that this set contains a while (2) implies that this set does not contain b. This contradicts $a \sim_S b$. Hence $S = E_1$ and we can construct E_1 by constructing S. The result is shown.

Next we give, a graph-theoretic equivalence of the theorem but we first define a graph showing the relationship between two equivalence relations on a set.

Definition 4. *Let C and D be two equivalence relations on a set V. Form a bipartite graph $B(C, D) = (G, E)$, where the nodes G is such that $G = \{[u]_C \mid [u]_C \in C\} \cup \{[u]_D \mid [u]_D \in D\}$ and the edges E are such that*

$E = \{([u]_C, [v]_D) \mid \exists\, x \in V : x \in [u]_C \text{ and } x \in [v]_D\}$. *We call this the* **incidence graph** *of the pair* (C, D).

Theorem 2. *Let* V *be a finite set and let* $L_2 L_1 : \mathcal{P}(V) \to \mathcal{P}(V)$ *be a given fully defined operator on* $\mathcal{P}(V)$. *If there exists solutions* (E_1, E_2) *then the incidence graph of* E_1 *and* E_2, $B(E_1, E_2)$, *is such that there are no compete bipartite subgraphs as components other than edges.*

Proof. This is a direct translation of the previous theorem graph-theoretically. Suppose that the incidence graph of E_1 and E_2, $B(E_1, E_2)$, contains a complete bipartite subgraph as a component. Then the partition corresponding to E_2 violates Condition (i) of the theorem and the partition corresponding to E_1 violates condition (ii) of the theorem.

Corollary 3. *Let* V *be a finite set and* $L_2 L_1 : \mathcal{P}(V) \to \mathcal{P}(V)$ *be a given defined operator. If* (E_1, E_2) *is a unique solution for the operator then* $|E_1| < 2^{|E_2|}$ *and* $|E_2| < 2^{|E_1|}$.

Proof. This follows directly from the conditions since in the incidence graph of a unique solution (E_1, E_2), each equivalence class in E_1 is mapped to a unique non-empty subset of equivalence classes in E_2 and vice versa.

The next natural question is, without assuming conditions on the equivalence relations, are there instances when the algorithm produces a unique solution? Example 4.1 is an example of a unique decomposition of a given $L_2 L_1$ operator. So this leads naturally to the next question. What conditions result in a unique solution to a given $L_2 L_1$? Can we find characterising features of the pairs of equivalences relations which give a unique $L_2 L_1$ operator?

We note that the algorithm always produces a solution for a fully defined $L_2 L_1$ operator which has at least one solution. Hence, if there is a unique solution then these pairs of equivalence relations satisfy the conditions of Theorem 1. Recall that in Example 4.2, we were given an $L_2 L_1$ operator defined on $\mathcal{P}(V)$ for $V = \{a, b, c, d\}$ such that $L_2 L_1(X) = \emptyset$ for all $X \neq V$ and $L_2 L_1(V) = V$. This example shows us that in addition to a solution which would satisfy the conditions of the theorem, which applying the algorithm gives us; $E_1 = \{\{a, b, c, d\}\}$ and $E_2 = \{\{a, b, c, d\}\}$, we also have solutions of the form $E_1 = \{\{a, b\}, \{c, d\}\}$ and $E_2 = \{\{a, c\}, \{b, d\}\}$ or $E_1 = \{\{a, b\}, \{c, d\}\}$ and $E_2 = \{\{a, d\}, \{b, c\}\}$ amongst others. In Lemma 1, we showed that the solution given by the algorithm is the coarsest pair compatible with a given defined $L_2 L_1$ operator. We now try to find a condition such that after applying the algorithm, we may deduce whether or not the (S, R) solution is unique. This leads us to the next section.

4.1 Characterising Unique Solutions

Theorem 3. *Let* V *be a finite set and let* $L_2 L_1 : \mathcal{P}(V) \to \mathcal{P}(V)$ *be a fully defined operator on* $\mathcal{P}(V)$. *If* (S, R) *is returned by Algorithm 4.1, then* (S, R) *is the unique solution of the operator iff the following holds:*

(i) *For any $[x]_R \in R$, there exists $[z]_S \in S$ such that, $|[x]_R \cap [z]_S| = 1$.*
(ii) *For any $[x]_S \in S$, there exists $[z]_R \in R$ such that, $|[x]_S \cap [z]_R| = 1$.*

Proof. We prove $\Leftarrow$ direction first. So assume the conditions. We note that by Lemma 1, any other solutions, (E_1, E_2) to the given $L_2 L_1$ operator must be coarser than (S, R). Thus, if there is another solution to the given $L_2 L_1$ operator, (E_1, E_2) then at least one of $E_1 < S$, $E_2 < R$ must hold.

First we assume to get a contradiction that there exists a solution (E_1, E_2) which is such that $E_1 < S$. That is, E_1 contains a splitting of at least one of the equivalences classes of S, say $[a]_S$. Hence $|[a]_S| \geq 2$. By assumption there exists a $[z]_R \in R$ such that $|[a]_S \cap [z]_R| = 1$. Hence there is a $[z]_{E_2} \in E_2$ such that $|[a]_S \cap [z]_{E_2}| = 1$ since $E_2 \leq R$. Call the element in this intersection v say. We note that $[v]_{E_2} = [z]_{E_2}$. Now as $[a]_S$ is spilt into smaller classes in E_1, v must be in one of these classes, $[v]_{E_1}$. Consider the minimal pre-image of the minimal output set of $L_2 L_1$ which contains $[v]_R$. Call this set $Y_{(S,R)}$. For the solution (S, R), $Y_{(S,R)}$ contains all of $[a]_S$ since $v \in [a]_S$. But for the solution (E_1, E_2), the minimal pre-image of the minimal output set of $L_2 L_1$ which contains $[v]_R$, $Y_{(E_1, E_2)}$, is such that $Y_{(E_1, E_2)} = (Y_S - [a]_s) \cup [v]_{E_1} \neq Y_S$. Hence the output list for (E_1, S) is different from the given one which is a contradiction.

Next, suppose to get a contradiction there exists a solution (E_1, E_2) which is such that $E_2 < R$. That is, E_2 contains a splitting of at least one of the equivalences classes of R, say $[a]_R$. Hence $|[a]_R| \geq 2$. By assumption there exists a $[z]_S \in S$ such that $|[a]_R \cap [z]_S| = 1$. Hence there is a $[z]_{E_1} \in E_1$ such that $|[a]_R \cap [z]_{E_1}| = 1$ since $E_1 \leq S$. Call the element in this intersection v say. We note that $[v]_{E_1} = [z]_{E_1}$. Now as $[a]_R$ is spilt into smaller classes in E_2, v must be in one of these classes, $[v]_{E_2}$. Consider the set $[a]_R - [v]_{E_2}$. The minimal pre-image of the minimal output set which contains this set in the (S, R) solution, $Y_{(S,R)}$ contains $[v]_S$ since here the minimal output set which contains $([a]_R - [v]_{E_2})$, must contain all of $[a]_R$ which contains v. If (E_1, E_2) were the solution then the minimal pre-image of the minimal output set which contains $([a]_R - [v]_{E_2})$, $Y_{(E_1, E_2)}$, would not contain $[v_s]$ since $([a]_R - [v]_{E_2}) \cap [v]_S = \emptyset$. That is, $Y_{(E_1, E_2)} \neq Y_S$. Hence the output list for (E_1, E_2) is different from the given one which is a contradiction.

Now we prove $\Rightarrow$ direction. Suppose that (E_1, E_2) is the unique solution, and assume that the condition does not hold. By Theorem 1, $(E_1, E_2) = (S, R)$. Then either there exists an $[x]_R \in R$ such that for all $[y]_S \in S$ such that $[x]_R \cap [y]_S \neq \emptyset$ we have that $|[x]_R \cap [y]_S| \geq 2$ or there exists an $[x]_S \in S$ such that for all $[y]_R \in R$ such that $[x]_S \cap [y]_R \neq \emptyset$ we have that $|[x]_S \cap [y]_R| \geq 2$.

We consider the first case. Suppose that $[x]_R$ has non-empty intersection with n sets in S. We note that $n \geq 1$. Form a sequence of these sets; $S_1, ... S_n$. Since $|[x]_R \cap S_i| \geq 2$ for each i such that $i = 1, ... n$, let $\{a_{i1}, a_{i2}\}$ be in $[x]_R \cap S_i$ for each i such that $i = 1, ... n$. We split $[x]_R$ to form a finer E_2 as follows: Let $P = \{a_{i1} \mid i = 1, ... n\}$ and $Q = [x]_R - P$ be equivalence classes in E_2 and for the remaining equivalence classes in E_2, let $[y] \in E_2$ iff $[y] \in R$ and $[y]_R \neq [x]_R$. Now, $L_R L_S(X) = Y$ iff the union of S-equivalence classes in X contains the union of R-equivalence classes which is equal to Y. So, for the (S, E_2) solution, the only way that $L_{E_2} L_S(X)$ could be different from $L_R L_S(X)$ is if there is a union of

S-equivalence classes in X which contain P but not Q or which contain Q but not P (since P and Q always occur together as $[x]_R$ for the (S, R) solution). However, this is not the case as follows. Since P and Q exactly spilt all of the equivalence classes of S which have non-empty intersection with $[x]_R$, we have that $u_S(P) = u_S(Q)$. That is, P intersects exactly the same equivalence classes of S as Q. Therefore, P is contained by exactly the same union of equivalence classes in S as Q. Therefore, a union of S-equivalence classes in X contains P iff it contains Q iff its contains $[x]_R$. Hence, $L_R L_S(X) = L_{E_2} L_S(X)$ for all $X \in \mathcal{P}(V)$ and if (S, R) is a solution for the given vector, then so is (S, E_2) which is a contradiction of assumed uniqueness of (S, R).

We consider the second case. Suppose that $[x]_S$ has non-empty intersection with n sets in R. We note that $n \geq 1$. Form a sequence of these sets; $R_1, \ldots R_n$. Since $|[x]_S \cap R_i| \geq 2$ for each i such that $i = 1, \ldots n$, let $\{a_{i1}, a_{i2}\}$ be in $[x]_S \cap R_i$ for each i such that $i = 1, \ldots n$. We split $[x]_S$ to form a finer E_1 as follows: Let $P = \{a_{i1} \mid i = 1, \ldots n\}$ be one equivalence class and let $Q = [x]_R - P$ be another and for any $[y]_S \in S$ such that $[y]_S \neq [x]_S$, let $[y] \in E_1$ iff $[y] \in S$. Again, $L_R L_S(X) = Y$ iff the union of S-equivalence classes in X contains the union of R-equivalence classes which is equal to Y. So, for the (E_1, R) solution, the only way that $L_R L_{E_1}(X)$ could be different from $L_R L_S(X)$ is if (i) P is contained in $L_R L_S(X)$ while Q is not contained in $L_R L_S(X)$ or (ii) Q is contained in $L_R L_S(X)$ while P is not contained in $L_R L_S(X)$. Since P and Q spilt all of the equivalence classes of R which have non-empty intersection with $[x]_S$, this implies that $u_R(P) = u_R(Q)$. That is, P and Q intersect exactly the same equivalence classes of R. So if P is needed to contain an equivalence class in R for the (S, R) solution, then Q is also needed. In other words, if $L_2 L_1(X) = Y$, then for any minimal set such $Y_m \subseteq X$ such that $L_2 L_1(Y_m) = Y$, P is contained in Y_m iff Q is contained in Y_m iff $[x]_S$ is contained in Y_m. Hence, $L_R L_S(X) = L_R L_{E_1}(X)$ for all $X \in \mathcal{P}(V)$ and if (S, R) is a solution for the given vector, then so is (E_1, R) which is a contradiction of assumed uniqueness of (S, R).

The following theorem sums up the results of Theorems 1 and 3.

Theorem 4. *Let V be a finite set and let $L_2 L_1 : \mathcal{P}(V) \to \mathcal{P}(V)$ be a fully defined successive approximation operator on $\mathcal{P}(V)$. If (E_1, E_2) is a solution of the operator then it is the unique solution iff the following holds:*

(i) For each $[x]_{E_2}, [y]_{E_2} \in E_2$, if $[x]_{E_2} \neq [y]_{E_2}$ then $u_{E_1}([x]_{E_2}) \neq u_{E_1}([y]_{E_2})$,
(ii) For each $[x]_{E_1}, [y]_{E_1} \in E_1$, if $[x]_{E_1} \neq [y]_{E_1}$ then $u_{E_2}([x]_{E_1}) \neq u_{E_2}([y]_{E_1})$.
(iii) For any $[x]_{E_2} \in E_2$, there exists $[z]_{E_1} \in E_1$ such that, $|[x]_{E_2} \cap [z]_{E_1}| = 1$.
(iv) For any $[x]_{E_1} \in E_1$, there exists $[z]_{E_2} \in E_2$ such that, $|[x]_{E_1} \cap [z]_{E_2}| = 1$.

Remark 4.3. If an equivalence relation pair satisfies the conditions of Theorem 1, then the $L_2 L_1$ operator based on those relations would be such that if there exists other solutions then they would be finer pairs of equivalence relations. On the other hand, if an equivalence relation pair satisfies the conditions of Theorem 3, then the $L_2 L_1$ operator based on those relations would be such that if there exists other solutions then they would be coarser pairs of equivalence relations. Hence, if an equivalence relation pair satisfies the conditions of both Theorem 1 and Theorem 3, then the L_2L_1 operator produced by it is unique.

Corollary 4. *Let V be a finite set and let $L_2L_1 : \mathcal{P}(V) \to \mathcal{P}(V)$ be a fully defined successive approximation operator on $\mathcal{P}(V)$. If (S, R) is the solution returned by Algorithm 4.1, is such that it is the unique solution then following holds:*

For any $x \in V$ we have that;

(i) $[x]_S \not\supseteq [x]_R$ *unless* $|[x]_R| = 1$
(ii) $[x]_R \not\supseteq [x]_S$ *unless* $|[x]_S| = 1,$

Proof. This follows directly from the conditions in Theorem 3.

Example 4.1 (*revisited*): Consider again, the given output vector of Example 4.1. First we form the (S, R) pair using Algorithm 4.1. We get that $R = \{\{a, b\}, \{c, d\}, \{e\}\}$ and $S = \{\{a, c\}, \{b\}, \{d, e\}\}$. Since this is the pair produced from Algorithm 4.1, we know that it satisfies the conditions of Theorem 1. Now we need only to check if this pair satisfies the conditions of Theorem 3 to see if it is the only solution to do so. To keep track of which equivalence class a set belongs to, we will index a set belonging to either S or R by S or R respectively. Then we see that $|\{a, b\}_R \cap \{b\}_S| = 1$, $|\{c, d\}_R \cap \{a, c\}_S| = 1$ and $|\{e\}_R \cap \{d, e\}_S| = 1$. This verifies both conditions of Theorem 3 and therefore this is the unique solution of the given operator.

Proposition 9. *Let V be a finite set and $L_2L_1 : \mathcal{P}(V) \to \mathcal{P}(V)$ be a given defined operator. If (E_1, E_2) is a unique solution such that either $E_1 \neq Id$ or $E_2 \neq Id$ where Id is the identity equivalence relation on V then,*

(i) $E_1 \not\leq E_2,$
(ii) $E_2 \not\leq E_1.$

Proof. We first observe that if E_1 and E_2 are unique solutions and both of them are not Id then one of them cannot be equal Id. This is because if (E_1, Id) were solutions to a given L_2L_1 operator corresponding to L_1 and L_2 respectively then (Id, E_1) would also be solutions corresponding to L_1 and L_2 respectively and the solutions would not be unique. Hence, each of E_1 and E_2 contains at least one equivalence class of size greater than or equal to two.

Suppose that $E_1 \leq E_2$. Consider an $e \in E_2$ such that $|e| \geq 2$. Then e either contains a $f \in E_1$ such that $|f| \geq 2$ or two or more singletons in E_1. Then first violates the condition of Corollary 4 and the second violates the second condition of Theorem 1. Hence the solutions cannot be unique. Similarly, if we suppose that $E_2 \leq E_1$.

Corollary 5. *Let V be a finite set and $L_2L_1 : \mathcal{P}(V) \to \mathcal{P}(V)$ be a given defined operator. If there exists a unique solution (E_1, E_2) such that either $E_1 \neq Id$ or $E_2 \neq Id$ where Id is the identity equivalence relation on V then,*
(i) $k = \gamma(E_1, E_2) = \frac{|POS_{E_1}(E_2)|}{|V|} < 1$ *or* $E_1 \not\approx E_2$
(ii) $k = \gamma(E_2, E_1) = \frac{|POS_{E_2}(E_1)|}{|V|} < 1$ *or* $E_2 \not\approx E_1.$

Proof. This follows immediately from definitions.

Proposition 10. *Let V be a finite set and $L_2L_1 : \mathcal{P}(V) \to \mathcal{P}(V)$ be a given defined operator. If there exists a unique solution (E_1, E_2) then,*

(i) for any $[x]_{E_1} \in E_1$, $|POS_{E_2}([x]_{E_1})| \leq 1$
(ii) for any $[x]_{E_2} \in E_2$, $|POS_{E_1}([x]_{E_2})| \leq 1$.

Proof. This follows from the conditions in Theorem 4 and Corollary 4 which imply that for a unique pair solution (E_1, E_2), an equivalence class of one of the equivalence relations cannot contain any elements of size greater than one of the other relation and can contain at most one element of size exactly one of the other relation.

Corollary 6. *Let V be a finite set where $|V| = l$ and $L_2L_1 : \mathcal{P}(V) \to \mathcal{P}(V)$ be a given defined operator. If there exists a unique solution (E_1, E_2) such that $|E_1| = n$ and $|E_2| = m$ then,*

(i) $k = \gamma(E_1, E_2) = \frac{|POS_{E_1}(E_2)|}{|V|} \leq \frac{m}{l}$
(ii) $k = \gamma(E_2, E_1) = \frac{|POS_{E_2}(E_1)|}{|V|} \leq \frac{n}{l}$.

Proof. Let (E_1, E_2) be the unique solution of the given L_2L_1 operator. This result follows directly from the previous proposition by summing over all the elements in one member of this pair for taking its positive region with respect to the other member of the pair.

Corollary 7. *Let V be a finite set such that $|V| = n$ and $L_2L_1 : \mathcal{P}(V) \to \mathcal{P}(V)$ be a given defined operator. If there exists a unique solution (E_1, E_2) then,*

*(i) if the minimum size of an equivalence class in E_1, k_1 where $k_1 \geq 2$ then
$k = \gamma(E_1, E_2) = \frac{|POS_{E_1}(E_2)|}{|V|} = 0$.*
*(ii) if the minimum size of an equivalence class in E_2, k_2 where $k_2 \geq 2$ then
$k = \gamma(E_2, E_1) = \frac{|POS_{E_2}(E_1)|}{|V|} = 0$.*

Proof. Since no member of E_2 can contain any member of E_1 because E_1 has no singletons, we get that $\frac{|POS_{E_1}(E_2)|}{|V|} = 0$. Similarly for Part (ii).

Proposition 11. *Let V be a finite set and $L_2L_1 : \mathcal{P}(V) \to \mathcal{P}(V)$ be a given defined operator. If there exists a unique solution (E_1, E_2) such that $|E_1| = m$ and $|E_2| = n$ and S_1 is the number of singletons in E_1 and S_2 is the number of singletons in E_2, then,*

(i) $S_1 \leq n$
(ii) $S_2 \leq m$.

Proof. We note that the conditions in Theorem 4 imply that no two singletons in E_1 can be contained by any equivalence class in E_2 and vice versa. The result thus follows on application of the pigeonhole principle between the singletons in one equivalence relation and the number of elements in the other relation.

4.2   A Derived Preclusive Relation and a Notion of Independence

In [1], Cattaneo and Ciucci found that preclusive relations are quite useful for using rough approximations in information systems. In this direction, we will define a related notion of independence of equivalence relations from it.

Let V be a finite set and let $\mathfrak{C}_V$ be the set of all equivalence relations on V. Also, let $\mathfrak{C}_V^0 = \mathfrak{C}_V - Id_V$, where Id_V is the identity relation on V. From now on, where the context is clear, we will omit the subscript. We now define a relation on $\mathfrak{C}^o$, $\not\Rightarrow_{\mathfrak{C}^o}$, as follows:

Let E_1 and E_2 be in $\mathfrak{C}^o$. Let $L_2 L_1 : \mathcal{P}(V) \to \mathcal{P}(V)$ where L_1 and L_2 are lower approximation operators based on E_1 and E_2 respectively. Then,

$$E_1 \not\Rightarrow_{\mathfrak{C}^o} E_2 \text{ iff } L_2 L_1 \text{ is a unique approximation operator.}$$

That is, if for no other E_3 and E_4 in $\mathfrak{C}^o$ where at least one of $E_1 \neq E_3$ or $E_2 \neq E_4$ holds, is it the case that the operator $L_2 L_1 = L_3 L_4$, where L_3 and L_4 are lower approximation operators based on E_3 and E_4 respectively.

Definition 5. *Let V be a set and $E_1, E_2 \in \mathfrak{C}_V^0$. We say that E_1 is $\mathfrak{C}_V^0$-independent of E_2 iff $E_1 \not\Rightarrow_{\mathfrak{C}_V^0} E_2$. Also, if $\neg(E_1 \not\Rightarrow_{\mathfrak{C}_V^0} E_2)$, we simply write $E_1 \Rightarrow_{\mathfrak{C}_V^0} E_2$. Here, we say the E_1 is $\mathfrak{C}_V^0$-dependent of E_2 iff $E_1 \Rightarrow_{\mathfrak{C}_V^0} E_2$.*

Proposition 12. $\not\Rightarrow_{\mathfrak{C}_V^0}$ *is a preclusive relation.*

Proof. We recall that a preclusive relation is one which is irreflexive and symmetric. Let $E \in \mathfrak{C}^o{}_V$. Since $E \neq Id$, then by application of Proposition 4.2.3 (E, E) does not generate a unique $L_2 L_1$ operator and therefore $E \Rightarrow_{\mathfrak{C}_V^0} E$. Hence $\not\Rightarrow_{\mathfrak{C}_V^0}$ is irreflexive.

Now, suppose that $E_1, E_2 \in \mathfrak{C}^o{}_V$ are such that $E_1 \not\Rightarrow_{\mathfrak{C}_V^0} E_2$. Then (E_1, E_2) satisfies the conditions of Theorem 4. Since together, the four conditions of the theorem are symmetric (with conditions (i) and (ii) and conditions (iii) and (iv) being symmetric pairs), then (E_2, E_1) also satisfies the conditions of the theorem. Then by this theorem, we will have that $E_2 \not\Rightarrow_{\mathfrak{C}_V^0} E_1$. Hence, $\not\Rightarrow_{\mathfrak{C}_V^0}$ is symmetric.

Remark 4.4. From the previous proposition we can see that dependency relation $\Rightarrow_{\mathfrak{C}_V^0}$ is a similarity relation.

Proposition 13. *If $E_1 \Rightarrow E_2$ then $E_1 \Rightarrow_{\mathfrak{C}_V^0} E_2$.*

Proof. This follows from Corollary 9.

Proposition 14. *It is not the case that $E_1 \Rightarrow_{\mathfrak{C}_V^0} E_2$ implies that $E_1 \Rightarrow E_2$.*

Proof. In Example 4.2 we see (E_1, E_2) does not give a corresponding unique $L_2 L_1$ operator, hence $E_1 \Rightarrow_{\mathfrak{C}_V^0} E_2$ but $E_1 \not\Rightarrow E_2$.

Remark 4.5. From Propositions 13 and 14, we see that $\mathfrak{C}_V^0$-*dependency* is a more general notion of equivalence relation dependency that $\Rightarrow$ (or equivalently $\leq$). Similarly $\mathfrak{C}_V^0$-*independence* is a stricter notion of independence than $\not\Rightarrow$.

Theorem 5. *Let V be a finite set and E_1 and E_2 equivalence relations on V. Then*
$E_1 \not\approx_{\mathcal{E}_V^0} E_2$ *iff the following holds:*

(i) *For each $[x]_{E_2}, [y]_{E_2} \in E_2$, if $[x]_{E_2} \neq [y]_{E_2}$ then $\boldsymbol{u}_{E_1}([x]_{E_2}) \neq \boldsymbol{u}_{E_1}([y]_{E_2})$,*
(ii) *For each $[x]_{E_1}, [y]_{E_1} \in E_1$, if $[x]_{E_1} \neq [y]_{E_1}$ then $\boldsymbol{u}_{E_2}([x]_{E_1}) \neq \boldsymbol{u}_{E_2}([y]_{E_1})$.*
(iii) *For any $[x]_{E_2} \in E_2$, there exists $[z]_{E_1} \in E_1$ such that, $|[x]_{E_2} \cap [z]_{E_1}| = 1$.*
(iv) *For any $[x]_{E_1} \in E_1$, there exists $[z]_{E_2} \in E_2$ such that, $|[x]_{E_1} \cap [z]_{E_2}| = 1$.*

Proof. This follows directly from Theorem 4.

4.3 Seeing One Equivalence Relation Through Another

We will first give a proposition which will show a more explicit symmetry between conditions (i) and (ii) and conditions (iii) and (iv) in Theorem 4 for unique solutions.

Proposition 15. *Let V be a finite set and let E_1 and E_2 be two equivalence relations on V. Then;*
For any $[x]_{E_1} \in E_1$, $\exists [z]_{E_2} \in E_2$ such that, $|[x]_{E_1} \cap [z]_{E_2}| = 1$ iff it is not the case that $\exists Y, Z \in \mathcal{P}(V)$ such that $[x]_{E_1} = Y \cup Z$, $Y \cap Z = \emptyset$ and $\boldsymbol{u}_{E_2}(Y) = \boldsymbol{u}_{E_2}(Z) = \boldsymbol{u}_{E_2}([x]_{E_1})$.

Proof. We prove $\Rightarrow$ first. Let $[x]_{E_1} \in E_1$ and suppose that $\exists [z]_{E_2} \in E_2$ such that, $|[x]_{E_1} \cap [z]_{E_2}| = 1$. Then let $[x]_{E_1} \cap [z]_{E_2} = t$. Now for any spilt of $[x]_{E_1}$, that is for any $Y, Z \in \mathcal{P}(V)$ such that $[x]_{E_2} = Y \cup Z$ and $Y \cap Z = \emptyset$, t is in exactly one of these sets. Thus exactly one of $\boldsymbol{u}_{E_2}(Y)$, $\boldsymbol{u}_{E_2}(Z)$ contains $[t]_{E_2} = [z]_{E_2}$. Hence $\boldsymbol{u}_{E_2}(Y) \neq \boldsymbol{u}_{E_2}(Z)$.

We prove the converse by the contrapositive. Let $[x]_{E_1} \in E_1$ be such that for all $[z]_{E_2} \in E_2$ whenever $[x]_{E_1} \cap [z]_{E_2} \neq \emptyset$ (and clearly some such $[z]_{E_2}$ must exist), we have that $|[x]_{E_1} \cap [z]_{E_2}| \geq 2$. Suppose that $[x]_{E_1}$ has non-empty intersection with n sets in E_2. We note that $n \geq 1$. Form a sequence of these sets; $R_1, \ldots R_n$. Since $|[x]_{E_1} \cap R_i| \geq 2$ for each i such that $i = 1, \ldots n$, let $\{a_{i1}, a_{i2}\}$ be in $[x]_{E_1} \cap R_i$ for each i such that $i = 1, \ldots n$. Let $Y = \{a_{i1} \mid i = 1, \ldots n\}$ and let $Z = [x]_{E_1} - Y$. Then, $[x]_{E_1} = Y \cup Z$, $Y \cap Z = \emptyset$ and $\boldsymbol{u}_{E_2}(Y) = \boldsymbol{u}_{E_2}(Z) = \boldsymbol{u}_{E_2}([x]_{E_1})$.

Using the preceding proposition we obtain an equivalent form of Theorem 4.

Theorem 6. *Let V be a finite set and E_1 and E_2 equivalence relations on V. Then (E_1, E_2) produces a unique $L_2 L_1 : \mathcal{P}(V) \to \mathcal{P}(V)$ operator iff the following holds:*

(i) *For each $[x]_{E_2}, [y]_{E_2} \in E_2$, if $[x]_{E_2} \neq [y]_{E_2}$ then $\boldsymbol{u}_{E_1}([x]_{E_2}) \neq \boldsymbol{u}_{E_1}([y]_{E_2})$*
(ii) *For each $[x]_{E_1}, [y]_{E_1} \in E_1$, if $[x]_{E_1} \neq [y]_{E_1}$ then $\boldsymbol{u}_{E_2}([x]_{E_1}) \neq \boldsymbol{u}_{E_2}([y]_{E_1})$*
(iii) *For any $[x]_{E_2} \in E_2$, if $\exists Y, Z \in \mathcal{P}(V)$ such that $[x]_{E_2} = Y \cup Z$ and $Y \cap Z = \emptyset$ then $\boldsymbol{u}_{E_1}(Y) \neq \boldsymbol{u}_{E_1}(Z)$.*
(iv) *For any $[x]_{E_1} \in E_1$, if $\exists Y, Z \in \mathcal{P}(V)$ if $[x]_{E_1} = Y \cup Z$ and $Y \cap Z = \emptyset$ then $\boldsymbol{u}_{E_2}(Y) \neq \boldsymbol{u}_{E_2}(Z)$.*

Conceptual Translation of the Uniqueness Theorem. The conditions of the above theorem can be viewed conceptually as follows: (i) Through the eyes of E_1, no two equivalence classes of E_2 are the same; (ii) Through the eyes of E_2, no two equivalence classes of E_1 are the same; (iii) No equivalence class in E_2 can be broken down into two smaller equivalence classes which are equal to it through the eyes of E_1; (iv) No equivalence class in E_1 can be broken down into two smaller equivalence classes which are equal to it through the eyes of E_2. In other words we view set $V \bmod E_1$. That is, let $V \bmod E_1$ be the set obtained from V after renaming the elements of V with fixed representatives of their respective equivalence classes in E_1. Similarly let $V \bmod E_2$ be the set obtained from V after renaming the elements of V with fixed representatives of their respective equivalence classes in E_2. We then have the following equivalent conceptual version of Theorem 4.

Theorem 7. *Let V be a finite set and E_1 and E_2 equivalence relations on V. Then*

(E_1, E_2) *generate a unique L_2L_1 operator iff the following holds:*

(i) No two distinct members of E_2 are equivalent in V $\bmod E_1$.
(ii) No two distinct members of E_1 are equivalent in V $\bmod E_2$.
(iii) No member E_2 can be broken down into two smaller sets which are equivalent to it in V $\bmod E_1$.
(iv) No member E_1 can be broken down into two smaller sets which are equivalent to it in V $\bmod E_2$.

5   Decomposing U_2U_1 Approximations

We now investigate the case of double upper approximations. This is dually related to the case of double lower approximations because of the relationship between upper and lower approximations by the equation, $U(X) = -L(-X)$ (see property 10 in Sect. 2.1.1). The following proposition shows that the problem of finding solutions for this case reduces to the case in the previous section:

Proposition 16. *Let V be a finite set and let $U_2U_1 : \mathcal{P}(V) \to \mathcal{P}(V)$ be a given fully defined operator on $\mathcal{P}(V)$. Then any solution (E_1, E_2), is also a solution of $L_2L_1 : \mathcal{P}(V) \to P\mathcal{P}(V)$ operator where $L_2L_1(X) = -U_2U_1(-X)$ for any $X \in \mathcal{P}(V)$. Therefore, the solution (E_1, E_2) for the defined U_2U_1 operator is a unique iff the solution for the corresponding L_2L_1 operator is unique.*

Proof. Recall that $L_2L_1(X) = -U_2U_1(-X)$. Hence, if there exists a solution (E_1, E_2) which corresponds to the given U_2U_1 operator, this solution corresponds to a solution for the L_2L_1 operator which is based on the same (E_1, E_2) by the equation $L_2L_1(X) = -U_2U_1(-X)$. Similarly for the converse.

Algorithm: Let V be a finite set and let $U_2U_1 : \mathcal{P}(V) \to \mathcal{P}(V)$ be a given fully defined operator on $\mathcal{P}(V)$. To solve for a solution, change it to solving for a solution for the corresponding L_2L_1 operator by the equation $L_2L_1(X) =$

$-U_2U_1(-X)$. Then, when we want to know the L_2L_1 output of a set we look at the U_2U_1 output of its complement set and take the complement of that. Next, use Algorithm 4.2 and the solution found will also be a solution for the initial U_2U_1 operator.

5.1 Characterising Unique Solutions

Theorem 8. *Let V be a finite set and let $U_2U_1 : \mathcal{P}(V) \to \mathcal{P}(V)$ be a given fully defined operator on $\mathcal{P}(V)$. If (E_1, E_2) is a solution then, it is unique iff the following holds:*

(i) for each $[x]_{E_2}, [y]_{E_2} \in E_2$, if $[x]_{E_2} \neq [y]_{E_2}$ then $u_{E_1}([x]_{E_2}) \neq u_{E_1}([y]_{E_2})$,
(ii) for each $[x]_{E_1}, [y]_{E_1} \in E_1$, if $[x]_{E_1} \neq [y]_{E_1}$ then $u_{E_2}([x]_{E_1}) \neq u_{E_2}([y]_{E_1})$.
(iii) For any $[x]_{E_2} \in E_2$, there exists $[z]_{E_1} \in E_1$ such that, $|[x]_{E_2} \cap [z]_{E_1}| = 1$.
(iv) For any $[x]_{E_1} \in E_1$, there exists $[z]_{E_2} \in E_2$ such that, $|[x]_{E_1} \cap [z]_{E_2}| = 1$.

Proof. This follows from Proposition 16 using Theorem 4.

6 Decomposing U_2L_1 Approximations

For this case, we observe that $U_2L_1(X) = -L_2(-L_1(X)) = U_2(-U_1(-X))$. Since we cannot get rid of the minus sign between the Ls (or Us), duality will not save us the work of further proof here like it did in the previous section. In this section, we will see that U_2L_1 approximations are tighter than L_2L_1 (or U_2U_1) approximations. For this decomposition we will use an algorithm that is very similar to Algorithm 4.1, however notice the difference in step 2 where it only requires the use of minimal sets with respect to $\subseteq$ instead of minimum sets (which may not necessarily exist).

Algorithm 4.2: For Partial Decomposition of Double Successive Lower Approximations
Let V be a finite set. Given an input of a fully defined operator $U_2L_1 : \mathcal{P}(V) \to \mathcal{P}(V)$, if a solution exists, we can produce a solution (S, R), i.e. where L_1 and U_2 are the lower and upper approximation operators of equivalence relations S and R respectively, by performing the following steps:

1. Let J be the set of output sets of the given U_2L_1 operator. We form the relation R to be such that for $a, b \in V$, $a \sim_R b \iff (a \in X \iff b \in X)$ for any $X \in J$. It is clear that R is an equivalence relation.
2. For each $Y \neq \emptyset$ output set, find the minimal pre-image sets with respect to $\subseteq$, Y_m, such that $U_2L_1(Y_m) = Y$. Collect all these minimal sets in a set K. Note that we can always find these minimal sets since $\mathcal{P}(V)$ is finite.
3. Using K, we form the relation S to be such that for $a, b \in V$, $a \sim_S b \iff (a \in X \iff b \in X)$ for any $X \in K$. It is clear that S is an equivalence relation.

4. Form the operator $U_R L_S : \mathcal{P}(V) \to \mathcal{P}(V)$ generated by (S, R). If for all $X \in \mathcal{P}(V)$, the given $U_2 L_1$ operator is such that $U_2 L_1(X) = U_R L_S(X)$, then (S, R) is a solution proving that a solution exists (note that it is not necessarily unique). Return (S, R). Otherwise, discard S and R and return 0 signifying that no solution exists.

We will prove the claim in step 4 in this section.

Lemma 4. *Let V be a set and $U_2 L_1 : \mathcal{P}(V) \to \mathcal{P}(V)$ be a given fully defined operator on $\mathcal{P}(V)$ with L_1 and E_2 based on unknown E_1 and E_2 respectively. Let R and S be equivalence relations defined on V as constructed in Algorithm 4.3. Then $E_2 \leq R$ and $E_1 = S$.*

Proof. We first prove $E_2 \leq R$. Now the output set of a non-empty set in $\mathcal{P}(V)$ is obtained by first applying the lower approximation L_1 to it and and after applying the upper approximation, U_2 to it. Hence by definition of U_2, the non-empty output sets are unions of equivalence classes of the equivalence relation which corresponds to U_2. If a is in an output set but b is not in it then they cannot belong to the same equivalence class of E_2 i.e. $a \not\sim_R b$ implies that $a \not\sim_{E_2} b$. Hence $E_2 \leq R$.

Now, the minimal pre-image, X say, of a non-empty output set which is a union of equivalence classes in E_2, has to be a union of equivalence classes in E_1. For suppose it was not. Let $Y = \{y \in X \mid [y]_{E_1} \not\subseteq X\}$. By assumption, $Y \neq \emptyset$. Then $L_1(X) = L_1(X - Y)$. Hence $U_2 L_1(X) = U_2 L_1(X - Y)$ but $|X - Y| < |X|$ contradicting minimality of X. Therefore, if a belongs to the minimal pre-image of a non-empty output set but b does not belong to it, then a and b cannot belong to the same equivalence class in E_1 i.e. $a \not\sim_S b$ which implies that $a \not\sim_{E_1} b$. Hence $E_1 \leq S$.

We now prove the converse, that $S \leq E_1$. For suppose it was not. That is, $E_1 < S$. Then there exists at least one equivalence class in S which is split into smaller equivalence classes in E_1. Call this equivalence class $[a]_S$. Then there exists $w, t \in V$ such that $[w]_{E_1} \subset [a]_S$ and $[t]_{E_1} \subset [a]_S$. Now consider the pre-images of a minimal output sets of $U_2 L_1$, containing t. That is, X such that $U_2 L_1(X) = Y$ where Y is the minimal output set such that $t \in Y$ and for any $X_1 \subset X$, $U_2 L_1(X_1) \neq Y$. The following is a very useful observation.

Claim: For any $v \in \boldsymbol{u}_{E_1}([y]_{E_2})$, $[v]_S$ is a minimal set such that $U_2 L_1([v]_S) \supseteq [y]_{E_2}$. The above follows because (1) $U_2 L_1([v]_S) \supseteq [y]_{E_2}$ since $v \in \boldsymbol{u}_{E_1}([y]_{E_2})$ and (2) For any $Z \subset [v]_S$, $U_2 L_1(Z) = \emptyset$ since $L_1(Z) = \emptyset$.

Now for $U_2 L_1(X)$ to contain t, then it must contain $[t]_{E_2}$. Hence by the previous claim, $X = [t]_S$ is such a minimal pre-image of a set containing t. If L_1 is based on S, then $X = [t]_S = [a]_S$. However, if L_1 is based on E_1, then $X = [a]_S$ is not such a minimal set because $X = [t]_{E_1}$ is such that $U_2 L_1(X) = Y$ but $[t]_{E_1} \subset [a]_S$. Hence, $U_R L_S(X) \neq U_{E_2} L_{E_1}(X)$ for all $X \in \mathcal{P}(V)$ which is a contradiction to (E_1, E_2) also being a solution for the given $U_2 U_1$ operator. Thus we have that $E_1 = S$.

Lemma 5. *Let V be a finite set and $U_2L_1 : \mathcal{P}(V) \to \mathcal{P}(V)$ be a fully defined operator. If there exists equivalence pair solutions to the operator (E_1, E_2) which is such that there exists $[x]_{E_2}, [y]_{E_2} \in E_2$, such that $[x]_{E_2} \neq [y]_{E_2}$ and $u_{E_1}([x]_{E_2}) = u_{E_1}([y]_{E_2})$, then there exists another solution, (E_1, H_2), where H_2 is an equivalence relation formed from E_2 by combining $[x]_{E_2}$ and $[y]_{E_2}$ and all other elements are as in E_2. That is, $[x]_{E_2} \cup [y]_{E_2} = [z] \in H_2$ and if $[w] \in E_2$ such that $[w] \neq [x]_{E_2}$ and $[w]_{E_2} \neq [y]_{E_2}$, then $[w] \in H_2$.*

Proof. Suppose that (E_1, E_2) is a solution of a given U_2L_1 operator and H_2 is as defined above. Now, $U_2L_1(X) = Y$ iff the union of E_1-equivalence classes in X intersects the equivalence classes of E_2 whose union is equal to Y. So, in the (E_1, H_2) solution, the only way that $U_{H_2}L_{E_1}(X)$ could be different from $U_{E_2}L_{E_1}(X)$(which is $= U_2L_1(X)$) is if there some equivalence class of E_1 which either intersects $[x]_{E_2}$ but not $[y]_{E_2}$ or intersects $[y]_{E_2}$ but not $[x]_{E_2}$. However, this is not the case since we have that $u_{E_1}([x]_{E_2}) = u_{E_1}([y]_{E_2})$. Hence, $U_{E_2}L_{E_1}(X) = U_{H_2}L_{E_1}(X)$ for all $X \in \mathcal{P}(V)$ and therefore if (E_1, E_2) is a solution to the given operator then so is (E_1, H_2).

Next, we prove the claim in step 4 of Algorithm 4.2.

Theorem 9. *Let V be a finite set and $U_2L_1 : \mathcal{P}(V) \to \mathcal{P}(V)$ a fully defined operator. If there exists an equivalence relation pair solution, then there exists a solution (E_1, E_2), which satisfies,*

(i) for each $[x]_{E_2}, [y]_{E_2} \in E_2$, if $[x]_{E_2} \neq [y]_{E_2}$ then $u_{E_1}([x]_{E_2}) \neq u_{E_1}([y]_{E_2})$,
Furthermore $E_1 = S$ and $E_2 = R$, where (S, R) are the relations obtained by applying Algorithm 4.2 to the given U_2L_1 operator.

Proof. Suppose that there exists a solution (C, D). Then by Lemma 4, $C = S$, where S is produced by Algorithm 4.2. If (S, D) satisfies condition (i) of the theorem then take $(E_1, E_2) = (C, D)$. Otherwise, use repeated applications of Lemma 5 until we obtain a solution, (S, E_2) which satisfies the condition of the theorem. Since $\mathcal{P}(V)$ is finite this occurs after a finite number of applications of the lemma. Moreover, by Lemma 4, $E_2 \leq R$.

Consider the minimal sets in the output list of the given U_2L_1 operator. It is clear that these sets are union of one or more equivalence classes of E_2. Let $[y]_{E_2} \in E_2$ then for any $v \in u_{E_1}([y]_{E_2}))$, $U_2L_1([v]_S) \supseteq [y]_{E_2}$ (by the claim in Lemma 4).

Claim: (i) For any $[y]_{E_2} \neq [z]_{E_2} \in E_2$, there exists an output set, $U_2L_1(X)$ such that it contains at least of $[y]_{E_2}$ or $[z]_{E_2}$ both it does not contain both sets.

Suppose that $[y]_{E_2} \neq [z]_{E_2} \in E_2$. By the assumed condition of the theorem, then $u_{E_1}([y]_{E_2}) \neq u_{E_1}([z]_{E_2})$. Hence either (i) there exists $a \in V$ such that $a \in u_{E_1}([y]_{E_2})$ and $a \notin u_{E_1}([z]_{E_2})$ or (ii) there exists $a \in V$ such that $a \notin u_{E_1}([y]_{E_2})$ and $a \in u_{E_1}([z]_{E_2})$. Consider the first case. This implies that $[a]_S \cap [y]_{E_2} \neq \emptyset$ while $[a]_S \cap [z]_{E_2} = \emptyset$. Therefore, $U_2L_1([a]_S) \supseteq [y]_{E_2}$ but $U_2L_1([a]_S) \not\supseteq [z]_{E_2}$. Similarly, for the second case we will get that $U_2L_1([a]_S) \supseteq [z]_{E_2}$ but $U_2L_1([a]_S) \not\supseteq [y]_{E_2}$ and the claim is shown.

We recall that $a \sim_R b \iff (a \in X \iff b \in X)$ for each X in the range of the given U_2L_1. From the previous proposition we have that $E_2 \leq R$. From the above claim we see that if $[y]_{E_2} \neq [z]_{E_2}$ in E_2 then there is an output set that contains one of $[y]_{E_2}$ or $[z]_{E_2}$, but not the other. Hence, if $x \not\sim_{E_2} y$ then $x \not\sim_R y$. That is, $R \leq E_2$. Therefore we have that $R = E_2$.

6.1   Characterising Unique Solutions

Theorem 10. *Let V be a finite set and let $U_2L_1 : \mathcal{P}(V) \to \mathcal{P}(V)$ be a fully defined successive approximation operator on $\mathcal{P}(V)$. If (S, R) is returned by Algorithm 4.1, then (S, R) is the unique solution of the operator iff the following holds:*

(i) For any $[x]_R \in R$, there exists $[z]_S \in S$ such that, $|[x]_R \cap [z]_S| = 1$.

Proof. We prove $\Leftarrow$ direction first. So assume the condition holds. Then by Theorem 9 if there is a unique solution, it is (S, R) produced by Algorithm 4.2. We note that by Lemma 4, any other solution, (E_1, E_2) to the given U_2L_1 operator must be such that $E_1 = S$ and $E_2 \leq R$.

So, suppose to get a contradiction, that there exists a solution (E_1, E_2) which is such that $E_2 < R$. That is, E_2 contains a splitting of at least one of the equivalences classes of R, say $[a]_R$. Hence $|[a]_R| \geq 2$. By assumption there exists a $[z]_S \in S$ such that $|[a]_R \cap [z]_S| = 1$. Call the element in this intersection v say. We note that $[v]_S = [z]_S$. Now as $[a]_R$ is spilt into smaller classes in E_2, v must be in one of these classes, $[v]_{E_2}$. Now, $U_2L_1([v]_S)$ when U_2 is based on E_2, contains $[v]_{E_2}$ but does not contain $[a]_R$. This is because $[v]_S \cap ([a]_R - [v]_{E_2}) = \emptyset$. That is, $U_{E_2}L_S([v]_S) \not\supseteq [a]_R$ but $U_RL_S([v]_S) \supseteq [a]_R$. Hence $U_{E_2}L_S(X) \neq U_RL_S(X)$ for all $X \in \mathcal{P}(V)$. This is a contradiction to (S, E_2) also being a solution to the given U_2L_1 operator for which (S, R) is a solution. Hence we have a contradiction and so $E_2 = R$.

Now we prove $\Rightarrow$ direction. Suppose that (E_1, E_2) is the unique solution, and assume that the condition does not hold. By uniqueness, $(E_1, E_2) = (S, R)$. Then, there exists an $[x]_R \in R$ such that for all $[y]_S \in S$ such that $[x]_R \cap [y]_S \neq \emptyset$ we have that $|[x]_R \cap [y]_S| \geq 2$.

Suppose that $[x]_R$ has non-empty intersection with n sets in S. We note that $n \geq 1$. Form a sequence of these sets; $S_1, \ldots S_n$. Since $|[x]_R \cap S_i| \geq 2$ for each i such that $i = 1, \ldots n$, let $\{a_{i1}, a_{i2}\}$ be in $[x]_R \cap S_i$ for each i such that $i = 1, \ldots n$. We split $[x]_R$ to form a finer E_2 as follows: Let $P = \{a_{i1} \mid i = 1, \ldots n\}$ and $Q = [x]_R - P$ be two equivalence classes in E_2 and for the rest of E_2, for any $[y]_R \in R$ such that $[y]_R \neq [x]_R$, let $[y] \in E_2$ iff $[y] \in R$. Now, $U_RL_S(X) = Y$ iff the union of S-equivalence classes in X intersects equivalence classes of E_2 whose union is equal to Y. So, for the (S, E_2) solution, the only way that $L_{E_2}L_S(X)$ could be different from $L_RL_S(X)$ is if there is an equivalence class in S which intersects P but not Q or Q but not P. However, this is not the case because $\boldsymbol{u}_S(P) = \boldsymbol{u}_S(Q)$. Hence, $L_RL_S(X) = L_{E_2}L_S(X)$ for all $X \in \mathcal{P}(V)$ and if (S, R) is a solution for the given vector, then so is (S, E_2) which is a contradiction of assumed uniqueness of (S, R).

The following result sums up the effects of Theorems 9 and 10.

Theorem 11. *Let V be a finite set and let $U_2L_1 : \mathcal{P}(V) \to \mathcal{P}(V)$ be a given fully defined operator on $\mathcal{P}(V)$. Then there exists a unique pair of equivalence relations solution (E_1, E_2) iff the following holds:*

(i) for each $[x]_{E_2}, [y]_{E_2} \in E_2$, if $[x]_{E_2} \neq [y]_{E_2}$ then $\boldsymbol{u}_{E_1}([x]_{E_2}) \neq \boldsymbol{u}_{E_1}([y]_{E_2})$,
(iii) For any $[x]_{E_2} \in E_2$, there exists $[z]_{E_1} \in E_1$ such that, $|[x]_{E_2} \cap [z]_{E_1}| = 1$.

7 Decomposing L_2U_1 Approximations

For this case we observe that L_2U_1 is dual to the case previously investigated U_2L_1 operator. Due to the duality connection between L_2U_1 and U_2L_1, the question of unique solutions of the former reduces to the latter as the following proposition shows.

Proposition 17. *Let V be a finite set and let $L_2U_1 : \mathcal{P}(V) \to \mathcal{P}(V)$ be a given fully defined operator on $\mathcal{P}(V)$. Then any solution (E_1, E_2), is also a solution of $U_2L_1 : \mathcal{P}(V) \to P\mathcal{P}(V)$ operator where $U_2L_1(X) = -L_2U_1(-X)$ for any $X \in \mathcal{P}(V)$. Therefore, the solution (E_1, E_2) for the defined U_2U_1 operator is a unique iff the solution for the corresponding U_2L_1 operator is unique.*

Proof. Recall that $U_2L_1(X) = -L_2U_1(-X)$. Hence, if there exists a solution (E_1, E_2) which corresponds to the given U_2L_1 operator, this solution corresponds to a solution for the L_2U_1 operator which is based on the same (E_1, E_2) by the equation $L_2U_1(X) = -U_2L_1(-X)$. Similarly for the converse.

Algorithm: Let V be a finite set and let $L_2U_1 : \mathcal{P}(V) \to \mathcal{P}(V)$ be a given fully defined operator on $\mathcal{P}(V)$. To solve for a solution, change it to solving for a solution for the corresponding U_2L_1 operator by the equation $U_2L_1(X) = -L_2U_1(-X)$. Then, when we want to know the U_2L_1 output of a set we look at the L_2U_1 output of its complement set and take the complement of that. Next, use Algorithm 4.2 and the solution found will also be a solution for the initial L_2U_1 operator.

7.1 Characterising Unique Solutions

Theorem 12. *Let V be a finite set and let $L_2U_1 : \mathcal{P}(V) \to \mathcal{P}(V)$ be a given fully defined operator on $\mathcal{P}(V)$. If (E_1, E_2) is a solution, then it is unique iff the following holds:*

(i) for each $[x]_{E_2}, [y]_{E_2} \in E_2$, if $[x]_{E_2} \neq [y]_{E_2}$ then $\boldsymbol{u}_{E_1}([x]_{E_2}) \neq \boldsymbol{u}_{E_1}([y]_{E_2})$,
(ii) For any $[x]_{E_2} \in E_2$, there exists $[z]_{E_1} \in E_1$ such that, $|[x]_{E_2} \cap [z]_{E_1}| = 1$.

Proof. This follows from Proposition 17, Theorems 9 and 10.

8   Conclusion

We have defined and examined the consequences of double successive rough set approximations based on two, generally unequal equivalence relations on a finite set. We have given algorithms to decompose a given defined operator into constituent parts. Additionally, in Sects. 4.2 and 4.3 we have found a conceptual translation of the main results which is very much in the spirit of what Yao suggested in [15]. These types of links help to better connect and organise the notions in rough set theory. This aids us in finding discoveries of isomorphisms between seemingly unrelated areas/concepts and so helps us to build a clearer picture of the field.

This type of analysis can be seen as somewhat analogous to decomposing a wave into constituent sine and cosine waves using Fourier analysis. In our case, we work out the possibilities of what can be reconstructed if we know that a system has in-built layered approximations. It is possible for example that some heuristics of how the brain works can be modelled using such approximations and cognitive science is a possible application for the theory which we have begun to work out.

References

1. Cattaneo, G., Ciucci, D.: A quantitative analysis of preclusivity vs. similarity based rough approximations. In: Alpigini, J.J., Peters, J.F., Skowron, A., Zhong, N. (eds.) RSCTC 2002. LNCS (LNAI), vol. 2475, pp. 69–76. Springer, Heidelberg (2002). https://doi.org/10.1007/3-540-45813-1_9
2. Pawlak, Z.: Rough sets, decision algorithms and Bayes' theorem. Eur. J. Oper. Res. **136**(1), 181–189 (2002)
3. Zhong, N., Dong, J., Ohsuga, S.: Using rough sets with heuristics for feature selection. J. Intell. Inf. Syst. **16**, 199–214 (2001)
4. Hu, K., Lu, Y., Shi, C.: Feature ranking in rough sets. AI Commun. **16**(1), 41–50 (2003)
5. Zhou, L., Jiang, F.: A rough set approach to feature selection based on relative decision entropy. LNCS Rough Sets Knowl. Technol. **6954**, 110–119 (2011)
6. Mahajan, P., Kandwal, R., Vijay, R.: Rough set approach in machine learning: a review. Int. J. Comput. Appl. **56**(10), 1–13 (2012)
7. Hu, X., Cercone, N.: Learning in relational databases: a rough set approach. Comput. Intell. **11**(2), 323–338 (1995)
8. Grzymala-Busse, J., Sedelow, S., Sedelow Jr., W.: Machine learning and knowledge acquisition, rough sets, and the English semantic code. In: Lin, T.Y., Cercone, N. (eds.) Rough Sets and Data Mining, pp. 91–107. Springer, Heidelberg (1997). https://doi.org/10.1007/978-1-4613-1461-5_5
9. Tsumoto, S.: Rough sets and medical differential diagnosis. Intell. Syst. Ref. Libr.: Rough Sets Intell. Syst.-Profr. Zdzislaw Pawlak Mem. **42**, 605–621 (2013)
10. Tsumoto, S.: Mining diagnostic rules from clinical databases using rough sets and medical diagnostic model. Inf. Sci. **162**(2), 65–80 (2004)
11. Tripathy, B., Acharjya, D., Cynthya, V.: A framework for intelligent medical diagnosis using rough set with formal concept analysis. Int. J. Artif. Intell. Appl. **2**(2), 45–66 (2011)

12. Napoles, G., Grau, I., Vanhoof, K., Bello, R.: Hybrid model based on rough sets theory and fuzzy cognitive maps for decision-making. LNCS Rough Sets Intell. Syst. Parad. **8537**, 169–178 (2014)
13. Przybyszewski, A.: The neurophysiological bases of cognitive computation using rough set theory. LNCS Trans. Rough Sets IX **5390**, 287–317 (2008)
14. Pagliani, P., Chakraborty, M.K.: A Geometry of Approximation Rough Set Theory: Logic, Algebra and Topology of Conceptual Patterns. Trends in Logic, vol. 27. Springer, Heidelberg (2008). https://doi.org/10.1007/978-1-4020-8622-9
15. Yao, Y.Y.: The two sides of the theory of rough sets. Knowl.-Based Syst. **80**, 67–77 (2015)
16. Pawlak, Z.: Rough sets. Int. J. Comput. Inf. Sci. **11**, 341–356 (1982)
17. Pawlak, Z.: Rough Sets: Theoretical Aspects of Reasoning about Data. Theory and Decision Library D, vol. 9, 1st edn. Springer, Heidelberg (1991)

Dialectical Rough Sets, Parthood and Figures of Opposition-I

A. Mani$^{(\boxtimes)}$

Department of Pure Mathematics, University of Calcutta,
International Rough Set Society, 9/1B, Jatin Bagchi Road,
Kolkata (Calcutta) 700029, India
a.mani.cms@gmail.com
http://www.logicamani.in

Abstract. In one perspective, the main theme of this research revolves around the inverse problem in the context of general rough sets that concerns the existence of rough basis for given approximations in a context. Granular operator spaces and variants were recently introduced by the present author as an optimal framework for anti-chain based algebraic semantics of general rough sets and the inverse problem. In the framework, various sub-types of crisp and non-crisp objects are identifiable that may be missed in more restrictive formalism. This is also because in the latter cases concepts of complementation and negation are taken for granted - while in reality they have a complicated dialectical basis. This motivates a general approach to dialectical rough sets building on previous work of the present author and figures of opposition. In this paper dialectical rough logics are invented from a semantic perspective, a concept of dialectical predicates is formalized, connection with dialetheias and glutty negation are established, parthood analyzed and studied from the viewpoint of classical and dialectical figures of opposition by the present author. The proposed method become more geometrical and encompass parthood as a primary relation (as opposed to roughly equivalent objects) for algebraic semantics.

Keywords: Rough objects · Dialectical rough semantics ·
Granular operator spaces · Rough mereology · Polytopes of dialectics ·
Antichains · Dialectical rough counting ·
Axiomatic approach to granules · Constructive algebraic semantics ·
Figures of opposition · Unified semantics

1 Introduction

It is well known that sets of rough objects (in various senses) are quasi or partially orderable. Specifically in classical or Pawlak rough sets [1], the set of roughly equivalent sets has a quasi Boolean order on it while the set of rough and crisp objects is Boolean ordered. In the classical semantic domain or classical meta level, associated with general rough sets, the set of crisp and rough objects is

© Springer-Verlag GmbH Germany, part of Springer Nature 2019
J. F. Peters and A. Skowron (Eds.): TRS XXI, LNCS 10810, pp. 96–141, 2019.
https://doi.org/10.1007/978-3-662-58768-3_4

quasi or partially orderable. Under minimal assumptions on the nature of these objects, many orders with rough ontology can be associated - these necessarily have to do with concepts of discernibility. Concepts of rough objects, in these contexts, depend additionally on approximation operators and granulations used. Many generalizations of classical rough sets from granular perspectives have been studied in the literature. For an overview the reader is referred to [2]. These were part of the motivations for the invention of the concept of granular operator spaces by the present author in [3] and developed further in [4-6]. In the paper, it is shown that collections of mutually distinct objects (relative to some formulas) form antichains. Models of these objects, proposed in the paper, have been associated with deduction systems in [4]. Collections of antichains can be partially ordered in many ways. Relative to each of these partial orders, maximal antichains can be defined. Maximal antichains are sets of mutually distinct objects and any of its supersets is not an antichain. In one sense, this is a way of handling roughness. To connect these with other rough objects, various kinds of negation-like operations (or generalizations thereof) are of interest. Such operations and predicates are of interest when rough parthoods are not partial or quasi orders [5,7,8].

Given some information about approximations in a context, the problem of providing a rough semantics is referred to as an *inverse problem*. This class of problems was introduced by the present author in [9] and has been explored in [7,10] in particular. The solution for classical rough sets may be found in [11]. For more practical contexts, abstract frameworks like *rough Y-systems* RYS [7] are better suited for problem formulation. However, good semantics may not always be available for RYS. It is easier to construct algebraic semantics over granular operator spaces and generalizations thereof [4,5] as can be seen in the algebraic semantics of [4] that involve distributive lattices without universal negations. Therefore, these frameworks are better suited for solving inverse problems and the focus of this paper will be limited to these.

The square of opposition and variants, in modern interpretation, refer to the relation between quantified or modal sentences in different contexts [12, 13]. These have been considered in the context of rough sets in [14] from a set theoretical view of approximations of subsets (in some rough set theories). The relation of parthood in the context of general rough sets with figures has not been investigated in the literature. This is taken up in the present research paper (but from a semantic perspective). Connections with dialectical predication and other kinds of opposition are also taken up in the light of recent developments on connections of para-consistency and figures of opposition.

At another level of conception, in the classical semantic domain, various subtypes of objects relate to each other in specific ways through ideas of approximations (the origin of these approximations may not be known clearly). These ways are shown to interact in dialectical ways to form other semantics under some assumptions. The basic structural schema can be viewed as a generalization of ideas like the square and hexagon of opposition - in fact as a combination of dialectical oppositions involved. This aspect is explained and developed in detail in the present paper. These are also used for introducing new methods of counting and semantics in a forthcoming paper by the present author. In [15], a dialectical rough set theory was developed by the present author using a specific concept of dialectical contradiction that refers to both rough and classical objects. Related parthood relations are also explored in detail with a view to construct possible models (semantics).

All these are part of a unified whole - the inverse problem and possible solution strategies in all its generality. Granular operator spaces and variants [3,5] are used as a framework for the problem, and as these have been restricted by the conditions imposed on the approximations, it makes sense to speak of a *part of a unified whole. Importantly no easy negation like operators are definable/defined in the framework and this is also a reason for exploring/identifying dialectical negations.* So at one level the entire paper is a contribution to possible solutions of the inverse problem and usable frameworks for the same. This also involves the invention of a universal dialectical negation (or opposition) from a formal perspective on the basis of diverse philosophical and practical considerations.

Semantic domains (or domains of discourse) associated with rough objects, operations and predications have been identified by the present author in [2,4,7, 16,17]. The problem of defining rough objects that permit reasoning about both intensional and extensional aspects posed in [18] corresponds to identification of suitable semantic domains. This problem is also addressed in this research (see Sect. 4).

The questions and problems that are taken up in this research paper and solved to varying extents also include the following:

1. What may be a proper formalization of a dialectical logic and dialectical opposition?
   - What is the relation between dialetheias, truth gluts and dialectical contradiction?
   - Should the objects formalized by the logic be interpreted as state transitions?
   - How do dialetheias and dialectical contradiction differ?
2. Do paraconsistent logics constitute a proper formalization of the philosophical intent in Hegelian and Marxist dialectics?
3. What is the connection between parthood in rough contexts and possible dialectical contradictions?
4. What is a rough dialectical semantics and is every parthood based rough semantics a dialectical one?
5. How does parthood relate to figures of opposition?
6. What is a useful representation of rough objects that addresses the concerns of [18]?

This paper is structured as follows. The next subsection includes background material on rough concepts, posets, granules and parthood. In the next section, the superiority of granular operator spaces over property systems is explained. Dialectical negation and logics are characterized from a critical perspective in the third section. In the following section, dialectical rough logics are developed and related parthoods are explored. Many examples are provided in the context of the semantic framework used in the fifth section. The sixth section is about figures of dialectical opposition generated by few specific parthood related statements in rough sets and a proposal for handling pseudo gluts. Some directions are also provided in the section. The reader can possibly omit some of the philosophical aspects of the section on *dialectical negation* on a first reading. But a proper understanding of the section would be useful for applications.

1.1   Background

In quasi or partially ordered sets, sets of mutually incomparable elements are called *antichains*. Some of the basic properties may be found in [19,20]. The possibility of using antichains of rough objects for a possible semantics was mentioned in [17,21,22] and has been developed subsequently in [3,4]. The semantics invented in the paper is applicable for a large class of operator based rough sets including specific cases of rough Y- systems (RYS) [7] and other general approaches like [23–26]. In [23,24,27], negation like operators are assumed in general and these are not definable operations in terms of order related operations in a algebraic sense (Kleene negation is also not a definable operation in the situation).

For basics of rough sets, the reader is referred to [1,8,28]. Distinct mereological approaches to rough sets can be found in [4,8,29].

Many concepts of *lower* and *upper approximation operators* are known in the literature. Relevant definitions are fixed below.

If S is any set (in ZFC), then a *lower approximation operator* l over S is be a map $l : \wp(S) \longmapsto \wp(S)$ that satisfies:

$$(\forall x \in \wp(S)) \neg(x \subset x^l) \qquad \text{(non-increasing)}$$
$$(\forall x \in \wp(S)) \, x^l = x^{ll} \qquad \text{(idempotence)}$$
$$(\forall a, b \in \wp(S)) \, (a \subseteq b \longrightarrow a^l \subseteq b^l) \qquad \text{(monotonicity)}$$

In the literature on rough sets many variants of the above are also referred to as lower approximations because the concept has to do with what one thinks a *lower approximation* ought to be in the context under consideration. Over the same set, an *upper approximation operator* u shall be a map $u : \wp(S) \longmapsto \wp(S)$ that satisfies:

$$(\forall x \in \wp(S)) \, x \subseteq x^u \qquad \text{(increasing)}$$
$$(\forall a, b \in \wp(S)) \, (a \subseteq b \longrightarrow a^u \subseteq b^u) \qquad \text{(monotonicity)}$$

All of these properties hold (in a non-trivial sense) in proto-transitive rough sets [17], while weaker properties hold for lower approximations in esoteric rough sets [2,30]. Conditions involving both l and u have been omitted for simplicity.

In some practical contexts, lower and upper approximation operators may be partial and l (respectively u) may be defined on a subset $S \subset \wp(S)$ instead. In these cases, the partial operation may not necessarily be easily completed. Further the properties attributed to approximations may be a matter of discovery.

These cases fall under the general class of *inverse problems* [7,9] where the goal is to see whether the approximations originate or fit a rough evolution (process). More details can be found in the next section.

An element $x \in \wp S$ will be said to be *lower definite* (resp. *upper definite*) if and only if $x^l = x$ (resp. $x^u = x$) and *definite*, when it is both lower and upper definite. In general rough sets, these ideas of definiteness are insufficient as it can happen that upper approximations of upper approximations are still not upper definite.

The concept of a Rough Y-System RYS was introduced by the present author in [7,31] and refined further in [32] and her doctoral thesis [33] as a very general framework for rough sets from an axiomatic granular perspective. The concept is not used in an essential way in the present paper and the reader may skip the few remarks that involve them. In simplified terms, it is a model of any collection of rough/crisp objects with approximation operators and a binary parthood predicate **P** as part of its signature.

Possible concepts of rough objects considered in the literature include the following:

- Non definite subsets of S; formally, x is a rough object if and only if $x^l \neq x^u$.
- Pairs of definite subsets of the form (a, b) that satisfy $a \subseteq b$.
- Pairs of subsets of the form (x^l, x^u).
- Sets in an interval of the form (x^l, x^u); formally,
- Sets in an interval of the form (a, b) satisfying $a \subseteq b$ with a, b being definite subsets.
- Higher order intervals bounded by definite subsets [7].
- Non-definite elements in a RYS [7]; formally, those x satisfying $\neg \mathbf{P} x^u x^l$ are rough objects.

In general, a given set of approximations may be compatible with multiple concepts of definite and rough objects - the eventual choice depends on one's choice of semantic domain. A detailed treatment, due to the present author, can be found in [2,7]. In [18], some of these definitions of rough objects are regarded as imperfect for the purpose of expressing both multiple extensions of concepts and rough objects - this problem relates to representation (given a restricted view of granularity) within the classical semantic domain of [7].

Concepts of representation of objects necessarily relate to choice of semantic frameworks. But, in general, order theoretic representations are of interest in most contexts. In operator centric approaches, the problem is also about finding ideal representations.

In simple terms, *granules* are the subsets that generate approximations and *granulations* are the collections of all such granules in the context. In the present author's view, there are at least three main classes of granular computing (five classes have been considered by her in [2]). The three main classes are

- Primitive Granular Computing Processes (PGCP): in which the problem require-ments are not rigid, effort on formalization is limited, scope of abstraction become limited and concept of granules are often vague, though they maybe concrete or abstract (relative to all materialist viewpoints).

- Classical Granular Computing Paradigm (CGCP): The precision based *classical granular computing paradigm*, traceable to Moore and Shannon's paper [34], is commonly understood as the granular computing paradigm (The reader may note that the idea is vaguely present in [35]). CGCP has since been adapted to fuzzy and rough set theories in different ways. An overview is considered in [36]. In CGCP, granules may exist at different levels of precision and granulations (granules at specific levels or processes) form a hierarchy that may be used to solve the problem in question.
- Axiomatic Approach to Granularity (AAG): The axiomatic approach to granularity, initiated in [31], has been developed by the present author in the direction of contamination reduction in [7]. The concept of admissible granules, described below, was arrived in the latter paper and has been refined/simplified subsequently in [4,16,32]. From the order-theoretic algebraic point of view, the deviation is in a very new direction relative the precision-based paradigm. The paradigm shift includes a new approach to measures.

Historical details can be found in a section in [2,17].

Granular operator spaces, a set framework with operators introduced by the present author in [3], will be used as considerations relating to antichains will require quasi/partial orders in an essential way. The evolution of the operators need not be induced by a cover or a relation (corresponding to cover or relation based systems respectively), but these would be special cases. The generalization to some rough Y-systems RYS (see [7] for definitions), will of course be possible as a result.

Definition 1. *A Granular Operator Space [3] S is a structure of the form $S = \langle \underline{S}, \mathcal{G}, l, u \rangle$ with $\underline{S}$ being a set, $\mathcal{G}$ an admissible granulation(defined below) over S and l, u being operators : $\wp(\underline{S}) \longmapsto \wp(\underline{S})$ ($\wp(\underline{S})$ denotes the power set of $\underline{S}$) satisfying the following ($\underline{S}$ will be replaced with S if clear from the context. Lower and upper case alphabets will both be used for subsets):*

$$a^l \subseteq a \ \& \ a^{ll} = a^l \ \& \ a^u \subset a^{uu}$$

$$(a \subseteq b \longrightarrow a^l \subseteq b^l \ \& \ a^u \subseteq b^u)$$

$$\emptyset^l = \emptyset \ \& \ \emptyset^u = \emptyset \ \& \ \underline{S}^l \subseteq S \ \& \ \underline{S}^u \subseteq S.$$

Here, admissible granulations are granulations $\mathcal{G}$ that satisfy the following three conditions (t is a term operation formed from the set operations $\cup, \cap, {}^c, 1, \emptyset$):

$$(\forall a \exists b_1, \ldots b_r \in \mathcal{G}) \, t(b_1, b_2, \ldots b_r) = a^l$$

$$\text{and } (\forall a) \, (\exists b_1, \ldots b_r \in \mathcal{G}) \, t(b_1, b_2, \ldots b_r) = a^u, \qquad \text{(Weak RA, WRA)}$$

$$(\forall b \in \mathcal{G})(\forall a \in \wp(\underline{S})) \, (b \subseteq a \longrightarrow b \subseteq a^l), \qquad \text{(Lower Stability, LS)}$$

$$(\forall a, b \in \mathcal{G})(\exists z \in \wp(\underline{S})) \, a \subset z, \, b \subset z \ \& \ z^l = z^u = z, \qquad \text{(Full Underlap, FU)}$$

Remarks

- The concept of admissible granulation was defined for RYS in [7] using part-hoods instead of set inclusion and relative to RYS, $\mathbf{P} = \subseteq$, $\mathbb{P} = \subset$.
- The conditions defining admissible granulations mean that every approximation is somehow representable by granules in a set theoretic way, that granules are lower definite, and that all pairs of distinct granules are contained in definite objects.

On $\wp(\underline{S})$, the relation $\sqsubset$ is defined by

$$A \sqsubset B \text{ if and only if } A^l \subseteq B^l \ \& \ A^u \subseteq B^u. \tag{1}$$

The rough equality relation on $\wp(\underline{S})$ is defined via

$$A \approx B \text{ if and only if } A \sqsubset B \ \& \ B \sqsubset A.$$

Regarding the quotient $\wp(\underline{S})| \approx$ as a subset of $\wp(\underline{S})$, an order $\Subset$ can be defined as follows:

$$\alpha \Subset \beta \text{ if and only if } \alpha^l \subseteq \beta^l \ \& \ \alpha^u \subseteq \beta^u. \tag{2}$$

Here α^l is being interpreted as the lower approximation of α and so on. $\Subset$ will be referred to as the *basic rough order*.

Definition 2. *By a roughly consistent object will be meant a set of subsets of $\underline{S}$ with mutually identical lower and upper approximations respectively. In symbols H is a roughly consistent object if it is of the form $H = \{A; (\forall B \in H) A^l = B^l, A^u = B^u\}$. The set of all roughly consistent objects is partially ordered by the inclusion relation. Relative this maximal roughly consistent objects will be referred to as* rough objects. *By* definite rough objects, *will be meant rough objects of the form H that satisfy*

$$(\forall A \in H) A^{ll} = A^l \ \& \ A^{uu} = A^u.$$

However, this definition of rough objects will not necessarily be followed in this paper.

Proposition 1. $\Subset$ *is a bounded partial order on* $\wp(\underline{S})| \approx$.

Proof. Reflexivity is obvious. If $\alpha \Subset \beta$ and $\beta \Subset \alpha$, then it follows that $\alpha^l = \beta^l$ and $\alpha^u = \beta^u$ and so antisymmetry holds.

If $\alpha \Subset \beta$, $\beta \Subset \gamma$, then the transitivity of set inclusion induces transitivity of $\Subset$. The poset is bounded by $0 = (\emptyset, \emptyset)$ and $1 = (S^l, S^u)$. Note that 1 need not coincide with (S, S). $\qquad \square$

The concept of *generalized granular operator spaces* has been introduced in [4,5] as a proper generalization of that of granular operator spaces. The main difference is in the replacement of $\subset$ by arbitrary *part of* ($\mathbf{P}$) relations in the axioms of admissible granules and inclusion of $\mathbf{P}$ in the signature of the structure.

Definition 3. *A General Granular Operator Space (GSP) S is a structure of the form $S = \langle \underline{S}, \mathcal{G}, l, u, \mathbf{P} \rangle$ with $\underline{S}$ being a set, $\mathcal{G}$ an admissible granulation(defined below) over S, l, u being operators : $\wp(\underline{S}) \longmapsto \wp(\underline{S})$ and $\mathbf{P}$ being a definable binary generalized transitive predicate (for parthood) on $\wp(\underline{S})$ satisfying the same conditions as in Definition 1 except for those on admissible granulations (Generalized transitivity can be any proper nontrivial generalization of parthood (see [22]). $\mathbb{P}$ is proper parthood (defined via $\mathbb{P}ab$ if and only if $\mathbf{P}ab$ & $\neg\mathbf{P}ba$) and t is a term operation formed from set operations):*

$$(\forall x \exists y_1, \dots y_r \in \mathcal{G}) \, t(y_1, y_2, \dots y_r) = x^l$$
$$\text{and } (\forall x) \, (\exists y_1, \dots y_r \in \mathcal{G}) \, t(y_1, y_2, \dots y_r) = x^u, \qquad \text{(Weak RA, WRA)}$$
$$(\forall y \in \mathcal{G})(\forall x \in \wp(\underline{S})) \, (\mathbf{P}yx \longrightarrow \mathbf{P}yx^l), \qquad \text{(Lower Stability, LS)}$$
$$(\forall x, y \in \mathcal{G})(\exists z \in \wp(\underline{S})) \, \mathbb{P}xz, \, \& \, \mathbb{P}yz \, \& \, z^l = z^u = z, \qquad \text{(Full Underlap, FU)}$$

It is sometimes more convenient to use only sets and subsets in the formalism as these are the kinds of objects that may be observed by agents and such a formalism would be more suited for reformulation in formal languages. For this reason higher order variants of general granular operator spaces have been defined in [37] by the present author. A detailed account can be found in [2].

In a partially ordered set *chains* are subsets in which any two elements are comparable. *Antichains*, in contrast, are subsets in which no two elements are comparable. Singletons are both chains and antichains.

Rough Sets and Parthood. It is necessary to clarify the nature of parthood even in set-theoretic structures like granular operator spaces. The restriction of the parthood relation to the case when the first argument is a granule is particularly significant. The theoretical assumption that objects are determined by their parts, and specifically by granules, may not be reasonable when knowledge of the context is evolving. Indeed, in this situation:

- granulation can be confounded by partial nature of information and noise,
- knowledge of all possible granulations may not be possible and the chosen set of granules may not be optimal for handling partial information, and
- the process has strong connections with apriori understanding of the objects in question.

2 Types of Preprocessing and Ontology

Information storage and retrieval systems (also referred to as information tables, descriptive systems, knowledge representation system) are basically representations of structured data tables. These have often been referred to as *information systems* in the rough set literature - the term means *an integrated heterogeneous system that has components for collecting, storing and processing data* in closely allied fields like artificial intelligence, database theory and machine learning. So it makes sense to avoid referring to *information tables* as *information systems* [38].

Information tables are associated with only some instances of data in real life. In many cases, such association may not be useful enough in the first place. *It is also possible that data collected in a context is in mixed form with some information being in information table form and some in the form of* relevant approximations *or all the main data is in terms of approximations.* The *inverse problem*, introduced by the present author in [9] and subsequently refined in [7], seeks to handle these types of situations. Granular operator spaces and higher order variants studied by the present author in [3,4,37] are important structures that can be used for its formulation. In simple terms, the problem is a generalization of the duality problem which may be obtained by replacing the semantic structures with parts thereof. In a mathematical sense, this generalization may not be proper (or conservative) in general.

The basic problem is *given a set of approximations, similarities and some relations about the objects, find an information system or a set of approximation spaces that fits the available information according to a rough procedure.* In this formalism, a number of information tables or approximation systems along with rough procedures may fit in. Even when a number of additional conditions like lattice orders, aggregation and commonality operations are available, the problem may not be solvable in a unique sense. Negation-like operations and generalizations thereof can play a crucial role in possible formalisms. In particular, the ortho-pair approach [39] to rough sets relies on negations in a very essential way.

It is also necessary to concentrate on the evolution of negation like operations or predicates at a higher level - because it is always necessary to explain the choice of operations used in a semantics. The following example (Example 1) from [4] illustrates the inverse problem. Some ways of identifying generalized negation-like predicates are shown in Example 2.

Example 1. This example has the form of a narrative in [4] that gets progressively complex. It has been used to illustrate a number of computational contexts in the paper.

Suppose Alice wants to purchase a laptop from an on line store for electronics. Then she is likely to be confronted by a large number of models and offers from different manufacturers and sellers. Suppose also that the she is willing to spend less than €x and is pretty open to considering a number of models. This can happen, for example, when she is just looking for a laptop with enough computing power for her programming tasks.

This situation may appear to have originated from information tables with complex rules in columns for decisions and preferences. Such tables are not information systems in the proper sense. Computing power, for one thing, is a context dependent function of CPU cache memories, number of cores, CPU frequency, RAM, architecture of chip set, and other factors like type of hard disk storage.

Proposition 2. *The set of laptops* $\mathbb{S}$ *that are priced less than* €x *can be totally quasi-ordered.*

Proof. Suppose $\prec$ is the relation defined according to $a \prec b$ if and only if price of laptop a is less than or equal to that of laptop b. Then it is easy to see that $\prec$ is a reflexive and transitive relation. If two different laptops a and b have the same price, then $a \prec b$ and $b \prec a$ would hold. So $\prec$ may not be antisymmetric. □

Suppose that under an additional constraint like CPU brand preference, the set of laptops becomes totally ordered. That is under a revised definition of $\prec$ of the form: $a \prec b$ if and only if price of laptop a is less than that of laptop b and if the prices are equal then CPU brand of b must be preferred over a's.

Suppose now that Alice has more knowledge about a subset C of models in the set of laptops $\mathbb{S}$. Let these be labeled as *crisp* and let the order on C be $\prec_{|C}$. Using additional criteria, rough objects can be indicated. Though lower and upper approximations can be defined in the scenario, the granulations actually used are harder to arrive at without all the gory details.

This example once again shows that granulation and construction of approximations from granules may not be related to the construction of approximations from properties in a cumulative way.

In [4], it is also shown that the number of data sets, of the form mentioned, that fit into a rough scheme of things are relatively less than the number of those that do not fit. Many combinatorial bounds on the form of rough object distribution are also proved in the paper.

Examples of approximations that are not rough in any sense are common in misjudgments and irrational reasoning guided by prejudice. So solutions to the problem can also help in judging the irrationality of reasoning and algorithms in different contexts.

Example 2. Databases associated with women badminton players have the form of multi-dimensional information tables about performance in various games, practice sessions, training regimen and other relevant information. Video data about all these would also be available. Typically these are updated at specific intervals of time. Players tend to perform better under specific conditions that include the state of the game in progress. They may also be able to raise the level of their game under specific kinds of stress - this involves dynamic learning. Additional information of the form can be expressed in terms of approximations, especially when the associations are not too perfect. Thus, a statement like

Player A is likely to perform at least as well as player B in playing conditions C

can be translated into $B^l \leq A^l$ where the approximations refer the specific property under consideration. *But, this information representation has no obvious rough set basis associated* and falls under the inverse problem, where the problem is of explaining the approximations from a rough perspective/basis.

Consider a pair (a, b) with Updates to the database can also be described through generalized negation-like predicates in the context.

2.1 Granular Operator Spaces and Property Systems

Limitations of the property system approach are mentioned in this subsection

In general, data can be presented in real life partly in terms of approxima-tions and in the object-attribute-value way of representing things (For those that like statistics, the collection of instances of the sentence has nice statis-tical properties). Such contexts were never intended to be captured through property systems or related basic constructors (see [28,40,41]). In particular, the examples of [42] *are abstract and the possible problems with basic construc-tors (when viewed from the perspective of approximation properties satisfied) are issues relating to construction - empirical aspects are missed.*

Definition 4. *A property system* [28,42–44] Π *is a triple of the form*

$$\langle U, P, R \rangle$$

with U being a universe of objects, P a set of properties, and $R \subseteq U \times P$ being a manifestation relation *subject to the interpretation object a has property b if and only if $(a, b) \in R$. When $P = U$, then Π is said to be a square relational system and Π then can be read as a Kripke model for a corresponding modal system.*

On property systems, basic constructors that may be defined for $A \subseteq U$ and $B \subseteq P$ are

$$<i> : \wp(U) \longmapsto \wp(P); \ <i>(A) = \{h : (\exists g \in A)\,(g, h) \in R\} \qquad (3)$$

$$<e> : \wp(P) \longmapsto \wp(U); \ <e>(B) = \{g : (\exists h \in B)\,(g, h) \in R\} \qquad (4)$$

$$[i] : \wp(U) \longmapsto \wp(P); [i](A) = \{h : (\forall g \in U)((g, h) \in R \longrightarrow g \in A)\} \qquad (5)$$

$$[e] : \wp(P) \longmapsto \wp(U); [e](B) = \{g : (\forall h \in P)((g, h) \in R \longrightarrow h \in B)\} \qquad (6)$$

It is known that the basic constructors may correspond to approximations under some conditions [28]. It may hold under some other conditions. Property system are not suitable for handling granularity and many of the inverse problem contexts. The latter part of the statement requires some explanation because suitability depends on the way in which the problem is posed - this has not been looked into comprehensively in the literature.

If all the data is of the form

Object X is definitely approximated by $\{A_1, \dots A_n\}$,

with the symbols X, A_i being potentially substitutable by objects, then the data could in principle be written in property system form with the sets of A_is forming the set of properties P - in this situation the relation R attains a different meaning. This is consistent with the structure being not committed to tractability of properties possessed by objects. Granularity would also be obscure in the situation.

If all the data is of the form

Object X's approximations are included in $\{A_1, \dots A_n\}$,

then the property system approach comes under even more difficulties. Granular operator spaces and generalized versions thereof [5] in contrast can handle all this.

3 Dialectical Negation

In general rough sets, most relevant concepts of negation and implication are dialectical in some sense. A universal definition of *dialectical negation* is naturally of interest - at least one that works for the associated vague contexts. Since vagueness is everywhere, multiple concepts of *dialectical negation* in the literature need to be reconciled (to the extent possible) for the purpose of a universal definition. The main questions relating to possible definitions of dialectical negations or contradictions at the formal level arise from the following reasons (these are explained below):

- The consensus that dialectical logics must be logics that concern state transitions.
- The view that paraconsistent logics are essentially dialectical logics (see [45]).
- The view that dialectical negation cannot be reduced to classical negation (see [12]). Indeed, in rough sets many kinds of negations and partial negations have been used in the literature (see for example, [2,9,11,46–48], [30,31,39, 42,49–51] and these lead to many contradictions as in
  1. *contradictions* [46] which are not false but that represent topological boundaries;
  2. *contradictions* [47] which are not false but lie between an absolute and local false;
  3. *contradictions* [42] which lead to at least a paraconsistent and a paracomplete logic.
- The view that dialectical negation is glutty negation (example [52,53]) is an intermediate kind of negation.
- The view that only propositional dialectical logics are possible ([54]).
- The present author's position that dialectical contradiction must be represented by binary predicates or binary logical connectives in general [15,55]. This is arrived at in what follows.

The relationship of an object and its negation may belong to one of three categories (an extension of the classification in [56]):

1. Cancellation as in *the attributes do not apply*. The ethical category of *the negative*, as used in natural language is often about this kind of cancellation. It is easy to capture this in logics admitting different types of atomic variables (or formalized for instance by labeled deductive systems [57]). In these the concept of the Negative, is usually not an atomic category. Obviously this type of negation carries more information in being a Not This but Among Those kind of negation as opposed to the simple Not This, Not This and Something Else and weakenings thereof.
2. Complementation understood in the sense of classical logic.
3. Glutty variants understood as something intermediate between the two.

In general rough sets, if A is a subset of attributes and c is set complementation, then the value of more common *negations in a rough sense* include

A^{uc}, A^{cl}, A^{uuc}, A^{clu}. Each of these is a nonempty subset of A^c in general. Consequently, the corresponding negation is glutty in a set theoretic sense.

The concept of *dialectical negation* as a material negation in logics about states is a reasonable abstraction of the core concept in Hegelian and Marxist dialectics (though this involves rejection of Hegel's idealist position). The negation refers to concepts in flux and so a logic with regard to the behavior of states rather than static objects would be appropriate. According to Hegel, the world, thought and human reasoning are dynamic and even the idea of true concepts are dynamic in nature. Poorly understood concepts undergo refinement as plural concepts (with the parts being *abstract* in Hegel's sense) that assume many *H(Hegelian)-Contradictory* forms. After successive refinements the resulting forms become reconciled or united as a whole. So for Hegel, H-contradictions are essential for all life and world dynamics. But Hegel's idealist position permitted only a closed world scheme of things. In Marx's materialist dialectic, the world is an open-ended system and so recursive applications of dialectical contradiction need not terminate or be periodic. All this means that the glutty interpretation of Hegel's contradiction may well be correct modulo some properties, while Marx's idea of dialectical contradiction does not reduce to such an interpretation in general. The debate on endurantism and perdurantism is very relevant in the context of Hegelian dialectics because the semantic domain associated is restricted by Hegel's world view. *In rough semantics, especially granular ones, approximations may be seen as transitions and so the preconditions are met in a sense.*

Example 3. The identity of an apple on a table can be specified in a number of ways. For the general class of apples, a set of properties X can be associated. The specific apple in question would also be possessing a set of specific properties Z that include the distribution and intensity of colors. Obviously, many of the specific properties will not be true of apples in general. Further they would be in dialectical opposition to the general. Note that an agent can have multiple views of what Z and X ought to be and multiple agents would only contribute to the pluralism. In Hegel's perspective all of this dialectical contradiction must necessarily be resolved in due course (this process may potentially involve non materialist assumptions), while in the Marxist perspective a refined plural that may get resolved would be the result. Thus, the apple may be of the Alice variety of *Malus Domestica* (a domestic apple) with many other specific features. The schematics (for an agent) is illustrated in Fig. 1 - $\sqsupset$ is a binary predicate with the intended meaning of being *in dialectical opposition.*

To see how ideas of unary operations as dialectical negations can fail, consider the color of apples alone. If the collection of all possible colors of apples is known, then the set of all possible colors would be knowable. Negation of a white apple may be definable by complementation in this case. If on the other hand only a few possible sets of colors and some collections of colors are known, then the operational definition can fail (or lack in meaning). This justifies the use of a general binary predicate $\sqsupset$ for expressing dialectical contradiction.

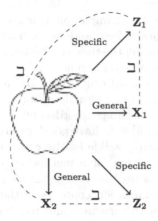

Fig. 1. Schematics of dialectics of identity

Da Costa et al. consider the heuristics of a possible dialectic logic in [45]. They seem to accept McGill's interpretation of unity of opposites [54] and restrict themselves to a propositional perspective on the basis of difficulties with formalizing four of the six principles. This results in a very weak dialectical logic. They are however of the view that formal logics based on Marxist and Hegelian dialectics intersect the class of paraconsistent logics and there is great scope for deviation and that it can be argued that paraconsistent logics represent a desired amendment of dialectics because of the latter's openness and non-rigid formalism.

The distinction between static and dynamic dialectical logics within the class of dialectical logics with dialectical contradiction as expressed with the help of an unary operation, may be attributed to Batens [58,59]. Adaptive logics in that perspective would appear to be more general than dynamic dialectical logics; the main idea is to interpret inconsistencies as consistently as is possible. Key to this class of logics is the concept of tolerance of contradictory statements that are not necessarily reduced in their level of contradictions by way of proof mechanisms. Through this one can capture parts of the thesis-antithesis-synthesis meta principle. All semantic aspects of adaptive logics are intended in a classical perspective as opposed to dialethic logics and these are very closely connected to paraconsistent logics as well. Some Hegelian approaches (see [60] for example) also seek to resolve universal contradictions.

Two of the most common misinterpretations or reductions of the concept of *dialectical opposition* relate to excluding the very possibility of formalizing it and the reduction of dialectical negation to simple negation or opposites. Examples of the latter type are considered in [61,62]. Some are of the view that Marx worked with normative ideas of concept and so introduction of related ideas in logic are improper. In modern terminology, Marx merely wanted concepts to be grounded in the material and was opposed to idealist positions that were designed for supporting power structures of oppression. This is reflected in Marx's position on Hegel's idealism and also, for example, on Wagner's position [63]. Marx and Hegel did not write about formal logic and the normative ideas of *concept* used by both and other authors during the time can be found in great numbers. From the point of view of less normative (or non normative) positions all these authors implicitly developed concepts at all times. It is also a reasonable position that Marxist methodologies should not be formalized independently of the normative restriction on possible ideas of concept afforded by actualization contexts. This is because it is always a good idea to have good grounding axioms for concepts to the extent that is permitted/possible by the context in question.

In the present author's opinion for a methodology or theory to qualify as *dialectic in Marx's sense* it is necessary that the idea of concepts used should be well grounded in the actualization contexts of the methodology or theory. In the context of reasoning with vagueness and rough sets, this may amount to requiring the approach be granular under specific conditions.

In general, formal versions of dialectical logics can be based on some of the following principles/heuristics.

A. Binary Logical predicates (that admit of universal substitution by propositional variables and well formed formulas) that are intended to signify binary dialectical contradiction are necessary for dialectical logics,
B. Unary logical connectives (that admit of universal substitution by propositional variables and well formed formulas) suffice for expressing dialectical contradiction,
C. The thesis-antithesis-synthesis principles must necessarily be included in the form of rules in any dialectical logic,
D. Higher order quantifiers must be used in the logical formalism in an essential way because dialectical contradictions happen between higher order concepts,
E. Dialectical logics should be realizable as fragments of first order predicate logic - this view is typically related to the position that higher order logics are superfluous.
F. Dialectical contradiction in whatever representation must be present at each stage of what is defined to constitute dialectical derivation - this abstraction is due to the present author and is not hard to realize.
G. All dialectic negations should be dialethic(*) in nature - this is a possibility explored in [56]. Dialethias are statements that can be both true and false simultaneously
H. A logic that permits expression of progression of knowledge states is a dialectical logic.
 I. a first order logic perspective
J. the point of view that dialectical contradiction can be expressed by binary predicates and not by unary operations
K. Dialectical logics as paraconsistent logics incorporating contradictions or as inconsistency adaptive logics.

Obviously many of these are mutually incompatible. [J] in particular is incompatible with [B] in the above. [H] leads to linear logics and variants. The meaning of dialectical logics that admit representation as a fragment of first order predicate logic will be naturally restricted. A few versions are known. Dynamic dialectical logics have been developed as inconsistency adaptive logics by Batens [64] in particular. In the present paper, [A] will be preferred as the binary predicate/predication cannot always be reduced to unary negations.

In general, it is obvious that given a dynamically changing subject, there will be at least a set of things which are dialectically contradictory to it in many ways. If a is in dialectical contradiction to b and c in two different senses, then it is perfectly possible that b is dialectically contradictory to c in some other sense. Further if X is dialectically contradictory to a conjunction of the form $Y \wedge Z$, then it is possible that X is dialectically contradictory to Y in some other sense and is virtually indifferent to Z.

3.1   Dialectical Contradiction and Contradiction

At a more philosophical level, the arguments of this section can be expressed in the language of functors [12]. However, a set-theoretic semantic approach is better suited for the present purposes. The concepts of contradiction and dialethic contradiction make essential use of negation operations (in some general sense), while that of dialectical contradiction when formulated on comparable terms does not necessarily require one. It is necessary to clarify the admissible negations in all this as many variations of the concept of logical negation are known and most are relevant for rough set contexts.

Let S be a partially ordered set with at least one extra partial unary operation f, a least element $\perp$ and a partial order $\leq$ ($\wedge$, $\vee$ being partial lattice infimum and supremum). In a partial algebra, two terms p and q, are weakly equal, ($p \overset{\omega}{=} q$), if and only if *whenever both terms are defined then they are equal*. Consider the following:

$$x \wedge f(x) \overset{\omega}{=} \perp \tag{N1}$$
$$(x \leq y \longrightarrow f(y) \leq f(x)) \tag{N2}$$
$$x \leq f^2(x) \tag{N3}$$
$$(x \leq f(y) \longrightarrow y \leq f(x)) \tag{N4}$$
$$f^n(x) = f^{n+m}(x) \text{ for some minimal } n, m \in N \tag{N5}$$
$$f(x \vee y) \overset{\omega}{=} f(x) \wedge f(y) \tag{N6}$$
$$x \wedge y = \perp \leftrightarrow y \leq f(x). \tag{N9}$$

Distinct combinations of these non-equivalent conditions can characterize negations. In [65], if an operation satisfies $N1$ and $N2$ over a distributive lattice, then it is taken to be a general negation. This is a reasonable concept for logics dealing with exact information alone as $N1$ does not hold in the algebras of vague logics, uncertain or approximate reasoning. For example it fails in the algebras of rough logic and generalizations thereof [7,11,17,30].

If $\forall x\, f^m(x) = f^n(x)$ holds, then the least n such that $m < n$ is called the *global period* of f, $s = n - m$, the *global pace* of f and (m, n), the index of f.

Theorem 1. *When the poset is a distributive lattice in the above context, then the following are separately true:*

1. *If f satisfies $N1$ and $N2$, then the index $(0, n)$ for $n > 2$ is not possible.*
2. *If f satisfies $N1$, $N2$, and $N3$, then $f(\bot) = T$ is the greatest element of the lattice and $f(T) = \bot$.*
3. *$N1$, $N2$, $N3$ together do not imply $N9$*
4. *$N9$ implies each of $N1$, $N2$ and $N3$.*
5. *An interior operator i on a poset is one that satisfies*
   - *$i(x) \leq x$,*
   - *$(a \leq b \longrightarrow i(a) \leq i(b))$*
   - *$i(i(x)) = i(x)$.*

If f is a regular negation (that is it satisfies the conditions $N1$, $N2$ and $N3$) and i an interior operator, then $g = if$ is a negation that satisfies $g^4 = g^2$.

Even at these levels of generality, the generalized negations can fail to express the appropriate concept of dialectical contradiction.

In a set-theoretical perspective, if a set of things A is dialectically opposed to another set B, then it may appear reasonable to expect A to contain the set of things dialectically opposed to B. Subject to this intent, the set of *all* things dialectically opposed to A would need to be expressed by $\sim A$. But in dialectical reasoning it will still be reasonable to say that A is dialectically opposed to some part of $\sim A$. For this the use of a unary $\sim$ can be glaringly insufficient. This is true not only from an algebraic system point of view (when working within a model) but also from perspectives generated by admissible sets of models. $N2$ is inherently incompatible with accepting f as a unary dialectical negation operator, especially when a lattice order is expected to be induced by some concept of logical equivalence from the order. $N3$ is perhaps the most necessary property of a dialectical negation operation.

Well-formed formulas of certain derived types alone may admit of a negation (in the sense of being equivalent to one of the negatives). Such a negation is partial. For instance, $\sim \sim x$ may not be defined in the first place, and some of $N1$–$N9$ may hold for such negations. Using such types of negation for expressing dialectical contradictions through compound constructions may be possible in adequately labeled deductive systems.

Dialethic logics are logics that tolerate contradictions and accept true contradictions. To be more specific a dialetheia is a pair of statements having the form $A \,\&\, \neg A$ with each of A and $\neg A$ being true. These statements may be interpreted sentences expressed in some language (that may be natural language or a language of thought, or anything). They can be used to formalize only some restricted cases of dialectical reasoning in which a unary dialectical contradiction operation is possible. It is also possible to reformulate some dialectical contexts as a dialethic process. Priest [66] had indicated the possibility of using dialethic logics as a base for dialectic logics. In [67], Priest develops a dialectical tense logic, where it is possible for a system to exist in both pre and post states during (at the instant of) state transitions. Zeleny [68] in particular has correctly pointed out (from a philosophical perspective) the possible shortcomings of a unary negation based approach. Though the issue of desiring incompatibility between classical logic and a possible dialectical logic is not a justified heuristic. The essential dialetheism principle is however usable in dialectical derivations. Such situations would allow dialectical opposition between proof patterns naturally.

The nature and meaning of negation in a dialethic logic is explained in [66,69]. From a philosophical meta perspective the negation of a formula is possibly a collection of formulas that may be represented by a single formula (from a logical perspective). It is with respect to such a *negation* that dialethic logics must tolerate contradictions and accept true ones. A survey of concepts of contradictions for dialetheism can also be found in [70]. Using any kind of universal paraconsistent system for describing inconsistencies is virtually shown to be an undesirable approach in [59]. Marxist dialectics is perceived from a dialetheistic perspective of things in [71]. It is claimed that dialetheias correspond to and realize the concept of historical contradiction. The methodological aspect of Marxist dialectics is also ignored by Priest (see [72]) to the point that *dialectic is a dialetheia*. There are no methodological strictures associated with dialetheias except for the requirement that they be real. This approach ignores

- the world view associated with Hegelian-Marxist dialectics,
- the principle of unity of opposites, and
- the basic problem with formalizing dialectical opposition with a unary operator - this is because the negation in dialectics is transient and dependent on the above two points.

In the present author's view dialetheias do exist in the real world and they may be the result of

- missing data/information
- a deliberate disregard for consistency. For example, a large number of people in the news media, religion and politics practice dialethic expression of a crude form and deceit. They may have their motivations for such actions, but those would not be justification for their dialetheias. It may also be possible to construct equivalent models or models with additional information that do not have true contradictions. For example, the statement X *has property* Q *and* $\neg Q$ may be replaced by the statements X *has property* Q *in state* A and X *has property* $\neg Q$ *in state* B. This amounts to interfering with the data and does not really change the state of affairs. Many religious functionaries have been convicted of sexual crimes and most were in harmony with their apparent dialethic behavior (religious texts may be full of contradictions, that only allows for prolonging the derivations).

The present author agrees with Priest's claims about dialetheias being not resolvable by revision of concepts and that they are better handled as such

[56, 72, 73]. However, she does not agree at all that the proper way of formalizing Hegelian-Marxist dialectics in all cases is through dialetheias. A mathematical formulation of the issue for many sub cases may be possible through rough sets. The *Cold vs Influenza* example considered in the subsection on examples of parthoods throws much light on the matter.

3.2   Dialectical Predication

At a philosophic level, dialectical predication is a relation between functors in the sense of [12]. At a formal model-theoretic semantic level, the best realization is through a binary dialectical predicate $\sqsupset$, that may have limited connections with negation operations (if any). The basic properties that are necessary (not sufficient) of the predicate are as follows (with $\oplus$ standing for aggregation):

$$\sqsupset(a, b) \longleftrightarrow \sqsupset(b, a) \qquad \text{(Commutativity)}$$
$$\neg \sqsupset(a, a) \qquad \text{(Anti-Reflexivity)}$$
$$\sqsupset(a, b) \longrightarrow \sqsupset(a \oplus c, b \oplus c) \qquad \text{(Aggregation)}$$

This predicate may be related to unary dialectical negation operations in a simple or complex way. One possibility (that leads to $N4$) is the following (for two predicates P and Q):

$$\sqsupset(a, b) \text{ iff } P(a) \implies \neg Q(b)$$

That the definitions are important is illustrated by the following example.

Let $\{x_n\}$ be a sequence of real numbers. In contexts where reasoning proceeds from the concrete to a general, let

- A be the statement that *Limit of the sequence is not conceivable*
- B be the statement that *Statement A is conceivable*
- C be the statement that *As B is true, the limit of the sequence is conceivable.*

A is dialectically opposed to C, but the scenario does not amount to a dialetheia if the entire context has enough information of the process (of B being true) being referred to by C.

Example 4 (Dialectics from Classification).
   The intent of this example is to show that

- strategies for classification of information can be dialectically opposed to each other and
- such information can fit into the rough set paradigm without the involvement of dialetheias.

From an abstract perspective, consider a general process or phenomena C described in abstract terms $A_1, \ldots, A_f$. Let every extension of the process have additional peculiar properties that lead to not necessarily independent classifications $C_1, C_2, \ldots, C_n$. Also let the categories $C_1^*, C_2^*, \ldots, C_r^*$ be formed by way of interaction between the members of said classes. This scenario leads to instances of parthood like $\mathbf{P}A_1C_1$ and $\mathbf{P}C_1C_1^*$ with $\mathbf{P}C_1C_1^*$ being in dialectical opposition to $\mathbf{P}A_1C_1$. Concrete instances of development over these lines are easy to find. In fact the historical development of any subject in the social sciences that has witnessed significant improvement over the last thirty years or so would fit under this schema. Two diverse contexts where such a dynamics may be envisioned are presented next.

Models and methods used for income estimation of rural agrarian households manage agrarian relations in different ways [74]. Long term ground level studies are required to clarify the nature of these relations. Suppose C is about estimating poverty in a village and A_1, A_2, A_3 are abstract categories based on volume of monetary transaction by farmers. Some economists may use this for estimating net income and as an indicator of absence of poverty, while in reality farmers may be having negative income or the sources may not be reliable.

Suppose, an improved classification C_1, C_2, C_3, C_4, C_5 has been arrived at based on estimates of investment, expenditure, consumption, exchange of labor and other factors through ground level studies. The resultant classification may need to be improved further to take non monetary transactions like barter of goods and resources into account. Thus, $C_1^*, C_2^*, C_3^*, C_4^*, C_5^*$ may be arrived at.

C_1 definitely takes A_1 into account and the latter is a causative factor for the former. This can be expressed by the parthood $\mathbf{P}A_1C_1$. C_1 is a much stronger causative factor of C_1^*. $\mathbf{P}C_1C_1^*$ then is dialectically opposed to $\mathbf{P}A_1C_1$. Such a relation can be used to track the context dynamics.

The subject of lesbian sexuality in particular has progressed significantly over the last few decades and can be expressed in similar abstract perspectives (see [75] and more recent literature). Women love women in different ways and this variation is significant for sub-classification. The parameters of classification relate to gender expression, sexual state variation, sexual performance, preferences in sexual interaction, routines, mutual communication, lifestyle choices, related social communication and more.

4   Dialectical Rough Sets

A dialectical approach to rough sets was introduced and a more general program was formulated in [15] by the present author. Multiple kinds of roughly equivalent objects and the dialectical relation between them are stressed in the approach. This is reflected in the two algebraic logics proposed in the mentioned paper. The entire universe is taken to be a dialectical relation in the second semantics and possible derivations revolve around it. The main intent was to include mixed kinds of objects in the semantics and so ideas of contamination apply differently. The essential content is repeated below (as [15] is a conference paper) and the nature of some possible parthoods involved is defined below.

A *pre-rough algebra* [11] is an algebra having the form

$$S = \langle \underline{S}, \sqcap, \sqcup, \Rightarrow, L, \neg, 0, 1 \rangle$$

of type $(2, 2, 2, 1, 1, 0, 0)$, which satisfies:

- $\langle \underline{S}, \sqcap, \sqcup, \neg \rangle$ is a De Morgan lattice.
- $\neg\neg a = a$; $L(a) \sqcap a = L(a)$.
- $L(a \sqcup b) = L(a) \sqcup L(b)$; $\neg L \neg L(a) = L(a)$.
- $LL(a) = L(a)$; $L(1) = 1$; $L(a \sqcap b) = L(a) \sqcap L(b)$; $\neg L(a) \sqcup L(a) = 1$.
- If $L(a) \sqcap L(b) = L(a)$ and $\neg L(\neg(a \sqcap b)) = \neg L(\neg a)$ then $a \sqcap b = a$. This is actually a quasi equation.
- $a \Rightarrow b = (\neg L(a) \sqcup L(b)) \sqcap (L(\neg a) \sqcup \neg L(\neg b))$.

A completely distributive pre-rough algebra is called a *rough algebra*. In all these algebras it is possible to define an operation $\diamond$ by setting $\diamond(x) = \neg L \neg(x)$ for each element x.

It should be noted that above definition has superfluous conditions. An equivalent definition that was used in [2] (based on [76]) is the following:

Definition 5. *An essential pre-rough algebra is an algebra of the form*

$$E = \langle \underline{E}, \sqcap, L, \neg, 0, 1 \rangle$$

that satisfies the following (with $\sqcup$ being a defined by $(\forall a, b)\, a \sqcup b = \neg(\neg a \sqcap \neg b)$ and $a \leq b$ being an abbreviation for $a \sqcap b = a$.)

$\langle \underline{E}, \sqcap, \sqcup, \neg, 0, 1 \rangle$ *is a quasi Boolean algebra.*

E1 $L1 = 1$
E2 $(\forall a)\, La \sqcap a = La$
E3 $(\forall a, b)\, L(a \sqcap b) = L(a) \sqcap L(b)$
E4 $(\forall a)\, \neg L \neg La = La$
E5 $(\forall a)\, \neg La \sqcap La = 0$
E6$(\forall a, b)\, (\neg L \neg a \leq \neg L \neg b\ \&\ La \leq Lb \longrightarrow a \leq b)$

An essential pre-rough algebra is said to be an essential rough algebra if $L(E)$ is also complete and completely distributive- that is it satisfies (for any subset X and element a)

$$a \sqcup \left(\bigsqcap X \right) = \bigsqcap \{a \sqcup x : x \in X\}\ \&\ a \sqcap \left(\bigsqcup X \right) = \bigsqcup \{a \sqcap x : x \in X\}$$

Essential pre-rough algebras are categorically equivalent to pre-rough algebras and essential rough algebras to rough algebras.

In this semantics explicit interaction between objects in the rough semantic domain and entities in the classical semantic domain is permitted. The requirement of explicit interaction is naturally tied to objects having a dual nature in the relatively *hybrid* semantic domain. Consequently an object's existence has dialectical associations. In application contexts, this approach can also be useful in enriching the interaction within the rough semantic domain with additional permissible information from the classical semantic domain.

Suppose S_1, S_2, S_3, and S_4 are four general approximation spaces. Suppose that the semantics of S_2 relative to S_1 and S_4 relative to S_3 are definable in a semantic domain. The question of equivalence of these *relativizations* is relevant. It may be noted that the essential problem is implicit in [7]. The hybrid dialectical approach is relevant in these contexts. But, of course, this approach is not intended to be compatible with contamination.

In *rough algebra semantics* it is not possible to keep track of the evolution of rough objects relative to the classical semantics suggested by the Boolean algebra with approximation operators. Conversely in the latter it is not possible to form rough unions and rough intersections relative to the *rough algebra* semantics. These are examples of relative distortions. The CERA semantics (*concrete enriched pre-rough algebra*), which is developed in the next subsection can deal with this, but distortions relative to *super rough semantics* [9] are better dealt with CRAD (*concrete rough dialectical algebra* introduced in the last subsection) like methods. For more on these considerations, the reader is referred to [7,77] and in the *three-valued perspective* to [46,47]. In [47], a three-valued sub domain and a classic sub domain formed by the union of the singleton granules of the classification are identified within the rough domain itself.

Jaskowski's discursive logic is an example of a subvaluationary approach in that it retains the non truth of contradictions in the face of truth-gluts. Connections with pre-rough, rough algebras and rough logics are well known (see [78]). In particular, Pawlak's five valued rough logic R_l (see [79]) and L_R [78] are not dialethic: though it is possible to know that something is roughly true and roughly false at the same time, it is taken to be *roughly inconsistent* as opposed to being just *true* or *roughly true*. This rejection definitely leads to rejection of other reasoning that leads to it as conclusion. Importantly a large set of logics intended for capturing rough semantics are paraconsistent and make use of skewed forms of conjunction and disjunction. It can be argued that the latter feature is suggestive of incomplete development of the logics due to inconsistencies in the application of the underlying philosophy (see [80] for example). The 4-valued DDT (see [65]) addresses some of these concerns with a justification for 3-valuedness in some semantics of classical RST. The NMatrix based logic [81] provides a different solution by actually avoiding conjunction and disjunction operations (it should, however, be noted that conjunctions and disjunctions are definable operations in the NMatrix based logic). In super rough semantics, due to the present author [9], the ability of objects to approximate is called in. These concerns become more acute in the semantics of more general rough sets.

In summary, the main motivations for the approach of this section are

- to provide a framework for investigating relative distortions introduced by different theories - this is in sharp contrast to the contamination reduction approach [7,9] of the present author,

- to improve the interface between rough and classical semantic domains in application contexts,
- to investigate *relativization of semantics* in the multi source general rough contexts (or equivalently in the general dynamic approximation contexts) - in [82] a distinct semantic approach to the problem is developed by the present author,
- address issues relating to truth and parthood at the semantic level,
- and develop a dialectical logic of rough semantics.

The nature of parthood was not considered in the context by the present author at the time of writing [15]. It is considered here to specify the nature of dialectical oppositions and potential diagrams of opposition.

4.1 Enriched Classical Rough Set Theory

Let $S = \langle \underline{S}, R \rangle$ be an approximation space with $\underline{S}$ being a set and R an equivalence. S will be used interchangeably with $\underline{S}$ and the intended meaning should be clear from the context. If $A \subset S$, $A^l = \bigcup\{[x]\,;\,[x] \subseteq A\}$ and $A^u = \bigcup\{[x]\,;\,[x] \cap A \neq \emptyset\}$ are the *lower* and *upper approximation* of A respectively. If $A, B \in \wp(S)$, then A is *roughly equal* to B ($A \approx B$) if and only if $A^l = B^l$ and $A^u = B^u$. $[A]$ shall be the equivalence class (with respect to $\approx$) formed by a $A \in \wp(S)$.

The proposed model may be seen as an extension of the pre-rough and rough algebra models in [11]. Here the base set is taken to be $\wp(S) \cup \wp(S)|\approx$ as opposed to $\wp(S)|\approx$ (used in the construction of a rough set algebra). The new operations $\oplus$, and $\odot$ introduced below are correspond to generalized aggregation and commonality respectively in the mixed domain. This is not possible in classical rough sets proper.

Definition 6. *On $Y = \wp(S) \cup \wp(S)|\approx$, the operations $\mathfrak{L}$, $\oplus$, $\odot$, $\blacklozenge$, $\rightsquigarrow$, $\rightarrow$, $\sim$ are defined as follows: (it is assumed that the operations $\cup, \cap, {}^c, {}^l, {}^u$ and $\sqcup, \sqcap, L, M, \neg, \Rightarrow$ are available on $\wp(S)$ and $\wp(S)|\approx$ respectively. Further*

$$\tau_1 x \Leftrightarrow x \in \wp(S) \text{ and } \tau_2 x \Leftrightarrow x \in \wp(S)|\approx.$$

-

$$\mathfrak{L}x = \begin{cases} x^l & \text{if } \tau_1 x \\ Lx & \text{if } \tau_2 x \end{cases}$$

-

$$\blacklozenge x = \begin{cases} x^u & \text{if } \tau_1 x \\ \neg L \neg x & \text{if } \tau_2 x \end{cases}$$

-

$$x \oplus y = \begin{cases} x \cup y & \text{if } \tau_1 x, \tau_1 y \\ \left[x \cup \left(\bigcup_{z \in y} z \right) \right] & \text{if } \tau_1 x, \tau_2 y \\ \left[\left(\bigcup_{z \in x} z \right) \cup y \right] & \text{if } \tau_2 x, \tau_1 y \\ x \sqcup y & \text{if } \tau_2 x, \tau_2 y \end{cases}$$

-

$$x \odot y = \begin{cases} x \cap y & \textit{if } \tau_1 x, \tau_1 y \\ \left[x \cap \left(\bigcap_{z \in y} z\right)\right] & \textit{if } \tau_1 x, \tau_2 y \\ \left[\left(\bigcap_{z \in x} z\right) \cap y\right] & \textit{if } \tau_2 x, \tau_1 y \\ x \sqcap y & \textit{if } \tau_2 x, \tau_2 y \end{cases}$$

-

$$\sim x = \begin{cases} x^c & \textit{if } \tau_1 x \\ \neg x & \textit{if } \tau_2 x \end{cases}$$

-

$$x \rightsquigarrow y = \begin{cases} x \cup y^c & \textit{if } \tau_1 x, \tau_1 y \\ \left[\bigcup_{z \in y}(x \cup z^c)\right] & \textit{if } \tau_1 x, \tau_2 y \\ x \implies y & \textit{if } \tau_2 x, \tau_2 y \\ \left[\bigcup_{z \in x}(z \cup y^c)\right] & \textit{if } \tau_2 x, \tau_1 y \end{cases}$$

-

$$x \twoheadrightarrow y = \begin{cases} [x \cup y^c] & \textit{if } \tau_1 x, \tau_1 y \\ \left[\bigcup_{z \in y}(x \cup z^c)\right] & \textit{if } \tau_1 x, \tau_2 y \\ x \implies y & \textit{if } \tau_2 x, \tau_2 y \\ \left[\bigcup_{z \in x}(z \cup y^c)\right] & \textit{if } \tau_2 x, \tau_1 y \end{cases}$$

It should be noted that $\odot$ is a very restrictive operation (because the commonality is over a class) when one of the argument is of type τ_1 and the other is of type τ_2. An alternative is to replace it with $\circ$ defined by the Eq. 7.

$$x \circ a = \begin{cases} x \cap a & \textit{if } \tau_1 x, \tau_1 a \\ \left[x \cap \left(\bigcup_{z \in a} z\right)\right] & \textit{if } \tau_1 x, \tau_2 a \\ \left[\left(\bigcup_{z \in x} z\right) \cap a\right] & \textit{if } \tau_2 x, \tau_1 a \\ x \sqcap a & \textit{if } \tau_2 x, \tau_2 a \end{cases} \tag{7}$$

Definition 7. *In the above context a partial algebra of the form*

$$W = \left\langle \wp(S) \cup \wp(S) | \approx, \neg, \sim, \oplus, \odot, \blacklozenge, \mathfrak{L}, 0, 1, \bot, \top \right\rangle$$

of type $(1, 1, 2, 2, 1, 1, 0, 0, 0, 0)$ *is a* concrete enriched pre-rough algebra *(CERA) if a pre-rough algebra structure is induced on* $\wp(S)| \approx$. Concrete enriched rough algebras *can be defined in the same manner. If the approximation space is* X, *then the derived CERA will be denoted by* $\mathfrak{W}(X)$. *Note that the two implication-like operations are definable in terms of other basic functions. A CERA in which* $\odot$ *has been replaced by* $\circ$ *is said to be a* soft *CERA.*

Proposition 3. *CERAs are well defined because of the representation theory of pre-rough algebras.*

Theorem 2. *A CERA satisfies all the following: (The first two conditions essentially state that the τ_is are abbreviations)*

$$(x \rightsquigarrow x = \top \longleftrightarrow \tau_1 x) \tag{type-1}$$

$$(\neg x = \neg x \longleftrightarrow \tau_2 x) \tag{type-2}$$

$$\sim\sim x = x; \ \mathfrak{LL}x = \mathfrak{L}x; \ \blacklozenge\mathfrak{L}x = \mathfrak{L}x \tag{ov-1}$$

$$\mathfrak{L}x \oplus x = x; \ \mathfrak{L}x \odot x = \mathfrak{L}x; \ \blacklozenge x \oplus x = \blacklozenge x; \ \blacklozenge x \odot x = x \tag{ov-2}$$

$$\mathfrak{L}\blacklozenge x = \blacklozenge x; \ x \oplus x = x; \ x \odot x = x \tag{ov-3}$$

$$(\tau_1 x \longrightarrow \sim x \oplus x = \top); \ (\tau_2 x \longrightarrow \sim \mathfrak{L}x \oplus \mathfrak{L}x = 1) \tag{qov-1}$$

$$\sim \bot = \top; \ \sim 0 = 1 \tag{qov-2}$$

$$x \oplus (x \oplus (x \oplus y)) = x \oplus (x \oplus y); \ x \odot (x \odot (x \odot y)) = x \odot (x \odot y) \tag{u1}$$

$$x \oplus y = y \oplus x; \ x \odot y = y \odot x \tag{u2}$$

$$(\tau_i x, \tau_i y, \tau_i z \longrightarrow x \oplus (y \oplus z) = (x \oplus y) \oplus z); i = 1, 2 \tag{ter(i1)}$$

$$(\tau_i x, \tau_i y, \tau_i z \longrightarrow x \oplus (y \odot z) = (x \oplus y) \odot (x \oplus z)); i = 1, 2 \tag{ter(i2)}$$

$$(\tau_i x, \tau_i y, \tau_i z \longrightarrow x \odot (y \odot z) = (x \odot y) \odot z); i = 1, 2 \tag{ter(i3)}$$

$$(\tau_i x, \tau_i y \longrightarrow x \oplus (x \odot y) = x, \sim (x \odot y) = \sim x \oplus \sim y); i = 1, 2 \tag{bi(i)}$$

$$(\tau_1 x, \tau_2 y, x \oplus y = y \longrightarrow \blacklozenge x \oplus y = y) \tag{bm}$$

$$(\tau_1 x, (1 \odot x = y) \vee (y = x \oplus 0) \longrightarrow \tau_2 y) \tag{hra1}$$

Definition 8. *An* abstract enriched pre-rough partial algebra *(AERA) will be a partial algebra of the form*

$$S = \langle \underline{S}, \neg, \sim, \oplus, \odot, \blacklozenge, \mathfrak{L}, 0, 1, \bot, \top \rangle$$

(of type $(1, 1, 2, 2, 1, 1, 0, 0, 0, 0)$) that satisfies:

RA dom$(\neg)$ *along with the operations* $(\oplus, \odot, \blacklozenge, \mathfrak{L}, \sim, 0, 1)$ *restricted to it and the definable* $\Rightarrow$ *forms a pre-rough algebra,*

BA $\underline{S} \setminus$ dom$(\neg)$ *with the operations* $(\oplus, \odot, \blacklozenge, \mathfrak{L}, \sim, \top, \bot)$ *restricted to it forms a topological Boolean algebra (with an interior and closure operator),*

IN *Given the definitions of type-1, type-2, all of u1, u2, ter(ij), bi(i), bm and hra hold for any i, j.*

Note that AERAs are actually defined by a set of quasi equations.

Theorem 3. *Every AERA S has an associated approximation space X (up to isomorphism), such that the derived CERA $\mathfrak{W}(X)$ is isomorphic to it.*

Proof. Given S, the topological Boolean algebra and the pre-rough algebra part can be isolated as the types can be determined with $\neg$, $\oplus$ and the 0-place operations. The representation theorems for the parts can be found in [83] and [11] respectively.

Suppose $\mathfrak{W}(Y)$ is a CERA formed from the approximation space Y (say) determined by the two parts. If Y is not isomorphic to X as a relational structure, then it is possible to derive an easy contradiction to the representation theorem of the parts.

Suppose $\mathfrak{W}(X)$ is not isomorphic to S, then given the isomorphisms between respective parts, at least one instance of $x \oplus' y \neq x \oplus y$ or $x \odot' y \neq x \odot y$ (for a type-1 x and a type-2 y with $'$ denoting the interpretation in $\mathfrak{W}(X)$). But, as type-1 elements can be mapped into type-2 elements (using 0 and $\oplus$), this will result in a contradiction to the representation theorem of parts. □

4.2   Dialectical Rough Logic

A natural dialectical interpretation can be assigned to the proposed semantics. A subset of the original approximation space has a dual interpretation in the classical and rough semantic domain. While an object in the latter relates to a set of objects in the classical semantic domain, it is not possible to transform objects in the rough domain to the former. For this reason, the universe is taken to be the set of tuples having the form $\{(x, 0 \oplus x) : \tau_1 x\} \cup \{(b, x) : \tau_2 b \,\&\, \tau_1 x \,\&\, x \oplus 0 = b\} = K$ (x and b being elements of a CERA). This universe is simply the described dialectical relation between objects in the two domains mentioned above. Other dialectical relations can also be derived from the specified one.

Definition 9. *A concrete rough dialectical algebra (CRAD) will be a partial algebra on K along with the operations $+, \cdot, \mathfrak{L}^*, \neg, \sim$ and 0-place operations $(\top, 1), (1, \top), (0, \bot), (\bot, 0)$ defined by (EUD is an abbreviation for Else Undefined)*

$$(a,b) + (c,e) = \begin{cases} (a \oplus c,\, b \oplus e) & \text{if } \tau_i a,\, \tau_i c \text{ if defined} \\ (a \oplus c,\, e \oplus a) & \text{if } \tau_1 a,\, \tau_2 c,\, (e \oplus a) \oplus 0 = a \oplus c,\ EUD \\ (a \oplus e,\, c \oplus b) & \text{if } \tau_2 a,\, \tau_1 c,\, (c \oplus b) \oplus 0 = a \oplus e,\ EUD \end{cases}$$

$$(a,b) \cdot (c,e) = \begin{cases} (a \odot c,\, b \odot e) & \text{if } \tau_i a,\, \tau_i c \text{ if defined} \\ (a \odot c,\, e \odot a) & \text{if } \tau_1 a,\, \tau_2 c,\, (e \odot a) \odot 0 = a \odot c,\ EUD \\ (a \odot e,\, c \odot b) & \text{if } \tau_2 a,\, \tau_1 c,\, (c \odot b) \odot 0 = a \odot e,\ EUD \end{cases}$$

$\mathfrak{L}^*(a, b) = (\mathfrak{L}a, \mathfrak{L}b)$ *if defined and* $\sim (a, b) = (\sim a, \sim b)$ *if defined.*

Illustrative Example

The following example is intended to illustrate key aspects of the theory invented in this section.

Let $S = \{a, b, c, e, f, q\}$ and R be the least equivalence relation generated by

$$\{(a, b),\ (b, c),\ (e, f)\}.$$

Under the conditions, the partition corresponding to the equivalence is

$$\mathcal{G} = \{\{a, b, c\},\ \{e, f\},\ \{q\}\}.$$

The quotient $S|R$ is the same as $\mathcal{G}$. In this example strings having the form ef are used as an abbreviation for $\{e, f\}$.

The set of triples having the form (x, x^l, x^u) for any $x \in \wp(S)$ is as below:

- $(a, \emptyset, abc)$, $(b, \emptyset, abc)$, $(c, \emptyset, abc)$, $(e, \emptyset, ef)$,
- $(f, \emptyset, ef)$, (q, q, q), $(ab, \emptyset, abc)$, $(ac, \emptyset, abc)$,
- $(ae, \emptyset, abcef)$, $(af, \emptyset, abcef)$, $(aq, q, abcq)$, $(bc, \emptyset, abc)$,
- $(be, \emptyset, abcef)$, $(bf, \emptyset, abcef)$, $(bq, q, abcq)$, $(ec, \emptyset, abcef)$,
- $(cf, \emptyset, abcef)$, (ef, ef, ef), (eq, q, efq), (fq, q, efq), (abc, abc, abc),
- $(abe, \emptyset, abcef)$, $(abf, \emptyset, abcef)$, $(abq, q, abcq)$, $(bce, \emptyset, abcef)$,
- $(bcf, \emptyset, abcef)$, $(bcq, q, abcq)$, $(ace, \emptyset, abcef)$, $(acf, \emptyset, abcef)$,
- $(acq, q, abcq)$, $(aef, ef, abcef)$, $(bef, ef, abcef)$, $(cef, ef, abcef)$,
- (aeq, q, S), (afq, q, S), (beq, q, S), (bfq, q, S), (ceq, q, S),
- (cfq, q, S), (efq, efq, efq), $(abce, abc, abcef)$, $(abcf, abc, abcef)$,
- $(abcq, abcq, abcq)$, $(abef, ef, abcef)$, $(abeq, q, S)$, $(abfq, q, S)$,
- $(bcef, ef, abcef)$, $(bceq, q, S)$, $(bcfq, q, S)$, $(aceq, q, S)$, $(acfq, q, S)$,
- $(acef, ef, abcef)$, $(aefq, efq, S)$, $(befq, efq, S)$, $(cefq, efq, S)$,
- $(abcef, abcef, abcef)$, $(abceq, abcq, S)$, $(abcfq, abcq, S)$,
  $(acefq, efq, S)$,
- $(bcefq, efq, S)$, (S, S, S).

From the values, it can be checked that the sets of roughly equivalent objects are

- $\{a, b, c, ab, ac, bc\}$, $\{e, f\}$, $\{q\}$ - the reader should note that elements belonging to the set are themselves sets.
- $\{ae, af, be, bf, ce, cf, abe, ace, acf, abf, bce, bcf\}$.
- $\{abq, acq, bcq, aq, bq, cq\}$, $\{abce, abcf\}$, $\{aef, bef, cef, abef, acef, bcef\}$.
- $\{eq, fq\}$, $\{abc\}$, $\{abcef\}$, $\{ef\}$, $\{abcq\}$, $\{efq\}$, S.
- $\{aeq, beq, ceq, afq, bfq, cfq, abeq, aceq, bceq, abfq, bcfq, acfq\}$.
- $\{aefq, befq, cefq, abefq, bcefq, acefq\}$ and $\{abceq, abcfq\}$.

The domain is taken to be $\wp(S) \cup \wp(S)|R$ in case of a CERA and interpretations of the unary operations are obvious. The nontrivial binary operations get interpreted as below:

$$bc \oplus [bf] = [bc \cup abcef] = [abcef]$$

$$b \oplus [f] = [bef] = \{aef, bef, cef, abef, acef, bcef\}$$

$$bc \odot [bf] = \left[bc \cap \bigcap \{ae, af, be, bf, ce, cf, abe, ace, acf, abf, bce, bcf\} \right] = [\emptyset]$$

$$b \odot [f] = [b \cap \bigcap \{e, f\}] = [\emptyset]$$

$$abcq \odot [q] = [q]$$

$$bc \rightsquigarrow [bf] = \left[\bigcup_{z \in [bf]} (bc \cup z^c) \right] = S$$

$$bc \rightsquigarrow [abceq] = \{abcef\}$$

$$bc \rightsquigarrow [S] = \{a, b, c, ab, bc, ac\}$$

$$[bf] \rightarrow bc = \left[\bigcup_{z \in [bf]} (z \cup bc^c) \right] = [S]$$

The universe of the CRAD associated with a CERA S is formed as the set of pairs having the form $(x, 0 \oplus x)$ and $(x \oplus 0, x)$ under the restriction that $\tau_1 x$ holds. So in the present example, some elements belonging to the universe are $(a, \{a, b, c, ab, ac, bc\})$, $(\{a, b, c, ab, ac, bc\}, bc)$, $(fq, \{eq, fq\})$.

$$(a, b) + (c, e) = \begin{cases} (a \oplus c,\ b \oplus e) & \text{if } \tau_i a,\ \tau_i c \\ (a \oplus c,\ e \oplus a) & \text{if } \tau_1 a,\ \tau_2 c,\ (e \oplus a) \oplus 0 = a \oplus c,\ \text{EUD} \\ (a \oplus e,\ c \oplus b) & \text{if } \tau_2 a,\ \tau_1 c,\ (c \oplus b) \oplus 0 = a \oplus e,\ \text{EUD} \end{cases}$$

To compute $(a, \{a, b, c, ab, ac, bc\}) + (\{eq, fq\}, fq)$ it is necessary to compute

- $a \oplus \{eq, fq\} = \{aefq, befq, cefq, abefq, bcefq, acefq\}$.
- $a \oplus fq = afq$.
- $afq + 0 = \{aeq, beq, ceq, afq, bfq, cfq, abeq, aceq, bceq, abfq, bcfq, acfq\}$.
- Clearly, $a \oplus \{eq, fq\} \neq afq + 0$.
- So $(a, \{a, b, c, ab, ac, bc\}) + (\{eq, fq\}, fq)$ is not defined.

Also note that $(a, \{a, b, c, ab, ac, bc\}) + (b, \{a, b, c, ab, ac, bc\})$ is defined, but $(a, \{a, b, c, ab, ac, bc\}) + (bc, \{a, b, c, ab, ac, bc\})$ is not.

Dialectical Negations in Practice. Real examples can be constructed (from the above example) by assigning meaning to the elements of S. An outline is provided below.

- Suppose $\{a, b, c, e, f, q\}$ is a set of attributes of lawn tennis players.
- For the above sentence to fit into the example context, it is necessary that they can be freely collectivized. This means that no combination of attributes should be explicitly forbidden.
- While sets of the form ec refer to players with specific attributes, roughly equal objects like $\{aef, bef, cef, abef, acef, bcef\}$ can be read as new class labels. Members of a class can be referred to in multiple ways.
- The operations defined permit aggregation, commonality and implications in novel ways. The $\oplus$ operation in particular can generate new classes that contain classes of roughly equal players and players with specific attributes in a mereological sense. For example, it can answer questions of the form: *What features can be expected of those who have a great backhand and have roughly equal performance on hard courts?*
- A number of dialectical negations can be defined in the situation. For example,

- An object x is in a sense dialectically opposed to a roughly equivalent set of objects H.
- An object x is in a sense dialectically opposed to $x \oplus H$.
- Likewise other operations defined provide more examples of opposition.

More generally, similar dialectical negation predicates can be defined over CRADs and new kinds of logical rules can be specified that concern transformation of one instance of dialectical negation into another, restricted introduction and inference rules. Related logic will appear separately.

4.3   Parthoods

In CERA related contexts, the universe is taken to be $W = \wp(S) \cup \wp(S) | \approx$ and the most natural parthoods are ones defined from the aggregation and commonality operators. Parthoods can also be based on information content and ideas of *consistent comparison*.

Definition 10. *The following parthoods can be defined in the mixed semantic domain corresponding to* CERA *on* W

$$\mathbf{P}_o ab \leftrightarrow [a] \leq [b] \qquad \text{(Roughly Consistent)}$$
$$\mathbf{P}_\oplus ab \leftrightarrow a \oplus b = b \qquad \text{(Additive)}$$
$$\mathbf{P}_\odot ab \leftrightarrow a \odot b = a \qquad \text{(Common)}$$

$\leq$ is the lattice order used in the definition of pre-rough algebras. Note that the operations $\oplus, \odot$ are not really required in the definitions of the last two parthoods which can equivalently be defined using the associated cases. This is significant as one of the goals is to *count the objects in specialized ways to arrive at semantics that make sense* [7].

In the definition of the base set K of CRAD, K is already a dialectic relation. Still definitions of parthoods over it make sufficient sense.

Definition 11. *The relation* $\mathbf{P}_\aleph$, *defined as below, will be called the* natural parthood *relation on* K:

$$\mathbf{P}_\aleph ab \leftrightarrow [e_1 a] \leq [e_1 b] \ \& \ [e_2 a] \leq [e_2 b],$$

where the operation e_i *gives the ith component for* $i = 1, 2$.

Admittedly the above definition is not internal to K as it refers to things that do not exist within K at the object level of reasoning.

5   General Parthood

Parthood can be defined in various ways in the framework of rough sets in general and granular operator spaces in particular. The rough inclusion defined earlier in the background section is a common example of parthood. Some others have been introduced in Definition 10 and in the illustrative example. The following are more direct possibilities that refer to a single non classical semantic domain (the parthoods of CERA are in the classical domain):

$$\mathbf{P}ab \longleftrightarrow a^l \subseteq b^l \qquad \text{(Very Cautious)}$$
$$\mathbf{P}ab \longleftrightarrow a^l \subseteq b^u \qquad \text{(Cautious)}$$
$$\mathbf{P}ab \longleftrightarrow a^l \subseteq b^u \setminus b^l \qquad \text{(Lateral)}$$
$$\mathbf{P}ab \longleftrightarrow a^u \subseteq b^u \qquad \text{(Possibilist)}$$
$$\mathbf{P}ab \longleftrightarrow a^u \subseteq b^l \qquad \text{(Ultra Cautious)}$$
$$\mathbf{P}ab \longleftrightarrow a^u \subseteq b^u \setminus b^l \qquad \text{(Lateral+)}$$
$$\mathbf{P}ab \longleftrightarrow a^u \setminus a^l \subseteq b^u \setminus b^l \qquad \text{(Bilateral)}$$
$$\mathbf{P}ab \longleftrightarrow a^u \setminus a^l \subseteq b^l \qquad \text{(Lateral++)}$$
$$\mathbf{P}ab \longleftrightarrow (\forall g \in \mathcal{G})(g \subseteq a \longrightarrow g \subseteq b) \qquad \text{(G-Simple)}$$

All these are valid concepts of parthoods that make sense in contexts as per availability and nature of information. Very cautious parthood makes sense in contexts with high cost of misclassification or the association between properties and objects is confounded by the lack of clarity in the possible set of properties. G-Simple is a version that refers granules alone and avoids references to approximations.

The above mentioned list of parthoods can be more easily found in decision making contexts in practice.

Example 5. Consider, for example, the nature of diagnosis and treatment of patients in a hospital in war torn Aleppo in the year 2016. The situation was characterized by shortage of medical personnel, damaged infrastructure, large number of patients and possibility of additional damage to infrastructure. Suppose patient B has bone fractures and a bullet embedded in their arm and patient C has bone fractures and shoulder dislocation due to a concrete slab in free fall, that only one doctor is on duty, and suppose that either of the two patients can be treated properly due to resource constraints. Suppose also that the doctor in question has access to some precise and unclear diagnostic information on medical conditions and that all of this data is not in tabular form.

In the situation, decision making can be based on available information and principles like

- Allocate resources to the patient who is definitely in the worst state - this decision strategy can be corresponded to *very cautious* parthoods,
- Allocate resources to the patient who seems to be in the worst state - this decision strategy can be corresponded to *cautious* parthoods,
- Allocate resources to the patient who is possibly in the worst state - this decision strategy can be corresponded to *possibilist* parthoods.
- Allocate resources to the patient who is likely to show more than the default amount of improvement - this decision strategy can be corresponded to the *bilateral* parthoods.

- If every symptom or unit complication that is experienced or certainly likely to be experienced by patient 1 is also experienced or is certainly likely to be experienced by patient 2, then prefer treating patient 2 over patient 1 - this decision strategy can be corresponded to the *g-simple* parthoods with symptoms/unit complications as granules.

Parthood can be associated with both dialectic and dialethic statements in a number of ways. Cautious parthood is consistent with instances having the form $\mathbf{P}ab$ and $\mathbf{P}ba$. It is by itself a dialectic relation within the same domain of discourse. Dialectics between parthoods in different semantic domains are of greater interest and will be considered in subsequent sections.

The *apparent parthood* relations considered in later sections of this paper typically arise from lack of clarity in specification of properties or due to imprecision (of fuzziness). This is illustrated in the next example.

Example 6 (Cold vs Influenza).
Detection of influenza within 48 h of *catching it* is necessary for effective treatment with anti-virals, but often patients fail to understand subtleties in distinguishing between cold and influenza. It is also not possible to administer comprehensive medical tests in a timely cost-effective way even in the best of facilities. So ground breaking insights even in restricted contexts can be useful.

The two medical conditions have similar symptoms. These may include *fever* - as indicated by elevated temperatures, *feverishness* - as indicated by personal experience (this may not be accompanied by fever), *sneezing, running nose, blocked nose, headache of varying intensity, cough and body pain.* Body pain is usually a lot more intense in case of flu (but develops after a couple of days).

Clearly, in the absence of confirmatory tests patients can believe both *instances of cold is apparently part of flu* and *instances of flu is apparently part of cold.* These statements are in dialectical contradiction to each other but no dialetheias are involved. Cold and flu are also in dialectical contradiction to each other. This is a useful and relevant formalism.

It is another matter that if glutty negations are permitted then *apparently part of* can as well be replaced by *part of.*

6   Figures of Dialectical Opposition

The scope of counting strategies and nature of possible models can be substantially improved when additional dialectical information about the nature of order-theoretic relation between rough and crisp objects is used. In the literature on generalizations of the square of opposition to rough sets, as in [14,84], it is generally assumed that realizations of such relations is the end result of semantic computations. This need not necessarily be so for reasons that will be explained below. In classical rough sets, a subset X of objects O results in a tri-partition of O into the regions $L(X)$ (corresponding to lower approximation of X), $B(X)$ (the boundary region) and $E(X)$ (complement of the upper approximation). These form a hexagon of opposition (see Fig. 2). In more general rough sets, this diagram generalizes to cubes of opposition [14].

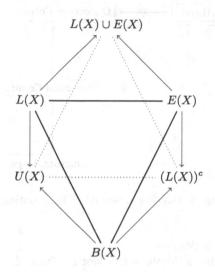

Fig. 2. Hexagon of opposition

The general strategy used in a forthcoming paper is illustrated in Fig. 3.

Some idea of parthood related ordering in the form of the following relations (a deeper understanding of ontology is essential for making sense of the vague usage (this is explored in [16,22])) can suffice in application contexts of any dialectical generalized scheme of the square or hexagon of opposition (examples have already been provided earlier):

AP Is Apparently Part of: understood from class, property, expected behavior, or some other perspective.
APN Is Apparently not Part of.
AP0 Is Apparently Neither Part of Nor Not Part of.
CP Is Certainly Part of.
CPN Is Certainly Not Part of.
CP0 Is Certainly Neither Part of Nor Not Part of. (This is intended to convey uncertainty.)
AI Is Apparently Indistinguishable from.
CI Is Certainly Indistinguishable from.
AW Is Apparently a Whole of
AWN Is Apparently Not a Whole of.
AW0 Is Apparently Neither a Whole of Nor Not a Whole of.
CW Is Certainly a Whole of.

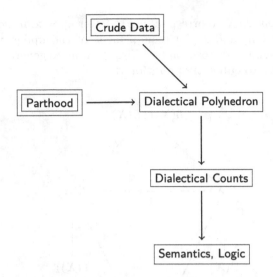

Fig. 3. Dialectical semantics by counting

CWN Is Certainly not a Whole of.
CW0 Is Certainly Neither a Whole of Nor Not a Whole of.

By the word apparently, the agent may be referring to the lack of models, properties possessed by the objects, relativized views of the same among other possibilities. For example, the word *apparently* can refer to the absence of any clear models about connections between diseases in data from a hospital chain in a single city or it can refer to problems caused by lack of data or relativizations about expected state of affairs relative to pre-existing models. Fuzzy and degree valuations of these perceptions are even less justified due to the use of approximate judgments. Predicates like AP0, CP0 are needed for handling indecision (which is likely to be happen often in practice).

As pointed out by a reviewer, AP, APN and AP0 form three-fourths of Belnap's *useful 4-valued logic* [85]. In this regard it should be noted that only those dialectical oppositions that can be sustained by inference procedures for progression of knowledge are relevant for logic.

The above set of predicates can be split into the following subsets of interest:

- Pure Apparence: AP, APN, AP0, AW, AWN, AW0
- Pure Certainty: CP, CPN, CP0, CW, CWN, CW0
- Mixed Apparence: AP, APN, AP0, AW, AWN, AW0, AI
- Mixed Certainty: CP, CPN, CP0, CW, CWN, CW0, CI
- Pure: AP, APN, AP0, AW, AWN, AW0, CP, CPN, CP0, CW, CWN, CW0
- Mixed: Union of all the above.

Before proceeding further it is necessary to fix the philosophical concepts of *Contradiction, Contrariety, Sub-contrariety and Sub-alternation* because the literature on these concepts is vast and there is much scope for varying their meaning (see for example [12,13]). One way of looking at the connection between truth value assignments and sentences, necessary for the diagram to qualify in the square of opposition (generalized) paradigm, is illustrated in Tables 1, 2, 3 and 4 below (NP means the assignment is not possible):

Table 1. Contrariety

α	β	$\mathbf{CY}(\alpha, \beta)$
T	T	NP
T	F	T
F	T	T
F	F	T

Table 2. Contradiction

α	β	$\mathbf{C}(\alpha, \beta)$
T	T	NP
T	F	T
F	T	T
F	F	NP

Table 3. Sub-contrariety

α	β	$\mathbf{SCY}(\alpha, \beta)$
T	T	T
T	F	T
F	T	T
F	F	NP

Table 4. SubAlternation

α	β	$\mathbf{AN}\,(\alpha, \beta)$
T	T	T
T	F	NP
F	T	T
F	F	T

The P^Q semantics [86] tries to take a simplified view of the situation. It may appear that the main problem with the proposal is a lack of suitable logical operators. But this drawback is not likely to be that significant for the counting based approach introduced in this paper, and developed further in a forthcoming paper. The P^Q approach in question is to look for answers to the questions:

- TT: Can the sentences be true together?
- FT: Can the sentences be false together?

After finding those answers, categories can be worked out according to Table 5.

Table 5. An opposition

TT	FT	
	1	0
1	Sub-alternation	Sub-contrariety
0	Contrariety	Contradiction

But dialectical contradiction requires additional categories that relate to the following questions:

- Dialethia: Can any one of the two statements be both true and false together? (Let $\delta(A)$ be the statement that A is both false and true together).
- Bi-Dialectic: Is either statement in dialectical opposition to the other? (Let $\sqsupset(A, B)$ be the statement that A is dialectically opposed to B).
- Dialectic: Is either statement a statement expressing dialectical opposition? (Let $\beta(A)$ be the statement that A expresses dialectical opposition with β being a particular associated predicate).

The above realization of the concept of dialetheia is pretty clear for implementation, but the latter two forms of dialectic depend on the choice of predicates and so many interpretations would be possible. In a typical concrete case, the parthood(s), the dialectical predicate and figure of opposition should be defined in order to obtain concrete answers.

The Question-Answer Semantic approach (QAS) of [13] constitutes a relatively more complete strategy in which the *sense* of a sentence α is an ordered set of questions $Q(\alpha) = \langle q_1(\alpha), \ldots, q_r(\alpha) \rangle$ and its *reference* is an ordered set of answers $A(\alpha) = \langle a_1(\alpha), \ldots, a_r(\alpha) \rangle$. These answers can be coerced to binary form (with possible responses being Yes or No).

If the dialectical approach of the present paper is extended to QAS approach, then the number of possible questions (like can question A and B be true together?) becomes very large and suitable subsets that are as efficient as the whole would be of interest. The possibilities are indicated in Table 6.

Table 6. Potential combinations

A	T	F	⊐	δ	δ	β	β	β	β	δ	δ	δ
B	T	F	⊐	δ	⊐	β	⊐	T	F	T	F	β

Connectives and operations can be involved in the definition of these predicates, but in the general case the meta concept of *suitable subsets* can only be roughly estimated and not defined unambiguously. Even apparent-parthood related statements can be handled by the answer set after fixing the necessary subsets of the question set. In a separate paper, it is shown by the present author that counting procedures can be initiated with the help of this strategy.

Relative to the nature of truth values, two approaches to the problem can be adopted:

1. Keep the concept of truth and falsity fixed and attempt suitable definitions of oppositions or
2. Permit variation of truth and falsity values. This in general would amount to deviating further from classical opposition paradigms.

6.1   Classical Case-1: Fixed Truth

When the concept of truth is not allowed to vary beyond the set $\{T, F\}$, then it is apparently possible to handle the cases involving apparent parthood without special external rules. A natural question that arises in the context of certain parthoods is about the admissibility of truth values. These aspects are considered in this subsection.

One instructive (but not exhaustive) way is to read CPab as all a with property $\pi(a)$ and none of properties in $\neg\pi(a)$ (in the domain of discourse) are part of any b with property $\pi(b)$. As a consequence CPNab is all a with property $\pi(a)$ and none of properties in $\neg\pi(a)$ (in the domain of discourse) are not part of any b with property $\pi(b)$.

In rough sets, the association of objects with properties happen only when mechanisms of associations are explicitly specified or are specifiable. There is much freedom to choose from among different mechanisms of associations in a abstract perspective. In praxis, these choices become limited but rarely do they ever become absent. Implicit in all this is the assumption of stable choice among possible mechanisms of associations. The stability aspect is an important research direction.

If truth tables for determining the nature of opposition is attempted using ideas of state resolution (instead of connectives) then perplexing results may happen. The choice of connectives is in turn hindered by an excess of choice. So the minimalist perspective based on two questions or the QAS-type approach should be preferred.

The following two theorems require interpretation.

Theorem 4. *The truth tables corresponding to two of the pairs formed from AP, APN, AP0 have the form indicated in Tables 7 and 8 ($P * Q$ is the resolution of the state relating to P and Q. This is abbreviated in the tables by $*$):*

Table 7. Contradiction?

APpq	APNpq	$*$
T	T	IN
T	F	T
F	T	T
F	F	IN

Table 8. Contradiction?

APpq	AP0pq	$*$
T	T	IN
T	F	T
F	T	T
F	F	IN

(IN is an abbreviation for indeterminate.)

Proof. The proof is direct. However, the interpretation is open and it is possible to read both tables as corresponding to contradiction. $\square$

Theorem 5. *The truth tables corresponding to the pairs formed from CP, CPN, CP0 and the pair CP, CI have the form indicated in Tables 9, 10, 11 and 12 (NP is an abbreviation for not possible):*

Table 9. Contradiction

CP*pq*	CPN*pq*	*
T	T	NP
T	F	T
F	T	T
F	F	NP

Table 10. Contrariety

CP*pq*	CP0*pq*	*
T	T	NP
T	F	T
F	T	T
F	F	T

Table 11. Contradiction

CPN*pq*	CP0*pq*	*
T	T	NP
T	F	T
F	T	T
F	F	NP

Table 12. Sub-alternation

CI*pq*	CP*pq*	*
T	T	T
T	F	NP
F	T	T
F	F	T

Proof. The proof consists in checking the possibilities by cases. In the table for CP and CP0, the last line is justified because no possibilities are covered by the last column.

From the safer (and questionable) two question framework, the above two theorems have the following form:

Theorem 6. *The answers to the two simultaneity (Sim) questions for the pairs formed from AP, APN, AP0 are in Tables 13 and 14.*

Table 13. Contradiction

AP*ab*	APN*ab*	Sim
T	T	NP
F	F	NP

Table 14. Sub-contrariety

AP*ab*	AP0*ab*	Sim
T	T	T
F	F	NP

(NP is an abbreviation for not possible.)

Proof. In the table for AP, AP0, TT discludes all possibilities and therefore yields T. Other parts are not hard to prove.

Theorem 7. *The simultaneity data corresponding to the pairs formed from CP, CPN, CP0 are in Table 15, 16, 17 and 18.*

Proof. In Table 16, for example, the question is can both CP*ab* be false and CP0*ab* be false? As the situation is impossible, NP is the result.

Table 15. Contradiction

CPab	CPNab	Sim
T	T	NP
F	F	NP

Table 16. Contradiction

CPab	CP0ab	Sim
T	T	NP
F	F	NP

Table 17. Contradiction

CPNab	CP0ab	Sim
T	T	NP
F	F	NP

Table 18. Sub-alternation

CIab	CPab	Sim
T	T	T
F	F	T

6.2 Case-2: Pseudo Gluts

A minimalist use of assumptions on possible grades of truth in the cases admitting apparent parthood leads to the following diagram of truth values. The figure is biased against falsity because in the face of contradiction agents are expected to be truth seeking - this admittedly is a potentially contestable philosophical statement.

Reading of truth tables in relation to state transition based conjunction is also relevant for dialectical interpretation. But these are not handled by the above tables and will be part of future work.

For the dialectical counting procedures introduced in the next section, the basic contexts are assumed to be very minimalist and possibly naive. In these some meta principles on aggregation of truth can be useful or natural. The states of truth mentioned in Fig. 4 relate to the following meta method of handling apparent truth. These will be referred as *Truth State Determining Rules* (TSR). In the rules α, β are intended in particular for formulas having the form $\mathbf{P}ab$ and variants.

Truth State Determining Rules

- If α is apparently true and β supports it, then α becomes more true.
- If α is apparently true and β opposes it, then α becomes less true.
- If α is less true (than true is supposed to be) and β opposes it, then α becomes even less true.
- If α is apparently false and β opposes it, then α becomes less false.
- In the figure, T denotes an intermediate truth value that can become stronger T_*, T^* or weaker $\mathsf{T}_\ominus$, $\mathsf{T}^\ominus$. This is because operators (apparently) like *less* and *even less* are available.

The above list of rules can be made precise using the distance between vertices in the graph and thus it would be possible to obtain truth values associated with combinations of sentences involving apparent parthood alone.

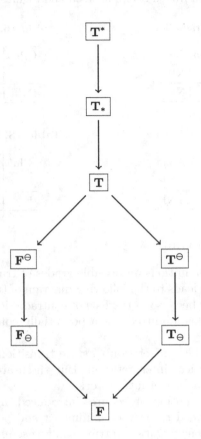

Fig. 4. Weak and strong truths

6.3   Counting Procedures and Dialectical Opposition

A brief introduction to the dialectical counting procedures and semantics invented by the present author is included in this subsection. The full version will appear in a separate paper.

New concepts of rough natural number systems were introduced in [7] from both formal and less-formal perspectives. These are useful for improving most rough set-theoretical measures in general rough sets and in the representation of rough semantics. In particular it was proved that the algebraic semantics of classical RST can be obtained from the developed dialectical counting procedures. In the counting contexts of [7], a pair of integers under contextual rules suffices to indicate the *number* associated with the element in an instance of the counting scheme under consideration. This is because those processes basically use a *square of discernibility* with the statements at vertices having the following form (Fig. 5):

- IS.NOT(a, b) meaning a is not b.
- IS(a, b) meaning a is identical with b.
- IND(a, b) meaning a is indiscernible from b.
- DIS(a, b) meaning a is discernible from b.

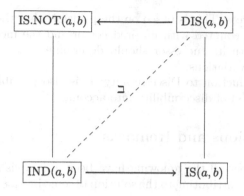

Fig. 5. Rough counting

A particular example of such a counting method is the *indiscernible predecessor based primitive counting* (IPC). The adjective *primitive* is intended to express a minimal use of granularity and related axioms. Let $S = \{x_1, x_2, \ldots, x_k, \ldots, \}$ be an at most countable set of elements in ZF that correspond to some collection of objects. If R is a binary relation on S that corresponds to *is weakly indiscernible from*, then IPC takes the following form:

Indiscernible Predecessor Based Primitive Counting (IPC). In this form of 'counting', the relation with the immediate preceding step of counting matters in a crucial way.

1. Assign $f(x_1) = 1_1 = s^0(1_1)$.
2. If $f(x_i) = s^r(1_j)$ and $(x_i, x_{i+1}) \in R$, then assign $f(x_{i+1}) = 1_{j+1}$.
3. If $f(x_i) = s^r(1_j)$ and $(x_i, x_{i+1}) \notin R$, then assign $f(x_{i+1}) = s^{r+1}(1_j)$.

For example, if $S = \{f, b, c, a, k, i, n, h, e, l, g, m\}$ and R is the reflexive and transitive closure of the relation

$$\{(a, b), (b, c), (e, f), (i, k), (l, m), (m, n), (g, h)\}$$

then S can be counted in the presented order as below.

$$\{1_1, 2_1, 1_2, 1_3, 2_3, 1_4, 2_4, 3_4, 1_5, 2_5, 1_6, 2_6\}, \tag{IPC}$$

Rough counting methods like the above have nice algebraic models and can also be used for representing semantics. The dialectical approaches of the present paper motivate generalizations based on the following:

- Counting by Dialectical Mereology: This method of counting is intended to be based on the principle that the mereological relation of the object being counted with its predecessors should determine its count and the enumeration should be on convex regular polygons (including squares and hexagons of opposition), polyhedrons or generalizations thereof (polytopes) of dialectical and classical opposition.

- Counting by Threes: This is based on the principle that the relation between the object being counted with its predecessor and the mereological relation of the object with its successor should determine its count. The approach admits of many variations.
- Counting by Reduction to Discernibility: It is also possible to count taking increasing scope(s) of discernibility into account.

Further Directions and Remarks

In this research paper all the following have been accomplished by the present author (apart from contributing to the solution of some inverse problem contexts)

- Formalization of possible concepts of dialectical contradiction has been done in one possible way using object level predicates.
- The difference between dialetheias and dialectical contradiction has been clarified. It has been argued that dialectical contradiction need not be reducible to dialetheias or be associated with glutty negation. But the latter correspond to Hegelian dialectical contradiction.
- A pair of dialectical rough semantics have been developed over classical rough sets. The nature of parthood is reexamined and used for counting based approaches to semantics.
- Opposition in the context of rough set and parthood related sentences is investigated and concepts of dialectical opposition and opposition are generated. Related truth tables show that the classical figures do not work as well for parthood related sentences. This extends previous work on figures of opposition of rough sets in new directions. A major contribution is in the use of dialectical negation predicates in the generation of figures of opposition.
- Possible diagrams of opposition are used for defining generalized counting process for constructive algebraic semantics. This builds on earlier work of the present author in [7].
- The antichain based semantics for general rough sets that has been developed in [3,4] has been supplemented with a constructive dialectical counting process and scope for using generalized negations in a separate paper. The foundations for the same has been laid in this paper.

Sub-classes of problems that have been motivated by the present paper relate to

- Formal characterization of conditions that would permit reduction of dialectical contradiction to dialetheia.
- Construction of algebraic semantics from dialectical counting using figures of opposition.
- Construction of general rough set algorithms from the antichain based dialectics.
- Algebraic characterization of parthood based semantics as opposed to rough object based approaches. This problem has been substantially solved by the present author in [4].

- Methods of property extraction by formal interpretation or translation across semantics - this aspect has not been discussed in detail in the present paper and will appear separately.
- Development of logics relating to glutty negation in the context of suitable rough contexts.

Acknowledgement. The present author would like to thank the referees for detailed remarks that led to improvement (especially of the readability) of the research paper.

References

1. Pawlak, Z.: Rough Sets: Theoretical Aspects of Reasoning About Data. Kluwer Academic Publishers, Dodrecht (1991)
2. Mani, A.: Algebraic methods for granular rough sets. In: Mani, A., Düntsch, I., Cattaneo, G. (eds.) Algebraic Methods in General Rough Sets. Trends in Mathematics, pp. 157–336. Birkhäuser, Basel (2018). https://doi.org/10.1007/978-3-030-01162-8_3
3. Mani, A.: Antichain based semantics for rough sets. In: Ciucci, D., Wang, G., Mitra, S., Wu, W.-Z. (eds.) RSKT 2015. LNCS (LNAI), vol. 9436, pp. 335–346. Springer, Cham (2015). https://doi.org/10.1007/978-3-319-25754-9_30
4. Mani, A.: Knowledge and consequence in AC semantics for general rough sets. In: Wang, G., Skowron, A., Yao, Y.Y., Ślęzak, D., Polkowski, L. (eds.) Thriving Rough Sets. SCI, vol. 708, pp. 237–268. Springer, Cham (2017). https://doi.org/10.1007/978-3-319-54966-8_12
5. Mani, A.: Pure rough mereology and counting. In: WIECON 2016, pp. 1–8. IEEXPlore (2016)
6. Mani, A.: On deductive systems of AC semantics for rough sets. ArXiv. Math (arXiv:1610.02634v1), pp. 1–12, October 2016
7. Mani, A.: Dialectics of counting and the mathematics of vagueness. In: Peters, J.F., Skowron, A. (eds.) Transactions on Rough Sets XV. LNCS, vol. 7255, pp. 122–180. Springer, Heidelberg (2012). https://doi.org/10.1007/978-3-642-31903-7_4
8. Polkowski, L.: Approximate Reasoning by Parts. Springer, Heidelberg (2011). https://doi.org/10.1007/978-3-642-22279-5
9. Mani, A.: Super rough semantics. Fundamenta Informaticae **65**(3), 249–261 (2005)
10. Mani, A.: Algebraic representation, duality and beyond. In: Mani, A., Düntsch, I., Cattaneo, G. (eds.) Algebraic Methods in General Rough Sets. Trends in Mathematics, pp. 459–552. Birkhäuser, Basel (2018). https://doi.org/10.1007/978-3-030-01162-8_6
11. Banerjee, M., Chakraborty, M.K.: Rough sets through algebraic logic. Fundamenta Informaticae **28**, 211–221 (1996)
12. Ioan, P.: Logic and Dialectics. Al. I. Cuza Universities Press (1998)
13. Schang, F.: Opposites and oppositions around and beyond the square of opposition. In: Beziau, J.Y., Jacquette, D., et al. (eds.) Around and Beyond the Square of Opposition. Studies in Universal Logic, vol. I, pp. 147–174. Birkhauser, Basel (2012)
14. Ciucci, D., Dubois, D., Prade, H.: Oppositions in rough set theory. In: Li, T., et al. (eds.) RSKT 2012. LNCS (LNAI), vol. 7414, pp. 504–513. Springer, Heidelberg (2012). https://doi.org/10.1007/978-3-642-31900-6_62

15. Mani, A.: Integrated dialectical logics for relativised general rough set theory. In: International Conference on Rough Sets, Fuzzy Sets and Soft Computing, Agartala, India, 6 p. (Refereed) (2009). http://arxiv.org/abs/0909.4876
16. Mani, A.: Ontology, rough Y-systems and dependence. Int. J. Comput. Sci. Appl. **11**(2), 114–136 (2014). Special Issue of IJCSA on Computational Intelligence
17. Mani, A.: Algebraic semantics of proto-transitive rough sets. In: Peters, J.F., Skowron, A. (eds.) Transactions on Rough Sets XX. LNCS, vol. 10020, pp. 51–108. Springer, Heidelberg (2016). https://doi.org/10.1007/978-3-662-53611-7_3
18. Chakraborty, M.K.: On some issues in the foundation of rough sets: the problem of definition. Fundamenta Informaticae **148**, 123–132 (2016)
19. Grätzer, G.: General Lattice Theory. Birkhauser, New York (1998)
20. Koh, K.: On the lattice of maximum-sized antichains of a finite poset. Algebra Universalis **17**, 73–86 (1983)
21. Mani, A.: Algebraic Semantics of Proto-Transitive Rough Sets, 1st edn. arXiv:1410.0572, July 2014
22. Mani, A.: Approximation dialectics of proto-transitive rough sets. In: Chakraborty, M.K., Skowron, A., Kar, S. (eds.) Facets of Uncertainties and Applications. Springer Proceedings in Math and Statistics, vol. 125, pp. 99–109. Springer, New Delhi (2015). https://doi.org/10.1007/978-81-322-2301-6_8
23. Ciucci, D.: Approximation algebra and framework. Fundamenta Informaticae **94**, 147–161 (2009)
24. Cattaneo, G., Ciucci, D.: Lattices with interior and closure operators and abstract approximation spaces. In: Peters, J.F., Skowron, A., Wolski, M., Chakraborty, M.K., Wu, W.-Z. (eds.) Transactions on Rough Sets X. LNCS, vol. 5656, pp. 67–116. Springer, Heidelberg (2009). https://doi.org/10.1007/978-3-642-03281-3_3
25. Yao, Y.Y.: Relational interpretation of neighbourhood operators and rough set approximation operators. Inf. Sci. **111**, 239–259 (1998)
26. Iwinski, T.B.: Rough orders and rough concepts. Bull. Pol. Acad. Sci. (Math.) **3–4**, 187–192 (1988)
27. Cattaneo, G.: Algebraic methods for rough approximation spaces by lattice interior-closure operations. In: Mani, A., Düntsch, I., Cattaneo, G. (eds.) Algebraic Methods in General Rough Sets. Trends in Mathematics, pp. 13–152. Birkhäuser, Basel (2018). https://doi.org/10.1007/978-3-030-01162-8_2
28. Pagliani, P., Chakraborty, M.: A Geometry of Approximation: Rough Set Theory: Logic, Algebra and Topology of Conceptual Patterns. Springer, Berlin (2008). https://doi.org/10.1007/978-1-4020-8622-9
29. Polkowski, L., Skowron, A.: Rough mereology: a new paradigm for approximate reasoning. Int. J. Approx. Reason. **15**(4), 333–365 (1996)
30. Mani, A.: Esoteric rough set theory: algebraic semantics of a generalized VPRS and VPFRS. In: Peters, J.F., Skowron, A. (eds.) Transactions on Rough Sets VIII. LNCS, vol. 5084, pp. 175–223. Springer, Heidelberg (2008). https://doi.org/10.1007/978-3-540-85064-9_9
31. Mani, A.: Choice inclusive general rough semantics. Inf. Sci. **181**(6), 1097–1115 (2011)
32. Mani, A.: Axiomatic granular approach to knowledge correspondences. In: Li, T., et al. (eds.) RSKT 2012. LNCS (LNAI), vol. 7414, pp. 482–487. Springer, Heidelberg (2012). https://doi.org/10.1007/978-3-642-31900-6_59
33. Mani, A.: Granular Foundations of the Mathematics of Vagueness, Algebraic Semantics and Knowledge Interpretation. University of Calcutta (2016)
34. Moore, E.F., Shannon, C.E.: Reliable circuits using less reliable relays-I. II. Bell Syst. Tech. J. **191–208**, 281–297 (1956)

35. Shannon, C.E.: A mathematical theory of communication. Bell Syst. Tech. J. **27**(379–423), 623–656 (1948)
36. Lin, T.Y.: Granular computing-1: the concept of granulation and its formal model. Int. J. Granular Comput. Rough Sets Int. Syst. **1**(1), 21–42 (2009)
37. Mani, A.: Approximations from anywhere and general rough sets. In: Polkowski, L., et al. (eds.) IJCRS 2017. LNCS (LNAI), vol. 10314, pp. 3–22. Springer, Cham (2017). https://doi.org/10.1007/978-3-319-60840-2_1
38. Ciucci, D.: Back to the beginnings: Pawlak's definitions of the terms information. In: Wang, G., Skowron, A., Yao, Y., Ślęzak, D., Polkowski, L. (eds.) Thriving Rough Sets. Studies in Computational Intelligence, vol. 708. Springer, Cham (2017). https://doi.org/10.1007/978-3-319-54966-8_11
39. Cattaneo, G., Ciucci, D.: Algebraic methods for orthopairs and induced rough approximation spaces. In: Mani, A., Düntsch, I., Cattaneo, G. (eds.) Algebraic Methods in General Rough Sets. Trends in Mathematics, pp. 553–640. Birkhäuser, Basel (2018). https://doi.org/10.1007/978-3-030-01162-8_7
40. Yao, Y.Y., Yao, B.: Covering based rough set approximations. Inf. Sci. **200**, 91–107 (2012)
41. Düntsch, I., Gediga, G.: Rough Set Data Analysis: A Road to Non-invasive Knowledge Discovery. Methodos Publishers, Bangor (2000)
42. Pagliani, P.: Covering rough sets and formal topology – a uniform approach through intensional and extensional constructors. In: Peters, J.F., Skowron, A. (eds.) Transactions on Rough Sets XX. LNCS, vol. 10020, pp. 109–145. Springer, Heidelberg (2016). https://doi.org/10.1007/978-3-662-53611-7_4
43. Sambin, G.: Intuitionistic formal spaces - a first communication. In: Skordev, D. (ed.) Mathematical Logic and Its Applications, pp. 187–204. Plenum Press, New York (1987)
44. Sambin, G., Gebellato, S.: A preview of the basic picture: a new perspective on formal topology. In: Altenkirch, T., Reus, B., Naraschewski, W. (eds.) TYPES 1998. LNCS, vol. 1657, pp. 194–208. Springer, Heidelberg (1999). https://doi.org/10.1007/3-540-48167-2_14
45. da Costa, N., Wolf, R.G.: Studies in paraconsistent logic-1: the dialectical principle of the unity of opposites. Philosophia - Philos. Q. Israel **15**, 497–510 (1974)
46. Pagliani, P.: Rough set theory and logic-algebraic structures. In: Orłowska, E. (ed.) Incomplete Information: Rough Set Analysis, pp. 109–190. Physica Verlag, Heidelberg (1998). https://doi.org/10.1007/978-3-7908-1888-8_6
47. Pagliani, P.: Local classical behaviours in three-valued logics and connected systems. Part 1. J. Multiple Valued Log. **5**, 327–347 (2000)
48. Cattaneo, G., Ciucci, D.: Algebras for rough sets and fuzzy logics. In: Peters, J.F., Andrzej, S., Grzymała-Busse, J.W., Kostek, B., Świniarski, R.W., Szczuka, M.S. (eds.) Transactions on Rough Sets. LNCS, vol. 3100, pp. 208–252. Springer, Heidelberg (2004)
49. Mani, A.: Algebraic semantics of similarity-based bitten rough set theory. Fundamenta Informaticae **97**(1–2), 177–197 (2009)
50. Ślęzak, D., Wasilewski, P.: Granular sets – foundations and case study of tolerance spaces. In: An, A., Stefanowski, J., Ramanna, S., Butz, C.J., Pedrycz, W., Wang, G. (eds.) RSFDGrC 2007. LNCS (LNAI), vol. 4482, pp. 435–442. Springer, Heidelberg (2007). https://doi.org/10.1007/978-3-540-72530-5_52
51. Cattaneo, G., Ciucci, D., Dubois, D.: Algebraic models of deviant modal operators based on De Morgan and Kleene lattices. Inf. Sci. **181**, 4075–4100 (2011)
52. Ficara, E.: Hegel's Glutty negation. Hist. Philos. Log. **36**, 1–10 (2014)

53. Brandom, R.: Between Saying and Doing. Oxford University Press, Oxford (2008)
54. McGill, V.P., Parry, W.T.: The unity of opposites - a dialectical principle. Sci. Soc. **12**, 418–444 (1948)
55. Mani, A.: Towards formal dialectical logics. Technical report (1999)
56. Priest, G.: Doubt Truth to be a Liar. Oxford University Press, Oxford (2008)
57. Gabbay, D.: Oxford Logic Guides, vol. 1, 1st edn, p. 33. Clarendon Press, Oxford (1996)
58. Batens, D.: Narrowing Down Suspicion in Inconsistent Premise Set (2006, preprint)
59. Batens, D.: Against global paraconsistency. Stud. Soviet Thought **39**, 209–229 (1990)
60. Apostol, L.: Logique et Dialectique. Gent (1979)
61. Hoffmann, W.C.: A formal model for dialectical psychology. Int. Log. Rev. 40–67 (1986)
62. Gorren, J.: Theorie Analytique de la Dialectique. South West Philos. Stud. **6**, 41–47 (1981)
63. Marx, K., Engels, F.: Marx and Engels: Collected Works, vol. 24. Progress Publishers, Delhi (1989)
64. Batens, D.: Dynamic dialectical logics. In: Paraconsistent Logic - Essays. Philosophia Verlag, Munich (1989)
65. Tzouvaras, A.: Periodicity of negation. Notre Dame J. Form. Log. **42**(2), 88–99 (2001)
66. Priest, G.: What not? A defence of a dialetheic theory of negation. In: Gabbay, D. (ed.) What is Negation?, pp. 101–120. Kluwer, Dordrecht (1999)
67. Priest, G.: To be and not to be: dialectical tense logic. Studia Logica **41**(2/3), 249–268 (1984)
68. Zeleny, J.: Paraconsistency and dialectical consistency. Log. Point View **1**, 35–51 (1994)
69. Priest, G.: Logicians setting together contradictories - a perspective on relevance, paraconsistency and dialetheism. In: Jacquette, D. (ed.) Blackwell Handbook to Philosophical Logic. Blackwell, Hoboken (2007)
70. Grim, P.: What is contradiction? In: Priest, G. (ed.) The Law of Non-contradiction. Oxford Universities Press, Oxford (2007)
71. Woods, J.: Dialectical Considerations on the Logic of Contradiction: Part I, II (2004, preprint)
72. Priest, G.: Dialectic and dialetheic. Sci. Soc. **53**(4), 388–415 (1990)
73. Priest, G.: Contradictory concepts. In: Weber, E., Wouters, D., Meheus, J. (eds.) Logic, Reasoning and Rationality. Interdisciplinary Perspectives from The Humanities and Social Sciences, vol. 5, pp. 197–216. Springer, Dordrecht (2014). https://doi.org/10.1007/978-94-017-9011-6_10
74. Swaminathan, M., Rawal, V.: A Study of Agrarian Relations. Tulika Books, New Delhi (2015)
75. Zimmerman, B. (ed.): Lesbian Histories and Cultures: An Encyclopedia. Garland Reference Library of the Social Sciences, vol. 1008. Garland Publishers, New York (2000)
76. Saha, A., Sen, J., Chakraborty, M.K.: Algebraic structures in the vicinity of pre-rough algebra and their logics II. Inf. Sci. **333**, 44–60 (2015)
77. Mani, A.: Contamination-free measures and algebraic operations. In: 2013 IEEE International Conference on Fuzzy Systems (FUZZ), pp. 1–8. IEEE (2013)

78. Bunder, M.W., Banerjee, M., Chakraborty, M.K.: Some rough consequence logics and their interrelations. In: Peters, J.F., Skowron, A. (eds.) Transactions on Rough Sets VIII. LNCS, vol. 5084, pp. 1–20. Springer, Heidelberg (2008). https://doi.org/10.1007/978-3-540-85064-9_1
79. Pawlak, Z.: Rough logic. Bull. Pol. Acad. Sci. (Tech.) **35**, 253–258 (1987)
80. Hyde, D., Colyvan, M.: Paraconsistent Vagueness-why not? Aust. J. Log. **6**, 207–225 (2008)
81. Avron, A., Konikowska, B.: Rough sets and 3-valued logics. Studia Logica **90**, 69–92 (2008)
82. Mani, A.: Towards logics of some rough perspectives of knowledge. In: Suraj, Z., Skowron, A. (eds.) Intelligent Systems Reference Library Dedicated to the Memory of Professor Pawlak ISRL, vol. 43, pp. 419–444. Springer, Heidelberg (2013). https://doi.org/10.1007/978-3-642-30341-8_22
83. Rasiowa, H.: An Algebraic Approach to Nonclassical Logics. Studies in Logic, vol. 78. North Holland, Warsaw (1974)
84. Ciucci, D.: Orthopairs and granular computing. Granular Comput. **1**(3), 159–170 (2016)
85. Belnap, N.D.: A useful four-valued logic. In: Dunn, J.M., Epstein, G. (eds.) Modern Uses of Multiple-valued Logic. Episteme, vol. 2, pp. 5–37. Springer, Dordrecht (1977). https://doi.org/10.1007/978-94-010-1161-7_2
86. Moretti, A.: Why the logical hexagon? Logica Univers **6**, 69–107 (2012)

A Logic for Spatial Reasoning
in the Framework of Rough Mereology

Lech Polkowski$^{(\boxtimes)}$

Department of Mathematics and Computer Science,
Chair of Mathematical Methods in Computer Science,
University of Warmia and Mazury in Olsztyn,
Słoneczna str. 54, 10–710 Olsztyn, Poland
polkow@pjwstk.edu.pl

To Professor Andrzej Skowron on the seventy fifth birthday.

Abstract. Spatial reasoning concerns a language in which spatial objects are described and argued about. Within the plethora of approaches, we single out the one set in the framework of mereology - the theory of concepts employing the notion of a part as the primitive one. Within mereology, we can choose between the approach based on part as the basic notion or the approach based on the notion of a connection from which the notion of a part is defined. In this work, we choose the former approach modified to the rough mereology version in which the notion of a part becomes 'fuzzified' to the notion of a part to a degree. The prevalence of this approach lies in the fact that it does allow for quantitative assessment of relations among spatial objects in distinction to only qualitative evaluation of those relations in case of other mereology based approaches.

In this work, we introduce sections on mereology based reasoning, covering part and connection based variants as well as rough mereology in order to provide the Reader with the conceptual environment we work in. We recapitulate shortly those approaches along with based on them methods for spatial reasoning. We then introduce the mereological approach in the topological context used in spatial reasoning, i.e., in collections of regular open or regular closed sets known to form complete Boolean algebras. In this environment, we create a logic for reasoning about parts and degrees of inclusion based on an abstract notion of a mass which generalizes geometric measure of area or volume and extends in the abstract manner the Lukasiewicz logical rendering of probability calculus. We give some applications, notably, we extend the relation of betweenness applied by us earlier in robot navigation and we give it the abstract characterization.

1 Introduction

In this section, we briefly render basic ideas and notions of theories mentioned in the abstract. This will allow the reader to fully grasp the environment for

© Springer-Verlag GmbH Germany, part of Springer Nature 2019
J. F. Peters and A. Skowron (Eds.): TRS XXI, LNCS 10810, pp. 142–168, 2019.
https://doi.org/10.1007/978-3-662-58768-3_5

our discussion of spatial reasoning. We discuss the part mereology, the rough mereology, the connection mereology. As our space of objects, we introduce the space of bounded regular open sets in a Euclidean n–space and we discuss the spatial reasoning in the ROM model. We recall basic notions of introduced by us earlier mereogeometry, in particular the notion of betweenness which we apply in navigation methods for teams of intelligent agents modeled as autonomous robots.

1.1   Mereology Based on the Notion of a Part

Mereology with part relation as the primitive notion was proposed in Leśniewski [16]. The interested reader may as well consult, e.g., Casati and Varzi [4] or Polkowski [33]. Given some collection U of things regarded as individuals in ontological sense, a *relation of a part* is a binary relation π on U which is required to be

M1 *Irreflexive: For each thing x, it is not true that $\pi(x,x)$.*
M2 *Transitive: For each triple x, y, z of things, if $\pi(x,y)$ and $\pi(y,z)$, then $\pi(x,z)$.*

The relation of a part does induce the relation of an *improper part* $\Pi(x,y)$, defined as

$$\Pi(x,y) \Leftrightarrow \pi(x,y) \vee x = y. \tag{1}$$

Basic properties of the relation Π which follow from M1, M2, are summed up in

Proposition 1. *1. $\Pi(x,x)$ for each thing x in the universe U.*
2. If $\Pi(x,y)$ and $\Pi(y,x)$ then $x = y$ for each pair x, y of things in U.
3. If $\Pi(x,y)$ and $\Pi(y,z)$ then $\Pi(x,z)$.

The relation Π is instrumental in definitions of other basic relations on the universe U. The relation of *overlapping*, $Ov(x,y)$ in symbols, is defined as follows

$$Ov(x,y) \Leftrightarrow \exists z.\Pi(z,x) \wedge \Pi(z,y). \tag{2}$$

Helped by the relation of overlapping, we introduce the third postulate for our model of mereology, which is as follows.

M3 *For each pair x, y of things, if for each thing z such that $\Pi(z,x)$ there exists a thing w such that $\Pi(w,y)$ and $Ov(z,w)$, then $\Pi(x,y)$.*

M3 is the standard reasoning tool in our discussion of mereology. The notion of overlapping is in turn instrumental in definition of the class operator in the sense of Leśniewski [16]. This operator assigns to each non-empty collection of things F in the universe (U, π) its class, $ClsF$ which is the thing satisfying the two conditions:

C1 *If $x \in F$, then $\Pi(x, ClsF)$, for each thing x in the universe U.*

C2 *If $\Pi(x, ClsF)$, then for each y with $\Pi(y, x)$ there exists $z \in F$ such that $Ov(y, z)$, for each thing x in the universe U.*

The immediate corollary from C1, C2, property 2 in Proposition 1 and M3 is the class uniqueness.

Proposition 2. *For each non-empty collection F of things in the universe U, if Cls_1F and Cls_2F satisfy conditions C1, C2, then $Cls_1F = Cls_2F$.*

As an example, we observe that each thing x is the class of its parts.

Proposition 3. *For each thing x, it is true that $x = Cls\{z : \Pi(z, x)\}$.*

For a universe (U, π) of things, we define the class V of all things.

$$V = Cls\{x : x \in U\}. \tag{3}$$

We call V the *universal thing*. We have

Proposition 4. *By C1, for each thing $x \in U$, it is true that $\Pi(x, V)$.*

We are now in a position to recall here two *fusion operators* due to Tarski [42]. These operators are the *sum* $x + y$ and the *product* $x \cdot y$ defined by means of

$$x + y = Cls\{z : \Pi(z, x) \vee \Pi(z, y)\} \tag{4}$$

and

$$x \cdot y = Cls\{z : \Pi(z, x) \wedge \Pi(z, y)\}. \tag{5}$$

Things x, y are *disjoint*, $dis(x, y)$ in symbols, whenever there is no thing z such that $\Pi(z, x)$ and $\Pi(z, y)$ (a fortiori, the product of x and y is not defined).

$$dis(x, y) \Leftrightarrow \neg Ov(x, y). \tag{6}$$

The *difference* $x - y$ is defined as follows

$$x - y = Cls\{z \in U : \Pi(z, x) \wedge \neg\Pi(z, y)\}. \tag{7}$$

It is well-known (see [42]) that the mereological universe (U, π) with the universal thing V and operations $+, \cdot, -$ is a complete Boolean algebra without the zero element, which we denote by the symbol $B(U, \pi)$. The complement $-x$ to a thing x in the universe (U, π) is the difference $V - x$.

$$-x = V - x. \tag{8}$$

We define the *mereological implication* denoted $x \rightharpoonup y$ as follows:

$$x \rightharpoonup y = -x + y. \tag{9}$$

Definition 1. *We declare $x \overset{\cdot}{\to} y$ true if and only if $-x + y = V$.*

We establish some facts on relationships among the notion of an improper part and facts of the Boolean algebra.

T1. $\Pi(x, y) \Leftrightarrow x \cdot y = x$.

Proof. Let $\Pi(x, y)$. For each thing z, if $\Pi(z, x)$ then by Proposition 1, 3., $\Pi(z, y)$ holds true, hence, $\Pi(z, x \cdot y)$. By M3, $\Pi(x, x \cdot y)$. That $\Pi(x \cdot y, x)$ follows from the definition (5). Finally, Proposition 1, 2. implies that $x = x \cdot y$.

T2. $\Pi(x, y) \Leftrightarrow x \overset{\cdot}{\to} y$.

Proof. $x \overset{\cdot}{\to} y$ true means $-x + y = V$, hence, for each thing z from $\Pi(z, x)$ it follows $\Pi(z, y)$ and thus by M3, $\Pi(x, y)$. Contrariwise, if $\Pi(x, y)$ holds true, then by T1, $x \cdot y = x$ and we have $x \overset{\cdot}{\to} y = -x + y = -(x \cdot y) + y = -x + -y + y = -x + V = V$, i.e. $x \overset{\cdot}{\to} y$ is true.

T3. $\Pi(x, y) \Leftrightarrow x \cdot y = x \Leftrightarrow x \overset{\cdot}{\to} y$.

1.2 Mereology Based on Connection

In Whitehead [44–46], a proposition of the notion of '*x extends over y*', appeared, dual to that of a part. Th. de Laguna [15] published a variant of the Whitehead scheme, which led Whitehead [47] to another version of his approach, based on the notion of '*x is extensionally connected to y*'. Connection Calculus based on the notion of a 'connection' was proposed in Clarke [5], which we outline here. The *predicate of being connected* $C(x, y)$ is subject to basic requirements:

CN1. $C(x, x)$ *for each thing x*.

CN2. *If* $C(x, y)$, *then* $C(y, x)$ *for each pair x, y of things*.

It follows that connection is reflexive and symmetric. This theory is sometimes called *Ground Topology* T, cf., Casati and Varzi [4]. The additional *extensionality* requirement:

CN3. *If* $\forall z.[C(z, x) \Leftrightarrow C(z, y)]$, *then* $x = y$,
produces the *Extensional Ground Topology* ET., see [4].
    Let us observe that the predicate C can be realized by taking $C = Ov$; clearly, CN1–CN3 are all satisfied with Ov. We call this model of connection mereology the *Overlap model*, denoted OVM. Also, letting $C(x, y)$ if and only if $x \cap y \neq \emptyset$, defines a connection relation on non-empty sets. In the universe endowed with C, satisfying CN1, CN2, one defines the notion of an improper part Π_C by letting:

$$\Pi_C(x, y) \Leftrightarrow \forall z.[C(z, x) \Rightarrow C(z, y)]. \tag{10}$$

The notion of a *C-part* π_C can be introduced as follows

$$\pi_C(x, y) \Leftrightarrow \Pi_C(x, y) \wedge x \neq y. \tag{11}$$

The predicate of *C–overlapping*, $Ov_C(x,y)$ is defined by means of

$$Ov_C(x,y) \Leftrightarrow \exists z.[\Pi_C(z,x) \wedge \Pi_C(z,y)]. \tag{12}$$

The notion of an *C–exterior things*, $extr_C(x,y)$ is defined by means of

$$extr_C(x,y) \Leftrightarrow \neg Ov_C(x,y). \tag{13}$$

A quasi-topological character of the relation $C(x,y)$ allows for notions having spatial relevance. A new notion is *C–external connectedness*, EC, defined as follows

$$EC(x,y) \Leftrightarrow C(x,y) \wedge extr(x,y). \tag{14}$$

It is easy to see that in the model OVM, EC is a vacuous notion.

The notion of a *tangential part* $T\Pi_C(x,y)$ is defined by means of

$$T\Pi_C(x,y) \Leftrightarrow \Pi_C(x,y) \wedge \exists z.EC(z,x) \wedge EC(z,y). \tag{15}$$

It follows that if there is some thing externally connected to x, then x is its tangential ingredient. This fact shows that the notion of a tangential ingredient falls short of the idea of a boundary. Dually, in absence of things externally connected to y, no ingredient of y can be a tangential ingredient.

A thing y is a *non–tangential ingredient* of a thing x, $NT\Pi_C(y,x)$, in case it is an improper part but not any tangential part of x:

$$NT\Pi_C(y,x) \Leftrightarrow \neg T\Pi_C(y,x) \wedge \Pi_C(y,x). \tag{16}$$

In absence of externally connected things, each thing is a non–tangential ingredient of itself, hence, in the model OVM each object is its own non–tangential ingredient and it has no tangential ingredients. To produce models in which EC, $NT\Pi_C, T\Pi_C$ will be exhibited in a non-trivial way, we resort to topology, see Sect. 9.1.

1.3   The Model ROM for Connection Based Mereology

We refer the Reader to the Sect. 9, below, for a discussion of topological notions including regular open and regular closed sets. We define in the space $RO(X)$ of regular open sets which model spatial objects in a regular topological space X the connection C by demanding that (Cl stands for the closure operator)

$$C_{rom}(x,y) \Leftrightarrow Clx \cap Cly \neq \emptyset. \tag{17}$$

For simplicity sake, we assume that the regular space X is connected so the boundary of each set is non–empty.

Proposition 5. *The following are characterizations of spatial relations induced by the relation C of being connected.*

1. $\Pi_{C_{rom}}(x, y) \Leftrightarrow x \subseteq y$.
2. $Ov_{C_{rom}}(x, y) \Leftrightarrow x \cap y \neq \emptyset$.
3. $EC_{rom}(x, y) \Leftrightarrow Clx \cap Cly \neq \emptyset \wedge x \cap y = \emptyset$.
4. $T\Pi_{C_{rom}}(x, y) \Leftrightarrow x \subseteq y \wedge Clx \cap Bdy \neq \emptyset$ where $Bdy = Cly - y$ is the boundary of y.
5. $NT\Pi_{C_{rom}}(x, y) \Leftrightarrow Clx \subseteq y$.

For the proofs, please see Sect. 9.

RCC: Region Connection Calculus. As an important example of mereological spatial reasoning we introduce here the RCC Calculus (Region Connection Calculus), cf. Randell, Cui and Cohn [41], Cohn, Randell, Cui, Bennett [9], Cohn, Gooday, Bennett, and Gotts [7], Cohn [6], Cohn and Varzi [10]. It is a calculus on closed regular sets (regions) in a regular topological space, i.e., in the frame of ROM. RCC admits Clarke's connection postulates CN1–CN3 and follows same lines in defining basic predicates. To preserve the flavor of this theory we give these predicates in the RCC notation

1. $DISCONNECTED\ FROM(x)(y)\ DC(x, y) \Leftrightarrow \neg C(x, y)$.
2. $IMPROPER\ PART\ OF(x)(y) :\ P(x, y) \Leftrightarrow \forall z.[C(z, y) \rightarrow C(z, x)]$.
3. $PROPER\ PART\ OF(x)(y) :\ PP(x, y) \Leftrightarrow P(x, y) \wedge \neg P(y, x)$.
4. $EQUAL(x)(y) :\ EQ(x, y) \Leftrightarrow P(x, y) \wedge P(y, x)$.
5. $OVERLAP(x)(y) : Ov(x, y) \Leftrightarrow \exists.z.P(x, z) \wedge P(y, z)$.
6. $DISCRETE\ FROM(x)(y) : DR(x, y) \Leftrightarrow \neg Ov(x, y)$.
7. $PARTIAL\ OVERLAP(x)(y) : POv(x, y) \Leftrightarrow Ov(x, y) \wedge \neg P(x, y) \wedge \neg P(y, x)$.
8. $EXTERNAL\ CONNECTED(x)(y) : EC(x, y) \Leftrightarrow C(x, y) \wedge \neg Ov(x, y)$.
9. $TANGENTIAL\ PART\ OF(x)(y) : TPP(x, y) \Leftrightarrow PP(x, y) \wedge \exists z.EC(x, z) \wedge EC(y, z)$.
10. $NON - TANGENTIAL\ PART\ OF(x)(y) : NTPP(x, y) \Leftrightarrow PP(x, y) \wedge \neg TPP(x, y)$.

To each non–symmetric predicate X, RCC adds the inverse Xi (e.g., to $TPP(x, y)$ it adds $TPPi(y, x)$). The eight predicates: *DC, EC, PO, EQ, TPP, NTPP, TPPi, NTPPi* show the *JEPD property* (Jointly Exclusive and Pairwise Disjoint) and they form the fragment of RCC called RCC8.

Due to topological assumptions, RCC has some stronger properties than Clarke's calculus of C, where connection is simply the set intersection. Witness, the following properties:

1. If $\forall z.Ov(x, z) \leftrightarrow Ov(y, z)$, then $x = y$ (extensionality of overlapping). (If $x \neq y$, then, e.g., there is $z \in x - y$ and regularity of the space yields us an open neighborhood V of z such that $ClV \cap y = \emptyset$ and $Ov(V, x)$ negating the premise).
2. If $PP(x, y)$, then $\exists z.P(x, z) \wedge DR(y, z)$.
3. $\forall x.EC(x, -x)$.

1.4   Rough Mereology

A scheme of mereology, introduced into a collection of things, sets an exact hierarchy of things of which some are (exact) parts of others; to ascertain whether a thing is an exact part of some other thing is in practical cases often difficult if possible at all, e.g., a robot sensing the environment by means of a camera or a laser range sensor, cannot exactly perceive obstacles or navigation beacons. Such evaluation can be done approximately only and one can discuss such situations up to a degree of certainty only. Thus, one departs from the exact reasoning scheme given by decomposition into parts to a scheme which approximates the exact scheme but does not observe it exactly.

Such a scheme, albeit its conclusions are expressed in an approximate language, can be more reliable, as its users are aware of uncertainty of its statements and can take appropriate measures to fend off possible consequences. Imagine two robots using the language of connection mereology for describing mutual relations; when endowed with touch sensors, they can ascertain the moment when they are connected; when a robot has as a goal to enter a certain area, it can ascertain that it connected to the area or overlapped with it, or it is a part of the area, and it has no means to describe its position more precisely.

Introducing some measures of overlapping, in other words, the extent to which one thing is a part to the other, would allow for a more precise description of relative position, and would add an expressional power to the language of mereology. Rough mereology answers these demands by introducing the notion of a *part to a degree* with the degree expressed as a real number in the interval $[0, 1]$. Any notion of a part by necessity relates to the general idea of *containment*, and thus the notion of a part to a degree is related to the idea of *partial containment* and it should preserve the essential intuitive postulates about the latter.

The predicate of a part to a degree stems ideologically from and has as one of motivations the predicate of an element to a degree introduced by Zadeh as a basis for fuzzy set theory [48]; in this sense, rough mereology is to mereology as the fuzzy set theory is to the naive set theory. To the rough set theory, owes rough mereology the fundamental interest in concepts as things for analysis.

The primitive notion of rough mereology is the notion of a *rough inclusion* which is a ternary predicate $\mu(x, y, r)$ where x, y are *things* and $r \in [0, 1]$, read 'the thing x is a part to degree at least of r to the thing y'. Any rough inclusion is associated with a mereological scheme based on the notion of a part by postulating that $\mu(x, y, 1)$ is equivalent to $\Pi(x, y)$, where the improper part relation is defined by the adopted mereological scheme. Other postulates about rough inclusions stem from intuitions about the nature of partial containment; these intuitions can be manifold, a fortiori, postulates about rough inclusions may vary. In our scheme for rough mereology, we begin with some basic postulates which would provide a most general framework. When needed, other postulates, narrowing the variety of possible models, can be introduced.

Rough Inclusions. We have already stated that a rough inclusion is a ternary predicate $x \rightarrow_m y$. We assume that a collection of things is given, on which a part relation π is introduced with the associated relation Π. We thus apply inference schemes of mereology due to Leśniewski, presented above.

Predicates $\mu(x, y, r)$ were introduced in Polkowski and Skowron [39, 40]; they satisfy the following postulates, relative to a given part relation π and the induced by π relation Π of an improper part, on a set of things:

RINC1. $\mu(x, y, 1) \Leftrightarrow \Pi(x, y)$.

RINC2. $\mu(x, y, 1) \Rightarrow \forall z[\mu(z, x, r) \Rightarrow \mu(z, y, r)]$.

RINC3. $x \rightarrow_m y \wedge s < r \Rightarrow \mu(x, y, s)$.

From postulates RINC1–RINC3, and known properties of the improper part notion, some consequences follow:

1. $\mu(x, x, 1)$.
2. $\mu(x, y, 1) \wedge \mu(y, z, 1) \Rightarrow \mu(x, z, 1)$.
3. $\mu(x, y, 1) \wedge \mu(y, x, 1) \Leftrightarrow x = y$.
4. $x \neq y \Rightarrow \neg\mu(x, y, 1) \vee \neg\mu(y, x, 1)$.
5. $\forall z \forall r[\mu(z, x, r) \Leftrightarrow \mu(z, y, r)] \Rightarrow x = y$.

Property 5 may be regarded as an *extensionality postulate* in rough mereology.

By a *model* for rough mereology, we mean a quadruple

$$M = (V_M, \pi_M, \Pi_M, \mu_M)$$

where V_M is a set with a part relation $\pi_M \subseteq V_M \times V_M$, the associated relation $\Pi_M \subseteq V_M \times V_M$, and a relation $\mu_M \subseteq V_M \times V_M \times [0, 1]$ which satisfies RINC1–RINC3.

We now describe some models for rough mereology which at the same time give us methods by which we can define rough inclusions, see Polkowski [25–31], a detailed discussion may be found in Polkowski [32].

Rough Inclusions from T–Norms. We resort to *continuous t–norms* which are continuous functions $T : [0, 1]^2 \rightarrow [0, 1]$ which are 1. symmetric. 2. associative. 3. increasing in each coordinate. 4. satisfying boundary conditions $T(x, 0) = 0, T(x, 1) = x$, cf., Polkowski [32], Chaps. 4, 6, Hájek [14], Chap. 2. Classical examples of continuous t–norms are

1. $L(x, y) = max\{0, x + y - 1\}$ *(the Łukasiewicz t–norm)*.
2. $P(x, y) = x \cdot y$ *(the product t–norm)*.
3. $M(x, y) = min\{x, y\}$ *(the minimum t–norm)*.

The *residual implication* $\Rightarrow_T$ induced by a continuous t–norm T is defined as

$$x \Rightarrow_T y = max\{z : T(x, z) \le y\}. \tag{18}$$

One proves that $\mu_T(x, y, r) \Leftrightarrow x \Rightarrow_T y \ge r$ is a rough inclusion; it is easy to see that each of them takes on the value of 1 when $x \le y$. In case $x > y$, particular cases are

1. $\mu_L(x,y,r) \Leftrightarrow min\{1, 1 - x + y \geq r\}$ (*the Łukasiewicz implication*).
2. $\mu_P(x,y,r) \Leftrightarrow \frac{y}{x} \geq r$ (*the Goguen implication*).
3. $\mu_M(x,y,r) \Leftrightarrow y \geq r$ (*the Gödel implication*).

A particular case of continuous t–norms are *Archimedean t–norms* which satisfy the inequality $T(x,x) < x$ for each $x \in (0,1)$. It is well–known, see Ling [17], that each archimedean t–norm T admits a representation:

$$T(x,y) = g_T(f_T(x) + f_T(y)), \tag{19}$$

where the function $f_T : [0,1] \to [0,1]$ is continuous decreasing with $f_T(1) = 0$, and $g_T : R \to [0,1]$ is the *pseudo–inverse* to f_T, i.e., $g \circ f = id$. It is known, cf., e.g., Hájek [14], that up to an isomorphism there are two Archimedean t–norms: L and P. Their representations are:

$$f_L(x) = 1 - x; \ g_L(y) = 1 - y \tag{20}$$

and

$$f_P(x) = exp(-x); \ g_P(y) = -lny. \tag{21}$$

For an Archimedean t–norm T, we define the rough inclusion μ^T on the interval $[0,1]$ by means of

$$(ari) \ \mu^T(x,y,r) \Leftrightarrow g_T(|x - y|) \geq r, \tag{22}$$

equivalently,

$$\mu^T(x,y,r) \Leftrightarrow |x - y| \leq f_T(r). \tag{23}$$

It follows from (23), that μ^T satisfies conditions RINC1–RINC3 with Π as identity =.

To give a hint of proof: for RINC1: $\mu^T(x,y,1)$ if and only if $|x-y| \leq f_T(1) = 0$, hence, if and only if $x = y$. This implies RINC2. In case $s < r$, and $|x - y| \leq f_T(r)$, one has $f_T(r) \leq f_T(s)$ and $|x - y| \leq f_T(s)$.

Specific recipes are:

$$\mu^L(x,y,r) \Leftrightarrow |x - y| \leq 1 - r \tag{24}$$

and

$$\mu^P(x,y,r) \Leftrightarrow |x - y| \leq -ln \ r \tag{25}$$

Both residual and archimedean rough inclusions satisfy the *transitivity condition*, where μ denotes either μ_T or μ^T,

(Trans) *If $\mu(x,y)$ and $\mu(y,z,s)$, then $\mu(x,z,T(r,s))$.*

In the way of a proof, assume, e.g., $\mu^T(x,y,r)$ and $\mu^T(y,z,s)$, i.e., $|x - y| \leq f_T(r)$ and $|y - z| \leq f_T(s)$. Hence, $|x - z| \leq |x - y| + |y - z| \leq f_T(r) + f_T(s)$, hence, $g_T(|x - z|) \geq g_T(f_T(r) + f_T(s)) = T(r,s)$, i.e., $\mu^T(x,z,T(r,s))$. Other cases go on same lines. Let us observe that rough inclusions of the form (ari) are also *symmetric*.

Rough Inclusions in Information Systems (Data Tables). An important domain where rough inclusions will play a dominant role in our analysis of reasoning by means of parts is the realm of *information systems* of Pawlak [23], cf., Polkowski [32], Ch. 6. We will define information rough inclusions denoted with a generic symbol μ^I.

We recall that an *information system* (a *data table*) is represented as a pair (U, A) where U is a finite set of things and A is a finite set of *attributes*; each attribute $a : U \rightarrow V$ maps the set U into the *value set* V. For an attribute a and a thing v, $a(v)$ is the value of a on v.

For things u, v the *discernibility set* $DIS(u, v)$ is defined as

$$DIS(u, v) = \{a \in A : a(u) \neq a(v)\}. \tag{26}$$

For an (ari) μ_T, we define a rough inclusion μ_T^I by means of

$$(airi)\ \mu_T^I(u, v, r) \Leftrightarrow g_T\left(\frac{|DIS(u, v)|}{|A|}\right) \geq r. \tag{27}$$

Then, μ_T^I is a rough inclusion with the associated ingredient relation of identity and the part relation empty.

For the Łukasiewicz t–norm, the *airi* μ_L^I is given by means of the formula

$$\mu_L^I(u, v, r) \Leftrightarrow 1 - \frac{|DIS(u, v)|}{|A|} \geq r. \tag{28}$$

We introduce the set $IND(u, v) = A \setminus DIS(u, v)$. With its help, we obtain a new form of (28)

$$\mu_L^I(u, v, r) \Leftrightarrow \frac{|IND(u, v)|}{|A|} \geq r. \tag{29}$$

The formula (29) witnesses that the reasoning based on the rough inclusion μ_L^I is the probabilistic one which goes back to Łukasiewicz [18]. Each (airi)–type rough inclusion μ_T^I satisfies the transitivity condition (Trans) and is symmetric.

Rough Inclusions on Sets and Measurable Sets. Formula (29) can be abstracted to set and geometric domains. For finite sets A, B,

$$\mu^S(A, B, r) \Leftrightarrow \frac{|A \cap B|}{|A|} \geq r, \tag{30}$$

where $|X|$ denotes the cardinality of X, defines a rough inclusion μ^S. For bounded measurable sets X, Y in an Euclidean space E^n,

$$\mu^G(X, Y, r) \Leftrightarrow \frac{||X \cap Y||}{||X||} \geq r, \tag{31}$$

where $||A||$ denotes the area (the Lebesgue measure) of the region A, defines a rough inclusion μ^G. Both μ^S, μ^G are symmetric but not transitive.

We now introduce a new method for obtaining rough inclusions which is an abstraction from μ^S, μ^G as well as from the logic of probability in Łukasiewicz [18].

2 The Mass-Based Rough Mereology. Masses on a Mereological Universe

We introduce a new type of rough inclusions derived from a basic notion of a mass $m(x)$ assigned to each thing x in the mereological universe U endowed with a part relation π and the derived relation Π of an improper part. The notion of mass in science is most often attributed to physical objects or linguistic category of mass expressions in dealing with which mereological tools are involved by some authors cf. Nicolas [20]. Here, we introduce mass as an attribute of things which may admit various interpretations depending on the specific context of usage.

We already know that the Tarski operators $+$ and $\cdot$ along with the universal thing V and the complement operator $-$ introduce in a mereological universe (U, π) the structure of the complete Boolean algebra without the null element $B(U, \pi)$ which satisfies T1; we denote by θ the null element of $B(U, \pi) = $ the empty thing, not in the mereological universe U.

The notion of a mass m in what follows should satisfy the following demands:

MS1. $m(x)$ *is a positive real number in the interval* $(0, 1]$ *for each thing x in U*.
MS2. $m(\theta) = 0$, *where θ denotes the empty thing, on the outside of the universe* U.

2.1 Mass-Based Logic (mRM-Logic)

We begin with axioms of mRM-logic.

A1. $x = V \Leftrightarrow m(x) = 1$.

A2. $x = \theta \Leftrightarrow m(x) = 0$.

A3. $(x \rightharpoonup y) \Rightarrow [m(y) = m(x) + m(-x \cdot y)]$.

We prove theses below, which delineate properties of the mass assignment m.

T4. $(x = y) \Rightarrow [m(x) = m(y)]$.

Proof. $x = y$ implies $x \rightharpoonup y$ and $y \hookrightarrow x$, hence by A3, $m(y) = m(x) + m((-x) \cdot y)$ so $m(y) \geq m(x)$ by MS1 and similarly $m(x) = m(y) + m((-y) \cdot x)$, i.e., $m(x) \geq m(y)$ and finally $m(x) = m(y)$.

T5. $m(x + y) = m(x) + m((-x) \cdot y)$.

Proof. Substitution $x/x + y$ in A3 yields $x \rightharpoonup (x + y)$ true so by A3, $m(x + y) = m(x) + m((-x) \cdot (x + y)) = m(x) + m((-x) \cdot y)$.

T6. $(x \cdot y = \theta) \Rightarrow [m(x + y) = m(x) + m(y)]$.

Proof. As $x \cdot y = \theta$, $(-x) \cdot y = y$ and T5 implies $m(x + y) = m(x) + m(y)]$.

T7. $m(x) + m(-x) = 1$.

Proof. As $x \cdot (-x) = \theta$, T6 implies that $m(x) + m(-x) = m(x + (-x)) = m(V) = 1$.

T8. $m(y) = m(x \cdot y) + m((-x) \cdot y)$.

Proof. As $x + (-x) = V$ and $x \cdot (-x) = \theta$, we have $y = y \cdot V = y \cdot (x + (-x))$, hence, $y = (y \cdot x) + (y \cdot (-x))$ so $m(y) = m(x \cdot y) + m((-x) \cdot y)$.

T9. $m(x + y) = m(x) + m(y) - m(x \cdot y)$.

Proof. $m(x + y) = m(x) + m((-x) \cdot y)$ by T5 and $m(y) = m(x \cdot y) + m((-x) \cdot y)$, hence, T9 follows.

T10. $\Pi(x, y) \Rightarrow [m(x) \leq m(y)]$.

Proof. $\Pi(x, y)$ is equivalent by T2 to $x \rightharpoonup y$ so by A3, $m(y) = m(x) + m((-x) \cdot y)$ and by MS1, $m(y) \geq m(x)$.

It follows that mass assignment observing A1-A3 is additive by T6 and monotone by T10.

T11. $m(x + y) = m(x) + m(y)] \Rightarrow (x \cdot y = \theta)$.

Proof. By T9, $m(x \cdot y) = 0$, hence by A2, $x \cdot y = \theta$.

T12. $\Pi(x, y) \Leftrightarrow m(x \rightharpoonup y) = 1)$.

Proof. $\Pi(x, y)$ is equivalent to $x \rightharpoonup y = V$ is equivalent by A1 to $m(x \rightharpoonup y) = 1$.

T13. $\Pi(y, x) \rightarrow y \cdot (-x) = \theta$.

Proof. $\Pi(y, x)$ is equivalent by T3 to $y \cdot x = y$, hence, $y \cdot (-x) = (y \cdot x) \cdot (-x) = y \cdot (x \cdot (-x)) = y \cdot \theta = \theta$.

T14. $\Pi(y, x) \Rightarrow m(x \rightharpoonup y) = 1 - m(x) + m(y)$.

Proof. As by T13 y and $-x$ are disjoint, $m(x \rightharpoonup y) = m(-x + y) = m(-x) + m(y)$ by T6 and by T7 we have that $m(x \rightharpoonup y) = 1 - m(x) + m(y)$.

T15. $m(x \rightharpoonup y) = 1 - m(x - y)$.

Proof. $m(x \rightharpoonup y) = m(-x + y) = m(V - (x - y))$. Also, $V = (x - y) + (V - (x - y))$, hence, $1 = m(x - y) + m(V - (x - y))$, and thus, $m(x \rightharpoonup y) = m(V - (x - y)) = 1 - m(x - y)$.

Comment. T15 is a general formula which specializes to T13 and T14 under appropriate assumptions.

The Notion of Independence.

Definition 2. *We say that things* x, y *are* independent *when* $m(x \cdot y) = m(x) \cdot m(y)$.

T16. $m(x \cdot y) = m(x) \cdot m(y) \Leftrightarrow m(-x \cdot y) = m(-x) \cdot m(y)$.

Proof. $m(y) = m(x \cdot y) + m(-x \cdot y)$ by T8, hence by Definition 1, $m(y) = m(x) \cdot m(y) + m(-x \cdot y)$, i.e., $m(-x \cdot y) = m(y) \cdot [1 - m(x)] = m(y) \cdot m(-x)$ by T7.

Comment. Things x, y are independent if and only if things $-x, y$ are independent. For the sequel, we denote the fact of independence of x and y with the symbol $I(x, y)$.

3   Rough Inclusions in Mass-Based Mereological Universe

We define a relation $\mu_m(x, y, r)$ on the Cartesian product $U \times U \times [0, 1]$.

Definition 3. $\mu_m(x, y, r) \Leftrightarrow \frac{m(x \cdot y)}{m(x)} \geq r$.

It remains to prove that μ_m is a rough inclusion by checking that properties RINC1-RINC3 are satisfied.

Proposition 6. *The relation $\mu_m(x, y, r)$ is a rough inclusion.*

Proof. $\mu_m(x, y, 1)$ is equivalent to $m(x \cdot y) = m(x)$ which, as $\Pi(x \cdot y, x)$ and by T10, $m(x \cdot y) \leq m(x)$ implies that $x \cdot y = x$ i.e. by T1, $\Pi(x, y)$. The converse is obvious. RINC1 is proved. For RINC2, assume that $\mu_m(x, y, 1)$ and $\mu_m(z, x, r)$. Hence, $x \cdot y = x$ and $\frac{m(z \cdot x)}{m(z)} \geq r$. As $z \cdot x = (z \cdot y) \cdot x$, we have $m(z \cdot x) \leq m(z \cdot y)$ and thus $\mu_m(z, y, r)$ holds true proving RINC. RINC3 is manifest.

We define in addition the function $x \rightarrow_m y : U^2 \rightarrow [0, 1]$ returning the maximal degree of inclusion of x into y.

Definition 4. $x \rightarrow_m y = \frac{m(x \cdot y)}{m(x)}$.

Let us observe that the mirror relation $\mu * (x, y, r)$ can be defined:

Definition 5. $\mu * (x, y, r) \Leftrightarrow \frac{m(x \cdot y)}{m(y)} \geq r$,

along with the function $x \rightarrow^*_m y$:

Definition 6. $x \rightarrow^*_m y = \frac{m(x \cdot y)}{m(y)}$.

Remark 1. While Definitions 3 and 6, reflect the point of view of mereology, determining what part of x is in y, Definitions 5 and 6 reflect the probabilistic point of view determining what part of y is in x, i.e., the conditional inclusion of x with respect to y. Clearly, $x \rightarrow^*_m y = y \rightarrow_m x$, hence, properties of $x \rightarrow^*_m y$ are properties of $x \rightarrow_m y$. Yet, we propose to apply $\rightarrow^*_m$ in order to separate the two understandings of partial inclusion: mereological vs. probabilistic.

We prove some theses on relationships of the function μ to the relation Π and we establish some of its properties. Alongside those theses we point to mirrored theses on $\rightarrow^*_m$.

T17. $\Pi(x, y) \Rightarrow x \rightarrow_m y = 1$.
T17*. $\Pi(y, x) \Rightarrow x \rightarrow^*_m y = 1$.

Proof. By T1 and Definition 3, Definition 5.

T18. $(x \rightarrow_m y = 1) \Rightarrow \Pi(x, y)$.
T18*. $(x \rightarrow^*_m y = 1) \Rightarrow \Pi(y, x)$.

Proof. As $(x \cdot y) \rightharpoonup x$, we have by A3 that $m(x) = m(x \cdot y) + m(x - x \cdot y) = m(x) + m(x - x \cdot y)$, hence, $m(x - x \cdot y) = 0$ and $x = x \cdot y$, and, this implies $\Pi(x, y)$. The proof in the mirror case goes along same lines.

T19. $\Pi(x, y) \Leftrightarrow x \to_m y = 1 \Leftrightarrow x \to y$.
T19*. $\Pi(y, x) \Leftrightarrow y \to_m^* x = 1 \Leftrightarrow y \to x$.

Proof. By T12, T17, T18.Proof in the mirror case goes by starry counterparts.

T20. $x \to_m -y) = 1 - x \to_m y$.

Proof. As by T8, $m(x) = m(x \cdot y) + m(x \cdot (-y))$, hence, $m(x \cdot (-y)) = m(x) - m(x \cdot y)$ and thus $\mu(x, -y) = 1 - x \to_m y$.

3.1 The Bayes Theorem in Mass-Based Rough Mereology

It is desirable to fully introduce probabilistic aspect into spatial reasoning by proving the Bayes theorem which allows for Bayesian form of reasoning. We begin with a simple form of the Bayes theorem whose proof is standard and with its consequences.

T22. $x \to_m y = \frac{m(y) \cdot y \to_m x}{m(x)}$.

T23. $\frac{x \to_m y}{y \to_m x} = \frac{m(y)}{m(x)}$.

T24. $\Pi(x, y) \Rightarrow \mu(y, x) = \frac{m(x)}{m(y)}$.

T25. $\frac{x \to_m y}{y \to_m x} \cdot \frac{y \to z}{z \to y} = \frac{x \to_m z}{z \to x}$.

T25 is the transitivity property for μ. It follows from T23. The general form of the Bayes theorem is as follows, cf. Łukasiewicz [18].

T26. $(+_{i \neq j} y_i \cdot y_j = \theta) \wedge (+_i y_i = V) \Rightarrow z \to_m x = \frac{m(x) \cdot x \to_m z}{\sum_{i=1}^{k} m(y_i) \cdot y_i \to_m z}$.

Proof. Let $Y = \{y_i : i \leq k\}$ be a maximal set of pairwise disjoint things in the universe U. Then $+Y = V$. Given a thing z, we prove the lemma.

Lemma 1. $z = Cls\{z \cdot y_i : i \leq k\}$.

Proof. Let $\Pi(x, z)$; there exist y, w such that $\Pi(y, x)$, $\Pi(y, w)$, $w = z \cdot y_i$ for some $i \leq k$, hence, $Ov(y, z \cdot y_i)$, and, by M3, $\Pi(x, Cls\{z \cdot y_i : i \leq k\}$. Conversely, let $\Pi(x, Cls\{z \cdot y_i : i \leq k\}$. BY C2 in the class definition, there exist things u, w such that $\Pi(u, x)$, $\Pi(u, w)$, $w = z \cdot y_j$ for some y_j. Hence, $\Pi(w, z)$ and by M3, $\Pi(Cls\{z \cdot y_i : i \leq k\}, z)$ implying by Proposition 1 that $z = Cls\{z \cdot y_i : i \leq k\}$.

In the same vein, we prove

Lemma 2. $Cls\{z \cdot y_i : i \leq k\} = +_{i=1}^{k} z \cdot y_i$.

T27. $z = +_{i=1}^{k} z \cdot y_i$.

Proof. By Lemmas 1 and 2.

T28. $m(z) = \sum_{i=1}^{k} m(z \cdot y_i) = \sum_{i=1}^{k} m(y_i) \cdot y \to_m z = \sum_{i=1}^{k} m(y_i) \cdot z \to_m^* y_i$.

Proof. By additivity of the mass m.

Comment. The reader has observed that in fact we have proved the dual to the classical Bayes theorem and in the star denoted form the exact rendition of the probabilistic Bayes theorem in the abstract framework of rough mereology: the reason is that in the classical probability, the conditional probability $P(A|B)$ is defined as part of B belonging in A whereas in mereology we are interested in part of x contained in y. The true Łukasiewicz version of the Bayes theorem is obtained with the mirror form involving $\mu*$ (see next section).

The mRM Logic vs. the Łukasiewicz Logic of Probability. Łukasiewicz [18] considered indefinite formulae on a finite universe U. For each formula $q(x)$, he assigned the numerical weight $w(q)$ defined as follows.

$$w(q) = \frac{|\{u \in U : q(u) \text{ is true}\}|}{|U|}. \tag{32}$$

$|U|$ denotes the number of elements in U. With this idea, he obtained the logical form of the calculus of probabilities. We define a translation of the mRM logic into the Łukasiewicz logic of probabilities by the function tr: $tr(x, a)$, $tr(y, b)$, $tr(m, w)$, $tr(x \to_m y, w_a(b))$, $tr(x \to_m^* y, w_b(a))$, $tr(\to, \Rightarrow)$, and for phrases of the form $\alpha \circ \beta$, define $tr(\alpha \circ \beta), (tr(\alpha) \circ tr(\beta))$, where $\circ$ denotes the concatenation operation.

We obtain via the function tr, the formulae of the logic of probability [18].

4   Spatial Reasoning in Mass Based Mereology: Betweenness

The notion of betweenness relation due to Tarski [43], modified by van Benthem [2] and adapted by us to the needs of data analysis and behavioral robotics cf. [32] will acquire here an abstract formulation in the framework of the mass mereology.

We introduce first the notion of distance $\delta(x, y)$ between two things x, y in the universe U based on the mass assignment m. We let

$$\delta(x, y) = m(x - y) + m(y - x) = m(x) + m(y) - 2 \cdot m(x \cdot y). \tag{33}$$

Clearly, $m(x, y) = 0$ if and only if $x = y$.

We say that the thing z *is between* things x, y, in symbols $Btw(z, x, y)$, when the following condition is satisfied   (B) *For each thing w not identical to z, and for each thing u such that $\Pi(u, w)$ there exists a thing t such that $\Pi(t, z)$ and either $\delta(t, x) < \delta(u, x)$ or $\delta(t, y) < \delta(u, y)$.* We have the following proposition

Proposition 7. *For each pair x, y of things, the sum $x + y$ is the only thing between x and y.*

Proof. Assume that $w \neq x + y$ so w does not satisfy the class definition for $x + y$, hence, there exists a thing u such that $\Pi(u, w)$ and u does overlap neither with x nor with y. Letting v as x or y, we have: $\delta(u, x) = m(u) + m(x) > 0$ while $\delta(x, x) = 0$, similarly in case of y.

The condition (B) as well as the notion of betweenness can be extended for finite sets of things to the notion $GBtw(z, T)$, where T is a finite set of things, of the *generalized betweenness relation* which holds when the codition (GB) is satisfied.

(GB) *For each thing $w \neq z$ and each thing u with $\Pi(u, w)$ there exist a thing v such that $\Pi(v, z)$ and a thing $t \in T$ such that $\delta(u, t) > \delta(v, t)$.*

Remark 2. In particular cases, important for applications, the mereological sum acquires specific renditions in the context of betweenness. In behavioral robotics, when mobile autonomous planar robots, or more generally intelligent agents, are modeled as planar rectangles, the mereological sum of two robots a, b is the extent $ext(a, b)$, i.e., the smallest rectangle containing a and b cf. Polkowski and Ośmiałowski [36, 37]. Hence, the notion of the mereological sum should be modified: for a given context C, the sum $x + y$ of two things satisfying the context C is the smallest thing satisfying the context C and such that it contains each thing being an improper part of either x or y (smallest, containment, are understood in terms of the relation Π). In the case of information/decision systems, where things are represented by means of their information sets, i.e., for a system with attributes in the set A and with the values of attributes in the set V, the information set for a thing u is the set $Inf(u) = \{a(u) : a \in A\}$, the mereological sum of things u and v relative to a partition $P = \{A_1, A_2\}$ is a thing w such that $Inf(w) = \{a(u) : a \in A_1\} \cup \{a(v) : a \in A_2\}$ cf. Polkowski [34].

5   Spatial Reasoning in Mass Based Mereology: The Environments of Regular Open and Regular Closed Sets

We denote by Cl the closure operator, already introduced in Sect. 9.1, and the symbol Int will denote the interior operator. A set x in a topological space (X, τ) is *regular open* if it is of the form $IntCly$ for some set y; then, by the property

$$IntClIntClz = IntClz \tag{34}$$

valid for each set $z \subseteq X$, we have that $x = IntClx$ which can be taken as the defining property of regular open sets. Dually, a set y is regular closed when $y = ClInty$.

It is shown in Sect. 9 that regular open sets form a complete Boolean algebra under operators

1. $x + y = IntCl(x \cup y)$.
2. $x \cdot y = x \cap y$.
3. $-x = X \setminus Clx$.

and with constants $\mathbf{0} = \emptyset, \mathbf{1} = X$.

Dually, regular closed sets form a complete Boolean algebra under operators

1. $x + y = x \cup y$.
2. $x \cdot y = ClInt(x \cap y)$.
3. $-x = X \setminus Intx$.

and with constants $\mathbf{0} = \emptyset, \mathbf{1} = X$.

We consider a mass assignment m on RO(X) respectively on RC(X) and we consider the distance function δ already applied in the betweenness discussion in Sect. 4.

$$\delta(x, y) = m(x - y) + m(y - x) \tag{35}$$

along with the distance function Δ

$$\Delta(x, y) = 1 - min\{x \to_m y, y \to_m x\} \tag{36}$$

so $\Delta(x, y) = 1 - m(x \cdot y) \cdot min\{\frac{1}{m(x)}, \frac{1}{m(y)}\}$.

In a discussion of planning and navigation for spatial things, we assume that those things are modeled as squares of fixed and constant size with area, say, C. As boundaries of squares are closed nowhere dense, hence, of measure zero, we can assume that $x \cdot y = x \cap y$ in case of squares.

We adopt the area measure $||x||$ as the mass assignment $m(x)$, hence, we have

$$\delta(x, y) = 2 \cdot C - 2 \cdot m(x \cap y) \tag{37}$$

and

$$\Delta(x, y) = 1 - \frac{m(x \cap y)}{C} \tag{38}$$

so it follows that

$$\delta(x, y) = 2 \cdot C \cdot \Delta(x, y), \tag{39}$$

i.e., both distance functions are equivalent: qualitative conclusions from both are identical. In particular, our result on betweenness relation in Sect. 4 is valid for Δ distance function.

6   Spatial Reasoning in Mass Based Mereology: Mereological Potential Fields

Classical methodology of potential fields works with integrable force field given by formulas of Coulomb or Newton which prescribe force at a given point as inversely proportional to the squared distance from the target; in consequence, the potential is inversely proportional to the distance from the target. The basic property of the potential is that its density (=force) increases in the direction toward the target. We observe this property in our construction.

In our construction of the potential field, region will be squares: robots are represented by squares circumscribed on them (simulations were performed with disk–shaped Roomba robots, the intellectual property of iRobot.Inc.), see Polkowski and Ośmiałowski [24, 36, 37], Ośmiałowski [21, 22].

Geometry induced by means of a rough inclusion can be used to define a generalized potential field: the force field in this construction can be interpreted as the density of squares that fill the workspace and the potential is the integral of the density. We present now the details of this construction. We construct the potential field by a discrete construction. The idea is to fill the free workspace of a robot with squares of fixed size C in such a way that the density of the square field (measured, e.g., as the number of squares intersecting the disc of a given radius r centered at the target) increases toward the target.

To ensure this property, we fix a real number – the field growth step in the interval (0, square edge length); in our exemplary case shown above the parameter field growth step is set to 0.01. The collection of squares grows recursively with the distance from the target by adding to a given square in the $(k + 1) - th$ step all squares obtained from it by translating them by $k \times$ field growth step with respect to a mass based distance Δ in basic eight directions: N, S, W, E, NE, NW, SE, SW (in the implementation of this idea, the *floodfill algorithm* with a queue has been used, see Ośmiałowski [21]. Once the square field is constructed, the path for a robot from a given starting point toward the target is searched for.

The idea of this search consists in finding a sequence of *way–points* which delineate the path to the target. Way–points are found recursively as centroids of unions of squares mereologically closest to the square of the recently found way–point.

In order to define the field force of the rough mereological field, let us consider how many generations of squares will be centered within the distance r from the target. Clearly, we have

$$d + 2d + ... + kd \leq r \tag{40}$$

where d is the field growth step, k is the number of generations. Hence,

$$k^2 d \leq \frac{k(k + 1)}{2} d \leq r \tag{41}$$

and thus

$$k \leq (\frac{r}{d})^{\frac{1}{2}} \tag{42}$$

The force $F(r)$ can be taken as $\sim r^{\frac{1}{2}}$. Hence, the force decreases with the distance r from the target slower than traditional Coulomb force. It has advantages of slowing the robot down when it is closing on the target. Parameters of this procedure are: the field growth step set to 0.01, and the size of squares which in our case is 1.5 times the diameter of the Roomba robot used in simulation.

A robot should follow the path proposed by planner shown in Fig. 1.

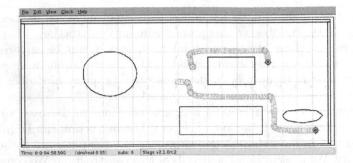

Fig. 1. Planned paths of Roomba robots to their targets.

7 Spatial Reasoning in Mass Based Mereology: Planning for Teams of Robots

Many authors attacked problems of route planning for autonomous agents/robots by extending methods elaborated for a single robot; helpful in those attempts were studies of behavior of migrating birds flying in 'boids', cf., Reynolds [38] which brought forth elementary behaviors like collision–avoidance, velocity adjustment, leader–following, flock–centering, transferred into robot milieu , e.g., in Mataric [19], Agah [1], which provided elementary robot behaviors like wandering, homing, following, avoidance, aggregation, dispersion. For details of the exposition which we give now, please consult Ośmiałowski [22] and Ośmiałowski and Polkowski [24]. We recall that agents are represented as squares in the plane and all things are rectangles with sides parallel to coordinate axes. For robots x, y, z, we say that a robot y is *between robots x and z*, in symbols

$$(\text{between } y \ x \ z) \tag{43}$$

in case the square y is contained in the extent of squares x, z, i.e., in the smallest rectangle containing x, z, see Sect. 4.

This notion can be generalized to the notion of *partial betweenness* which models in a more realistic manner spatial relations among x, y, z; we say in this case that y is *between x and z to a degree of at least r*, in symbols,

$$(\text{between–deg} r \ y \ x \ z \ ) \tag{44}$$

if and only if

$$y \to_m ext(x, z) \geq r, \tag{45}$$

where $ext(x, z)$ denotes the extent of x, z.

For a team of robots, $T(r_1, r_2, ..., r_n) = \{r_1, r_2, ..., r_n\}$, an *ideal formation IF* on $T(r_1, r_2, ..., r_n)$ is a betweenness relation on the set $T(r_1, r_2, ..., r_n)$.

In implementations, ideal formations are represented as lists of expressions of the form

$$(\text{between } r_0 \ r_1 \ r_2) \tag{46}$$

along with a list of expressions of the form

$$(\text{not–between } r_0 \ r_1 \ r_2). \tag{47}$$

To account for dynamic nature of the real world, in which due to sensory perception inadequacies, dynamic nature of the environment etc., we allow for some deviations from ideal formations by allowing that the robot which is between two neighbors can be between them to a degree in the sense of (44). This leads to the notion of a real formation.

For a team of robots, $T(r_1, r_2, ..., r_n) = \{r_1, r_2, ..., r_n\}$, a *real formation RF* on $T(r_1, r_2, ..., r_n)$ is a betweenness to degree relation on the set $T(r_1, r_2, ..., r_n)$ of robots.

Real formations are given as lists of expressions of the form,

$$(\text{between–deg } \eta \ r_0 \ r_1 \ r_2), \tag{48}$$

indicating that the thing r_0 is to degree of η in the extent of r_1, r_2, for all triples in the relation (between–deg ....), along with a list of expressions of the form,

$$(\text{not–between } r_0 \ r_1 \ r_2), \tag{49}$$

indicating triples which are not in the given betweenness relation.

In order to describe formations, the language derived from LISP–like s–expressions can be applied. Typically, LISP lists are hierarchical structures that can be traversed using recursive algorithms. We apply the restriction that top–level list (a root of whole structure) contains only two elements where the first element is always a formation identifier (a name). For instance

Example 1. (formation1 (some_predicate *param*1 ... *paramN*))

For each thing on a list (and for a formation as a whole) an extent can be derived and in facts, in most cases only extents of those things are considered. We have defined two possible types of things

1. *Identifier: robot or formation name (where formation name can only occur at top–level list as the first element);*

2. *Predicate: a list in LISP meaning where first element is the name of given predicate and other elements are parameters; number and types of parameters depend on given predicate.*

Minimal formation should contain at least one robot. For example

Example 2. (formation2 roomba0)

To help understand how predicates are evaluated, we need to explain how extents are used for computing relations between things. Suppose we have three robots (*roomba0*, *roomba1*, *roomba2*) with *roomba0* between *roomba1* and *roomba2*. We can draw an extent of this situation as the smallest rectangle containing the union *roomba1* ∪ *roomba2*. This extent can be embedded into bigger structure: it can be treated as a thing that can be given as a parameter to predicate of higher level in the list hierarchy. For example:

Example 3. (formation3 (between (between roomba0 roomba1 roomba2)
roomba3 roomba4))

Typical formation description may look like below, see Ośmiałowski [22]

Example 4. (cross
    (set
            (max–dist 0.25 roomba0 (between roomba0 roomba1 roomba2))
            (max–dist 0.25 roomba0 (between roomba0 roomba3 roomba4))
            (not–between roomba1 roomba3 roomba4)
            (not–between roomba2 roomba3 roomba4)
            (not–between roomba3 roomba1 roomba2)
            (not–between roomba4 roomba1 roomba2)
)
)

This is a description of a formation of five Roomba robots arranged in a cross
shape. The *max–dist* relation is used to bound formation in space by keeping all
robots close one to another.

We show a screen–shot of robots in the initial formation of cross–shape in
a crowded environment, see Fig. 2. These behaviors witness the flexibility of
our definition of a robot formation: first, robots can change formation, next, as
the definition of a formation is relational, without metric constraints on robots,
the formation can manage an obstacle without losing the prescribed formation
(though, this feature is not illustrated in figures in this chapter). The details of
implementation are discussed in Ośmiałowski [22].

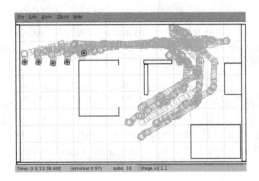

Fig. 2. Trails of robots moving in the line formation through the passage

8   Spatial Reasoning on the Boolean Algebra of Regular Open/Closed Sets with a Mass Assignment: Potential Issues

The point of our exposition in this work is to give theoretical foundations for
mass based spatial reasoning on regular sets be they open or closed. Given an

environment E for intelligent agents/robots of diameter r, with the free space $W = E - O$, where O is the space taken by obstacles, we say that two regions x, y are *r-connected* if and only if

$$x \rightarrow_m y \geq r \wedge y \rightarrow_m x \geq r. \tag{50}$$

An *r-path* from a region x to a region y is a sequence of regions $x_0 = x, x_1, x_2, \ldots, x_k = y$ such that x_i, x_{i+1} are r-connected for $i = 0, 1, \ldots, k-1$.

A set of regions R is *r-Connected* if and only if each pair x, y of regions in R are r-connected.

A set of regions R is *r-Acyclic* if and only if there is no r-path in it which is an *r-cycle*, i.e., the path with the beginning region identical to the ending region.

A set of regions R is an *r-Tree* if and only if it is r-Connected and r-Acyclic. The following is a simple conclusion.

Proposition 8. *In order for an agent a located, respectively, in a region x to be able to move without obstacles to a region y, it is sufficient and necessary that regions x, y be connected by an r-path, a fortiori, they be nodes in an r-Tree.*

9   Appendix: Topological Context of Spatial Reasoning

We denote as before with the symbol Cl the topological operator of closure and with the symbol Int the topological operator of interior. The essential property of those operators is

$$x \cap Cl y \subseteq Cl(x \cap y) \tag{51}$$

for each set y and each open set x in a topological space (X, τ). We focus on regular open and regular closed sets, already introduced in Sect. 9.1. A very basic property of regular sets is that they form complete Boolean algebras; regular open sets form the Boolean algebra, denoted $RO(X)$ and dually, regular closed sets form the complete Boolean algebra $RC(X)$. In order to justify this claim, we let

$$x^* = X \setminus Cl x.$$

The set x is regular open if and only if $x = x^{**}$. Indeed, $x^{**} = X \setminus Cl(X \setminus Cl x) = Int Cl x$. Properties of the operation x^* are (in proofs, one uses (51)

1. If $x \subseteq y$, then $y^* \subseteq x^*$.
2. If x is an open set, then $x \subseteq x^{**}$.
3. If x is an open set, then $x^* = x^{***}$, hence, $x^{**} = x^{****}$.
4. If x, y are open sets, then $(x \cap y)^{**} = x^{**} \cap y^{**}$.
5. $(x \cup y)^* = x^* \cup y^*$.
6. If x is an open set, then $(x \cup x^*)^{**} = X$.

Now, we define in the family $RO(X)$ of regular open sets operations $+, \cdot, -$:

1. $x + y = (x \cup y)^{**} = IntCl(x \cup y)$.
2. $x \cdot y = x \cap y$.
3. $-x = x^* = X \setminus Clx$.

and constants $\mathbf{0} = \emptyset, \mathbf{1} = X$.

All operations listed above give regular open sets by properties of $(.)^*$. It remains to check that axioms of Boolean algebra are satisfied. Commutativity laws $x + y = y + x, x \cdot y = y \cdot x$ are satisfied. The laws $x + \mathbf{0} = x, x \cdot \mathbf{1} = x$ are also manifest. We have $x \cdot -x = x \cap x^* = x \setminus Clx = \emptyset = \mathbf{0}$ as well as $x + (-x) = (x \cup x^{**})^{**}) = X = \mathbf{1}$. The distributive laws $x + (y \cdot z) = (x + y) \cdot (x + z)$ as well as $x \cdot (y + z) = (x \cdot y) + (x \cdot y)$ hold by Property 5.

It is easy to see that for a collection $\mathcal{A} \subseteq RO(X)$, the set $s(\mathcal{A}) = IntCl \bigcup \mathcal{A}$ is the least upper bound of $\mathcal{A}$.

By duality applied to the family $RC(X)$ of regular closed sets in X, we obtain a dual proposal that $RC(X)$ is a complete boolean algebra under operations $+, \cdot, -$ defined as follows:

1. $x + y = x \cup y$.
2. $x \cdot y = ClInt(x \cap y)$.
3. $-x = X \setminus Intx$.

and with constants $\mathbf{0} = \emptyset, \mathbf{1} = X$.

9.1   An Application: The Model ROM for Connection

We define in the space $RO(X)$ of regular open sets in a regular space X the connection C by demanding that

$$C(x, y) \Leftrightarrow Clx \cap Cly \neq \emptyset.$$

For simplicity sake, we assume that the regular space X is connected, so the boundary of each set is non–empty.

Ingredient in ROM. First, we investigate what Π_C means in ROM. By definition IC in (10), for $x, y \in RO(X)$,

$$\Pi_C(x, y) \Leftrightarrow \forall z \in RO(X).Clz \cap Clx \neq \emptyset \Rightarrow Clz \cap Cly \neq \emptyset.$$

This excludes the case when $x \setminus Cly \neq \emptyset$ as then we could find a $z \in RO(X)$ with

$$z \cap x \neq \emptyset = Clz \cap Cly$$

(as our space X is regular). It remains that $x \subseteq Cly$, hence, $x \subseteq IntCly = y$. It follows finally that in model ROM, $\Pi_C(x, y) \Leftrightarrow x \subseteq y$.

Overlap in ROM. Now, we can interpret overlapping in ROM. For $x, y \in RO(X)$, $Ov_C(x, y)$ means that there exists $z \in RO(X)$ such that $z \subseteq x$ and $z \subseteq y$ hence $z \subseteq x \cap y$, hence,

$$x \cap y \neq \emptyset.$$

This condition is also sufficient by regularity of X. We obtain that in ROM,

$$Ov_C(x, y) \Leftrightarrow x \cap y \neq \emptyset.$$

External Connectedness in ROM. The status of EC in ROM is

$$EC(x, y) \Leftrightarrow Clx \cap Cly \neq \emptyset \wedge x \cap y = \emptyset.$$

This means that closed sets Clx, Cly do intersect only at their boundary points.

Tangential Ingredient in ROM. We can address the notion of a tangential ingredient: $T\Pi_C(x, y)$ means the existence of $z \in RO(X)$ such that

$$Clz \cap Clx \neq \emptyset \neq Clz \cap Cly$$

and

$$z \cap x = \emptyset = z \cap y$$

along with $x \subseteq y$. In case

$$Clx \cap (Cly \setminus y) \neq \emptyset$$

letting $z = X \setminus Cly$ we have

$$Clz = Cl(X \setminus Cly)$$

and

$$Bdz = Clz \setminus z = Cl(X \setminus Cly) \setminus (X \setminus Cly)$$

which in turn is equal to

$$Cl(X \setminus Cly) \cap Cly = Cl(X \setminus y) \cap Cly = Bdy.$$

Hence, $Cly \setminus y \subseteq Clz$, and $Clz \cap Clx \neq \emptyset$; a fortiori, $Cly \cap Clz \neq \emptyset$. As $z \cap y = \emptyset$, a fortiori $z \cap x = \emptyset$ follows.

We know, then, that

$$Clx \cap (Cly \setminus y) \neq \emptyset \Rightarrow T\Pi_C(x, y).$$

Was to the contrary, $Clx \subseteq y$, from $z \cap Clx \neq \emptyset$ it would follow that $z \cap y \neq \emptyset$, negating $EC(x, y)$.

It follows finally that in ROM, $T\Pi_C(x, y)$ if and only if $x \subseteq y$ and $Clx \cap (Cly \setminus y) \neq \emptyset$, i.e.,

$$T\Pi_C(x, y) \Leftrightarrow x \subseteq y \wedge Clx \cap Bdy \neq \emptyset.$$

From this analysis we obtain also that $NT\Pi_C(x, y)$ if and only if $Clx \subseteq Inty$.

References

1. Agah, A.: Robot teams, human workgroups and animal sociobiology. A review of research on natural and artificial multi-agent autonomous systems. Adv. Robot. **10**, 523–545 (1997)
2. van Benthem, J.: The Logic of Time. Reidel. Dordrecht (1983)

3. Cao, Y.U., Fukunaga, A.S., Kahng, A.B.: Cooperative mobile robotics: antecedents and directions. Auton. Robot. **4**, 7–27 (1997)
4. Casati, R., Varzi, A.C.: Parts and Places. The Structures of Spatial Representation. MIT Press, Cambridge (1999)
5. Clarke, B.L.: A calculus of individuals based on connection. Notre Dame J. Form. Log. **22**(2), 204–218 (1981)
6. Cohn, A.G.: Calculi for qualitative spatial reasoning. In: Calmet, J., Campbell, J.A., Pfalzgraf, J. (eds.) AISMC 1996. LNCS, vol. 1138, pp. 124–143. Springer, Heidelberg (1996). https://doi.org/10.1007/3-540-61732-9_54
7. Cohn, A.G., Gooday, J.M., Bennett, B., Gotts, N.M.: A logical approach to representing and reasoning about space. In: Calmet, J., Campbell, J.A., Pfalzgraf, J. (eds.) Artificial Intelligence and Symbolic Mathematical Computation. Lecture Notes in Computer Science, vol. 1138, pp. 124–143. Springer, Heidelberg (1996). https://doi.org/10.1007/978-94-015-8994-9_8
8. Cohn, A.G., Gotts, N.M.: Representing spatial vagueness: a mereological approach. In: Proceedings of the 5th International Conference on Principles of Knowledge Representation and Reasoning, KR 1996, pp. 230–241. Morgan Kaufmann, San Francisco (1996)
9. Cohn, A.G., Randell, D., Cui, Z., Bennett, B.: Qualitative spatial reasoning and representation. In: Carrete, N., Singh, M. (eds.) Qualitative Reasoning and Decision Technologies, Barcelona, pp. 513–522 (1993)
10. Cohn, A.G., Varzi, A.C.: Connections relations in mereotopology. In: Prade H. (ed.) Proceedings of ECAI 1998 13th European Conference on Artificial Intelligence, pp. 150–154. Wiley, Chichester (1998)
11. Egenhofer, M.J.: Reasoning about binary topological relations. In: Gunther, O., Schek, H.(eds.) Proceedings of Advances in Spatial Databases, SSD 1991, Berlin, pp. 143–160 (1991)
12. Gotts, N.M., Gooday, J.M., Cohn, A.G.: A connection based approach to commonsense topological description and reasoning. Monist **79**(1), 51–75 (1996)
13. Gotts, N.M., Cohn, A.G.: A mereological approach to representing spatial vagueness. In: Working papers. The Ninth International Workshop on Qualitative Reasoning, QR 1995 (1995)
14. Hájek, P.: Metamathematics of Fuzzy Logic. Kluwer, Dordrecht (1998)
15. de Laguna, T.: Point, line and surface as sets of solids. J. Philos. **19**, 449–461 (1922)
16. Leśniewski, S.: Foundations of the General Theory of Sets (in Polish). Moscow (1916)
17. Ling, C.-H.: Representation of associative functions. Publ. Math. Debr. **12**, 189–212 (1965)
18. Łukasiewicz, J.: Die Logischen Grundlagen der Wahrscheinlichkeitsrechnung. Kraków, 1913. Cf. Borkowski, L. (ed.) Selected Works. North Holland-PWN, Amsterdam-Warszawa, pp. 16–63 (1970)
19. Matarić M.: Interaction and intelligent behavior. Ph.D. dissertation. MIT EECS Department (1994)
20. Nicolas, D.: The logic of mass expressions. In: Stanford Enc. Phil. https://plato.stanford.edu/entries/logic-masseapress/
21. Ośmiałowski, P.: On path planning for mobile robots: introducing the mereological potential field method in the framework of mereological spatial reasoning. J. Autom. Mob. Robot. Intell. Syst. (JAMRIS) **3**(2), 24–33 (2009)

22. Osmialowski P.: Planning and navigation for mobile autonomous robots. Ph.D. dissertation. Polkowski, L. Supervisor, Polish-Japanese Academy IT. PJAIT Publishers, Warszawa (2011)
23. Pawlak, Z.: Rough Sets: Theoretical Aspects of Data Analysis. Kluwer, Dordrecht (1992)
24. O'smiaiowski, P., Polkowski, L.: Spatial reasoning based on rough mereology: a notion of a robot formation and path planning problem for formations of mobile autonomous robots. In: Peters, J.F., Skowron, A., Słowiński, R., Lingras, P., Miao, D., Tsumoto, S. (eds.) Transactions on Rough Sets XII. LNCS, vol. 6190, pp. 143–169. Springer, Heidelberg (2010). https://doi.org/10.1007/978-3-642-14467-7_8
25. Polkowski, L.: Rough Sets. Mathematical Foundations. Springer, Heidelberg (2002)
26. Polkowski, L.: A rough set paradigm for unifying rough set theory and fuzzy set theory. Fundam. Inform. **54**, 67–88 (2003)
27. Polkowski, L.: Toward rough set foundations. Mereological approach. In: Tsumoto, S., Słowiński, R., Komorowski, J., Grzymała-Busse, J.W. (eds.) RSCTC 2004. LNCS (LNAI), vol. 3066, pp. 8–25. Springer, Heidelberg (2004). https://doi.org/10.1007/978-3-540-25929-9_2
28. Polkowski, L.: Formal granular calculi based on rough inclusions. In: Proceedings of IEEE 2005 Conference on Granular Computing GrC 2005, Beijing, China, pp. 57–62. IEEE Press (2005)
29. Polkowski, L.: Granulation of knowledge in decision systems: the approach based on rough inclusions. The method and its applications. In: Kryszkiewicz, M., Peters, J.F., Rybinski, H., Skowron, A. (eds.) RSEISP 2007. LNCS (LNAI), vol. 4585, pp. 69–79. Springer, Heidelberg (2007). https://doi.org/10.1007/978-3-540-73451-2_9
30. Polkowski, L.: A unified approach to granulation of knowledge and granular computing based on rough mereology: a survey. In: Pedrycz, W., Skowron, A., Kreinovich, V. (eds.) Handbook of Granular Computing, pp. 375–400. Wiley, Chichester (2008)
31. Polkowski, L.: Granulation of knowledge: similarity based approach in information and decision systems. In: Meyers, R.A. (ed.) Springer Encyclopedia of Complexity and System Sciences, pp. 1464–1487. Springer, Heidelberg (2009). https://doi.org/10.1007/978-1-4614-1800-9_94
32. Polkowski, L.: Approaimate Reasoning by Parts. An Introduction to Rough Mereology. Springer, Heidelberg (2011). https://doi.org/10.1007/978-3-642-22279-5
33. Polkowski, L.: Mereology in engineering and computer science. In: Calosi, C., Graziani, P. (eds.) Mereology and the Sciences. SL, vol. 371, pp. 217–291. Springer, Cham (2014). https://doi.org/10.1007/978-3-319-05356-1_10
34. Polkowski, L.: From Leśniewski, Łukasiewicz, Tarski to Pawlak: enriching rough set based data analysis. A retrospective survey. Fundam. Inform. **154**(1–4), 343–358 (2017)
35. Polkowski, L.: The counterpart to the Bayes theorem in mass-based rough mereology. In: Proceedings CS&P 2018. Humboldt Universität zu Berlin, September 2018. Informatik-Berichte series. Informatik-Bericht 248, pp. 47–56 (2018). http://ceur-ws.org/Vol-2240/paper4.pdf
36. Polkowski, L., Ośmiałowski, P.: Spatial reasoning with applications to mobile robotics. In: Aing-Jiang, J. (ed.): Mobile Robots Motion Planning. New Challenges. I-Tech, Vienna, pp. 433–453 (2008)
37. Polkowski, L., Ośmiałowski, P.: Navigation for mobile autonomous robots and their formations: an application of spatial reasoning induced from rough mereological geometry. In: Barrera, A. (ed.) Mobile Robots Navigation, pp. 339–354. In Tech, Zagreb (2010)

38. Reynolds, C.: Flocks, herds and schools. A distributed behavioral model. Comput. Graph. **21**(4), 25–34 (1987)
39. Polkowski, L., Skowron, A.: Rough mereology. In: Raś, Z.W., Zemankova, M. (eds.) ISMIS 1994. LNCS, vol. 869, pp. 85–94. Springer, Heidelberg (1994). https://doi.org/10.1007/3-540-58495-1_9
40. Polkowski, L., Skowron, A.: Rough mereology: a new paradigm for approaimate reasoning. Int. J. Approx. Reason. **15**(4), 333–365 (1997)
41. Randell D., Cui Z., Cohn A. G.: A spatial logic based on regions and connection. In: Proceedings of the 3rd International Conference on Principles of Knowledge Representation and Reasoning KR 1992. Morgan Kaufmann, San Mateo, pp. 165–176 (1992)
42. Tarski, A.: Zur Grundlegen der Booleschen Algebra I. Fund. Math. **24**, 177–198 (1935)
43. Tarski, A., Givant, S.: Symbolic logic. Bull **5**(2), 175–214 (1959)
44. Whitehead, A.N.: La théorie relationniste de l'espace. Revue de Métaphysique et de Morale **23**, 423–454 (1916)
45. Whitehead, A.N.: An Enquiry Concerning the Principles of Natural Knowledge. Cambridge University Press, Cambridge (1919)
46. Whitehead, A.N.: The Concept of Nature. Cambridge University Press, Cambridge (1920)
47. Whitehead, A.N.: Process and Reality: An Essay in Cosmology. Macmillan, New York (1929)
48. Zadeh, L.A.: Fuzzy sets. Inf. Control **8**, 338–353 (1965)

Compound Objects Comparators in Application to Similarity Detection and Object Recognition

Lukasz Sosnowski[✉]

Systems Research Institute, Polish Academy of Sciences,
ul. Newelska 6, 01-447 Warsaw, Poland
sosnowsl@ibspan.waw.pl

Abstract. This article presents similarity based reasoning approach for recognition of compound objects. It contains mathematical foundations for comparators theory as well as comparators network theory. It shows also three different practical applications in field of image recognition, text recognition and risk recognition.

1 Introduction

The functioning of the surrounding world is based on the ability to process information and to take decisions based on the said information. Information is presented in all ongoing processes. It may be saved in different ways on different types of media. Information exchange is connected with the ability to encode, decode and transmit information [82]. Meanwhile, decision-making is based on certain input information, which commonly comes from numerous sources, and also based on the processing of results of such input information in the form of synthesis, aggregation and decomposition.

It may be said that the world based on information consists of objects, which may be sources and receivers of information. Objects can be divided into two classes: compound objects and simple objects. Simple objects are called atomic, indivisible objects. They are ordinary entities with certain attributes, but not composed of other objects. Compound objects have their structure and may consist of other objects, either simple or compound. Both types are not direct participants of decision-making systems. They are converted to a representation, which is the most common description of a representative portion of an object. This representation can be different for the very object, depending on the type of processing, the purpose and context. Relationships between objects are important for the purposes of reasoning, in particular the relation of similarity, taxonomic order and granulation of compound objects [65].

In many areas of human life it is necessary to easily compare and classify objects. This requirement is driven by the necessity to detect objects' features, exclude certain events, prevent critical situations or to carry out identification based on similarity. Requests are typically generated by systems specialized in

© Springer-Verlag GmbH Germany, part of Springer Nature 2019
J. F. Peters and A. Skowron (Eds.): TRS XXI, LNCS 10810, pp. 169–300, 2019.
https://doi.org/10.1007/978-3-662-58768-3_6

information retrieval, monitoring, recommendation, identification, moderation and many others. Their main point of interest is a compound object and its relationships with certain patterns. In many cases the compound objects, which are processed in order to achieve a certain goal, are dealt with. The analysis of occurrences, characteristics and specificity of compound objects reveals that their processing is related to similarity (resemblance), comparison of features, structures of objects and their internal relations. There are certain domains where resemblance and compound objects play a special role. Below you will find several examples described in short:

- **CBR (Case-based reasoning)** - systems based on accumulated historical knowledge [61] (cases solved in the past). New solutions are produced by searching for cases which had previously been successfully completed. The basic steps of the decision-making process in CBR are the following: (a) to find the right match for the specified (new) problem with historical cases, (b) to adapt previous solutions to the current problem, (c) to solve the problem and save the result. In this case compound objects are represented by an input problem and reference elements. Both are used in comparative analysis.
- **Multimedia Database** - a database system allowing the storage of multimedia data [13], such as photographs, movies, documents, music, etc. Such databases are designed to process, search and aggregate information coded in those objects (not only to store). This is ensured by effective processing of objects and the provision of tools to communicate with the user as well meeting the user's expectations in the best possible way. This area is a direct extension of classical relational databases (but not exclusively), where the concept of a compound object was identified by means of tuples (relation described with certain attributes). In the case of a multimedia database, metadata come from objects (closed structures) and are saved there. However, database engines have to perform the same operations as a standard relational database, e.g. comparing, ranking, etc.
- **CEP (Complex event processing)** - systems for monitoring real-time events used in industry. These systems analyse data from various sources, compare them and detect patterns, trends and exceptions related to events analysed [47]. Many undesirable situations can be detected before they happen because appropriate decisions had been made on the basis of information received from such systems. In this area, compound objects are represented by configurations of observations as well as patterns and trends which are used for comparing acquired information.
- **RMS (Risk management systems)** - risk is a very complex phenomenon. A risk management system is an implementation of a set of procedures defined in a given area, e.g. information retrieval, fire rescue actions, etc. All these systems process compound objects. They are in a form of information, actions, processes and many other concepts and their instances, where risk can be identified. The desired information is the level of risk, which is a simple integer value. The problem is that the procedure of calculating these values is

based on many different conditions and concepts. The ability to compare such complex problems is one of the tools to produce these solutions in an efficient way.

- **RS (Recommender systems)** - one of the decision-making systems able to model user preferences and predict their needs, e.g. select the next element of resources available. Solutions of this kind are very popular in e-commerce, entertainment, etc. These systems have one common purpose: to select the best subset of items for the current user. The selection depends on user behaviour, previous activities in the short and long term. Any data can be important in terms of the final decision. There are many ways to do it. The first way would involve the implementation by means of a framework of comparators. The basic idea would to designate certain clusters of preferences, clusters of users or resources [34]. In consequence, a system of this kind must efficiently compare similarities to clusters and (must) find the closest one. This is the obvious area of using the framework that is easy to apply.
- **Search engines** - engines [16] for searching information of various types. They work with noisy data in most cases. Queries are often encoded imprecisely and ambiguously. Systems based on these engines optimize the relevance of results retrieved for a given query, in order to provide the most desired information to the user. Compound objects take the form of indexed documents, such as texts, images, sounds, videos and any kind of structured documents (e.g. scientific publications) in this context. These objects are analysed to determine additional knowledge (as a result of analysing their structure or content) and to compute similarity.

The above-mentioned examples of systems deal with quite different problems. What they have in common is the processing of compound objects, though completely different ones in individual cases. These examples show that compound objects may be differentiated by type, application or other criteria. However, in spite of the differences related to objects, the same needs related to their processing are clear. First and foremost, it is necessary to process similarities to a certain point of reference. By achieving knowledge about similarity, it becomes possible to rank the processed object in the ranking of similar elements, which can indirectly lead to its classification or recognition.

Similarity can be measured in many different ways and with different tools. However, regardless of the differences on the level of compound objects or problems to be solved, the approach to similarity processing should be standardized. It should provide a common algorithm for handling different types of objects and evaluating similarity unambiguously by expressing results by means of values of the same scale. In addition, the scale should determine the degree of similarity between given objects at a given level of accuracy. Therefore, the desired solution should ensure the universality of methodology of the procedure and return results in a standardized form. The last feature is important due to its ability to combine different results in order to build more complex structures. At the same time, the very results of comparisons must be easy to compare with other results obtained with this method. In addition, the model should be well-described from the mathematical point of view to understand its functioning and properties more easily.

The target solution has to model similarity in an easy way, based on the features of objects. Accumulation of many features will result in a high degree of complexity of the model, so it is necessary to provide a mechanism for easy correction of the similarity measurement method by excluding certain cases from the model. The mechanism can be analogous to that observed in other fields (e.g. digital systems [113]), where the so-called discriminators occur, which allow the exclusion of certain solutions on the basis of specific input values.

One of the tools for achieving goals set out in this dissertation is the comparator of compound objects. The comparator compares a given object against a set of reference objects. It uses uniform scales of similarity. It generates results of comparison in a consistent manner. Comparators can create multi-layered unidirectional networks. These structures make it possible to construct multi-lattice solutions without losing their readability. Both comparators and networks of comparators may be subject to learning, which significantly increases their effectiveness. It is also worth noting that the development of techniques and methods of comparing objects in one field automatically widens the range of methods in other fields. By introducing a new measure for a given solution, it is worth adding the possibility of using it in other contexts and making it available for different types of compound objects. For these reasons, a measure catalogue has been created and grouped according to types of objects processed. Thanks to its successive update it is an easy way to widen and enrich the application of the proposed methodology in many important areas related to processing, and in particular in comparison of compound objects.

1.1 Plan of the Dissertation

This chapter is devoted to a brief introduction to the problem of similarity, comparison and prevalence of this problem in areas related to decision-making systems. Simultaneously, in this chapter one can find the motivation behind the start of the research as well as the underlying assumptions and requirements faced by the solution. The thesis of this dissertation has been formulated in this chapter, together with the objectives for implementation. At the same time, this is a list of basic issues that this dissertation deals with.

The second section contains the theoretical foundations from different fields; ones that are used when formulating solutions which constitute the research of the author. This chapter presents the fuzzy sets theory, which is one of the basic theories used by compound comparators set out in the thesis. The chapter also presents the extension of fuzzy sets and other models of uncertain information. It also includes the selected approaches to modelling similarities, key concepts behind this study, concepts related to recognition and identification of objects. In addition, it contains the overview of methods of information synthesis, which were directly included in the processing of objects through the proposed comparator networks.

Section 3 contains the mathematical foundations of compound object comparators as well as the description of components and algorithms applied. It also includes the description of selected comparator learning techniques, which allow considering this methodology an example of practical implementation of. computational intelligence. In addition, this chapter discusses the issues of information flow in a comparator using granular structures. An important part of the dissertation is also the specification of the required properties, which should characterize each comparator of compound objects.

Section 4 describes the creation of complex network structures based on dedicated components in the form of comparators, aggregators, translators, etc. In this chapter you will also find a mathematical description of the network and classifications of types of networks along with the possible applications of each type of the network. This chapter, along with chapter three, is a presentation of the method developed, is the author's output, which comes from basic research, reflecting the experience gained in many practical projects.

Section 5 presents the general methodology of dealing with the design of comparator-based solutions and describes the action model. Three examples of solving identification and recognition problems of compound objects by means of the network of comparators are also presented. The examples refer to three completely different types of applications and completely different types of compound objects. The first example deals with image processing in the form of contour maps, the second deals with the processing of text objects and the third one involves the processing of risks during rescue and firefighting operations.

Section 6 summarizes the research done and results achieved. Assumptions, studies and results are briefly described. The chapter also contains defined paths for further development, which can be implemented for the purposes of further development of the field.

The dissertation also contains an appendix which gathers detailed information on possible methods of measuring similarity (a catalogue of similarities) as well as other specific issues related to the implementation of the method.

1.2   Research Problem. Thesis and Objectives of the Dissertation

The work done as part of this dissertation is focused on the recognition and identification of compound objects that rely on similarity methods. The need to handle similarities for very diverse objects is an important argument for developing a universal methodology of procedure, supported by tools and methods for comparing compound objects. The above mentioned premises allow the formulation of the basic thesis of the dissertation:

It is possible to develop the mathematical and algorithmic bases of a compound object comparator, which, firstly, allows the ordering of a set or subset of reference objects relative to input objects, and secondly it would be possible to for the said comparator to connect with other comparators for modelling compound object structures and their dependencies.

In order to substantiate the thesis above, the following objectives were formulated in the work:

- description of the mathematical basis of the comparator,
- description and characterization of the comparator network,
- description of the universal network design methodology,
- description of a comparator-based model of recognition and identification as a complementary method to known techniques and methods for processing compound objects,
- performance of experiments verifying the effectiveness of using comparators in recognition and identification of different types of compound objects.

2   Introductory Information

This chapter describes the mathematical foundations used in the later part of the thesis to describe the theory of compound object comparators. This chapter also includes descriptions of methods and decision models to which this methodology is a complementary approach. Topics are divided into sections but presented in a common context related to the main topic of the work.

2.1   Fuzzy Sets

Fuzzy sets were developed as an extension of the classic set theory initiated by Cantor. In the classic set theory, only the crisp membership of the item is considered. An item is always entirely or not entirely an element of a given set. For a given set A, this dependence is defined by the characteristic function in the following form:

$$f(x) = \begin{cases} 1 & x \in A \\ 0 & x \notin A \end{cases} A \subseteq U \tag{1}$$

where U is the space of the universe of elements. This form of function enables the modelling of crisp concepts only. In reality, however, we use fuzzy notions that can model the imprecision and may provide basis for some approximate reasoning.

The approach proposed by Zadeh introduces a change in the meaning of the item belonging to a set [117]. In this theory, degrees of membership are considered, meaning that a given element may belong to the collection also in part. This significantly changes the approach to modelling phenomena and concepts which are ambiguous or imprecise. Therefore:

Definition 1. *A fuzzy set A in a non-empty considerations space (universe) U is defined as the set of the following pairs:*

$$A = \{(x, \mu_A(x)) : x \in U\}, \tag{2}$$

where $\mu_A : U \to [0,1]$ is a membership function of the fuzzy set A [36], while $\mu_A(x) \in [0,1]$ is a degree of membership of the element x in the set A.

The membership function is a reflection of the characteristic function in the classic set theory. This function models three cases:

1. $\mu(x) = 1$, when element x belongs to set A completely, which is analogous to classic sets,
2. $\mu(x) = 0$, which means the total absence of belonging to set A, which again is defined by the analogous characteristic function given by formula (1),
3. $\mu(x) \in (0, 1)$, which means partial belonging to the fuzzy set and forms a completely new modelling area as compared to classic sets.

If a set $A \subset U$ is defined on the space with finite number of elements in the form of $U = \{x_1, \ldots, x_n\}$, then according to the Zadeh's notion we may express it as:

$$A = \frac{\mu_A(x_1)}{x_1} + \ldots + \frac{\mu_A(x_n)}{x_n} = \sum_{i=1}^{n} \frac{\mu_A(x_i)}{x_i} \tag{3}$$

where the horizontal line is not a quotient but only the allocation of the degree of belonging to individual elements of the set. Individual components of formula (3) correspond with formula (2) respectively $(x_1, \mu_A(x_1)), (x_2, \mu_A(x_2))$, etc. [77].

A fuzzy set can be used for the modelling of imprecise concepts that people use for communication. Examples include terms used in natural language, e.g. large, small, medium, suitable, sufficient, etc. Each of the examples above can be represented by a fuzzy set specifying to what extent a particular value belongs to each of them.

2.1.1 Basic Definitions

The fuzzy sets theory is characterised by a number of definitions and properties. Selected definitions and properties will be presented in the further part of the section. The selection was based indirectly or directly on the relation with further parts of this work.

Definition 2. *Elements of the space U, which possess non-negative membership value in the set A, are a support of the fuzzy set A, denoted as $suppA$.*

This definition may be expressed as:

$$suppA = \{x \in U : \mu_A(x) > 0\} \tag{4}$$

Definition 3. *The height of the fuzzy set A is a supremum of the values of the function of membership in A. We denote these properties as $h(A)$. A fuzzy set A is called normal if its height is $h(A) = 1$.*

In other words, the height of the fuzzy set is the upper bound of the value of the membership function, which can be expressed as:

$$h(A) = sup_{x \in X} \mu_A(x) \tag{5}$$

Definition 4. *If for every element of the space U the membership value $\mu_A(x) = 0$, then it is an empty set denoted as $A = \emptyset$.*

Definition 5. *Two fuzzy sets A and B in certain space U are equal if and only if*

$$\mu_A(x) = \mu_B(x)$$

for each $x \in U$.

The inclusion of sets is defined in a similar way by individual values of the membership function for the same elements of space, which gives the following definitions:

Definition 6. *The inclusion of fuzzy set A into fuzzy set B $(A, B \subseteq U)$ is denoted as $A \subset B$, when for each element $x \in U$ the following inequality is fulfilled:*

$$\mu_A(x) \leq \mu_B(x)$$

Definition 7. *α-cut of the fuzzy set $A \subseteq U$ is called a fuzzy set whose elements have degrees of membership of at least α.*

This dependence can be expressed as:

$$A_\alpha = \{x \in U : \mu_A(x) \geq \alpha\}, \forall_{\alpha \in [0,1]} \tag{6}$$

Triangular Norms and Negatives

Triangular norms are certain generalized operators for which operations of determining the product and the sum of fuzzy sets are performed. There are two types of triangular norms in the form of T-norms and S-norms (or interchangeably T-conorms) [37]. Both T-norm and S-norm are functions of two variables respectively in the form:

$$T : [0,1] \times [0,1] \rightarrow [0,1] \tag{7}$$

$$S : [0,1] \times [0,1] \rightarrow [0,1], \tag{8}$$

where the function (7) satisfies the following properties $\forall a, b, c \in [0,1]$:

1. $T(a,1) = a, T(a,0) = 0$
2. it is monotonous, i.e. $a \leq b \rightarrow T(a,c) \leq T(b,c)$
3. it is commutative, i.e. $T(a,b) = T(b,a)$
4. it is associative, i.e. $T[a, T(b,c)] = T[T(a,b), c]$

and S-norm in form of formula (8) satisfies the following properties $\forall a, b, c \in [0,1]$:

1. $S(a,0) = a, S(a,1) = 1$
2. it is monotonous, i.e. $a \leq b \rightarrow S(a,c) \leq S(b,c)$
3. it is commutative, i.e. $S(a,b) = S(b,a)$
4. it is associative, i.e. $S[a, S(b,c)] = S[S(a,b), c]$.

There are many examples of T-norms and S-norms. This section lists the most popular of them. The first and most common example is the T-norm *minimum* in the form:

$$T(a,b) = a \wedge b = min(a,b) \qquad (9)$$

The second T-norm is a algebraic product in the form:

$$T(a,b) = a \cdot b \qquad (10)$$

The third form i.e. the Łukasiewicz T-norm is defined in the following way:

$$T(a,b) = max(0, a+b-1) \qquad (11)$$

As for the S-norms, the most common form is the S-norm *maximum* in the form:

$$S(a,b) = a \vee b = max(a,b) \qquad (12)$$

Another known form is *the Probabilistic product*, denoted as:

$$S(a,b) = a + b - ab \qquad (13)$$

The third widely-know form of the S-norm in the literature is *Łukasiewicz S-norm* denoted in the form:

$$S(a,b) = min(a+b, 1) \qquad (14)$$

In order to fully perform operations on fuzzy sets, a negation operation is still required, which is represented by the negation operator. There are three types of negation: negation, *strict* negation and *strong* negation. Non-increasing function

$$N : [0,1] \to [0,1], \qquad (15)$$

is called negation, if it fulfils $N(0) = 1$ and $N(1) = 0$. Negation (15) is a *strict* negation if it is continuous and decreasing. On the other hand it is a *strong* negation if it is *strict* negation and in addition it is an involution (i.e. $N(N(a)) = a$) [32].

The simplest and most well known negation is the negation of *Zadeh*, which is a *strong* negation. This function is denoted in the form:

$$N(a) = 1 - a \qquad (16)$$

An example of *strict* negation is the *Yager's* negation, described by the dependence:

$$N(a) = (1 - a^p)^{\frac{1}{p}}, p > 0 \qquad (17)$$

After defining triangular norms and negations, basic operations on fuzzy sets can be performed.

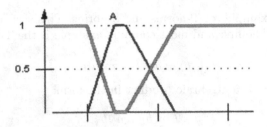

Fig. 1. Operation of the fuzzy set complement operator

Selected Operations on Fuzzy Sets. The *complement* of the fuzzy set $A \subseteq U$ is a fuzzy set $\neg A$ defined with a membership function in the form:

$$\mu_{\neg A}(x) = 1 - \mu_A(x), \forall x \in U \tag{18}$$

The geometrical interpretation of the complementary operator is shown in Fig. 1.

The intersection of two fuzzy sets $A, B \subseteq U$ is a fuzzy set $A \cap B$ defined by the membership function in the form:

$$\mu_{A \cap B}(x) = T(\mu_A(x), \mu_B(x)), \forall x \in U \tag{19}$$

Taking into account the previously defined T-norms, the minimum T-norm was used (see Sect. 2.1.1). The geometrical interpretation for this T-norm is shown in Fig. 2. The definition of intersection can be generalized to any finite number of fuzzy sets. In this situation, the result set will take the form:

$$\mu_{A_1 \cap A_2 \cap \ldots \cap A_n}(x) = T(\mu_{A_1}(x), \mu_{A_2}(x), \ldots, \mu_{A_n}(x)), \forall x \in U \tag{20}$$

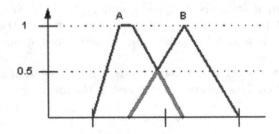

Fig. 2. Operators T-norm minimum for fuzzy sets A and B.

The *sum* of two fuzzy sets $A, B \subseteq U$ is a fuzzy set $A \cup B$ (or A + B), defined by the membership function in the form:

$$\mu_{A \cup B}(x) = S(\mu_A(x), \mu_B(x)), \forall x \in U \tag{21}$$

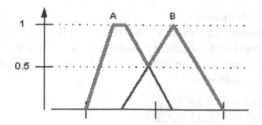

Fig. 3. S-norm maximum operator for fuzzy sets A and B.

The geometric interpretation for S-norm *maximum* is shown in Fig. 3. The sum definition can be generalized to any finite set of fuzzy sets. In this case, the result set will take the form:

$$\mu_{A_1 \cup A_2 \cup ... \cup A_n}(x) = S(\mu_{A_1}(x), \mu_{A_2}(x), \ldots, \mu_{A_n}(x)), \forall x \in U \qquad (22)$$

The Cartesian product of two fuzzy sets $A \subseteq U, B \subseteq V$ is denoted as $A \times B$ and defined for all $x \in U$ i $y \in V$ by the membership function in the form:

$$\mu_{A \times B}(x, y) = min(\mu_A(x), \mu_B(y)) \qquad (23)$$

or

$$\mu_{A \times B}(x, y) = \mu_A(x)\mu_B(y) \qquad (24)$$

Formula (23) can be generalized for any finite number of fuzzy sets $A_1 \subseteq X_1, A_2 \subseteq X_2, \ldots, A_n \subseteq X_n$, denoted as $A_1 \times A_2 \times \ldots \times A_n$ in the form:

$$\mu_{A_1 \times A_2 \times ... \times A_n}(x_1, x_2, \ldots, x_n) = min(\mu_{A_1}(x_1), \mu_{B_2}(x_2), \ldots, \mu_{A_n}(x_n)), \qquad (25)$$

and formula (24) in the form:

$$\mu_{A_1 \times A_2 \times ... \times A_n}(x_1, x_2, \ldots, x_n) = \mu_{A_1}(x_1)\mu_{B_2}(x_2) \ldots \mu_{A_n}(x_n) \qquad (26)$$

Both formulas (25) and (26) are fulfilled for all $x_1 \in U_1, x_2 \in U_2, \ldots, x_n \in U_n$.

Fuzzy Relations

Fuzzy relations are a key element in terms of application of fuzzy sets theory. They model relationships between two or more elements. One of the interpretations of these relationships is the concept of similarity. This relation determines the degree of similarity between the elements concerned. The fuzzy relation R between two nonempty non-fuzzy sets X and Y is a fuzzy set defined on the Cartesian product $X \times Y$ in the form:

$$R = \{((x, y), \mu_R(x, y))\}, \forall x \in X, \forall y \in Y, \qquad (27)$$

where $\mu_R : X \times Y \rightarrow [0,1]$ is a membership function to fuzzy relation R, $\mu_R(x, y) \in [0,1]$ is a degree in which element x is in relation with element y (e.g. similarity relation), and $X, Y \subset U$. Because the fuzzy relationship is a fuzzy set, the properties and operations are the same as discussed previously.

Fuzzy Similarity Relation

The dedicated extension of the fuzzy relation is called a *fuzzy similarity relation*. The relation represents similarity limited to a certain feature between two objects. This is the fuzzy relation defined with formula (27), which additionally fulfils the following conditions:

1. reflexivity: $\mu(x,x) = 1, \forall x \in U$
2. symmetry: $\mu(x,y) = \mu(y,x), \forall x,y \in U$
3. transitivity: $\mu(x,z) \geq max\{min\{\mu(x,y), \mu(y,z)\}\}, \forall x,y,z \in U$.

The definition of transitivity can be obviously reduced to the form: $\mu(x,z) \geq min\{\mu(x,y), \mu(y,z)\}$, $\forall x,y,z \in U$. This form can be further generalized to a T-norm, according to works [24, 29]. In practice, it is very difficult to ensure fulfilment of condition 3. In consequence, a bit simpler relation, called neighbourhood relation, is considered.

Neighbourhood Relation

Fuzzy relation known under a number of names, e.g. tolerance relation, proximity relation, relation of approximate equality, etc. It is a simplification of the relationship of the fuzzy similarity relation by excluding the transitivity condition. This is a fuzzy relation defined by formula (27) which additionally fulfils the following conditions:

1. reflexivity: $\mu(x,x) = 1, \forall x \in U$
2. symmetry: $\mu(x,y) = \mu(y,x), \forall x,y \in U$.

2.1.2   The Extension Principle

The principle developed by Zadeh, which permits the transfer of various functional dependencies of non-fuzzy sets into fuzzy sets. This is one of the strongest tools of fuzzy sets theory. Assuming that X is a Cartesian product of non-fuzzy sets $X_1 \times X_2 \times \ldots \times X_n$ and a non-fuzzy mapping is given in the form:

$$f : X_1 \times X_2 \times \ldots \times X_n \to Y \tag{28}$$

as well as certain fuzzy sets $A_1 \subseteq X_1, A_2 \subseteq X_2, \ldots, A_n \subseteq X_n$ then it follows from the extension principle that the resulting fuzzy set B, induced by the mapping f, takes the form:

$$B = f(A_1, A_2, \ldots, A_n) = \{(y, \mu_B(y)) : y = f(x_1, x_2, \ldots, x_n) \in Y\}, \tag{29}$$

where the membership function of the fuzzy set B takes the:

$$\mu_B(y) = \begin{cases} sup_{(x_1,\ldots,x_n \in f^{-1}(y))} T\{\mu_{A_1}(x_1), \ldots, \mu_{A_n}(x_n)\} & f^{-1}(y) \neq \emptyset \\ 0 & f^{-1}(y) = \emptyset \end{cases} \tag{30}$$

Example 21. *Let us assume that X is a Cartesian product of sets $X_1 = X_2 = \{3, 4, 5\}$ and A_1 is a fuzzy set of numbers close to 3 and A_2 is a fuzzy set of numbers close to 5 respectively in the form:*

$$A_1 = \frac{1}{3} + \frac{0.8}{4} \tag{31}$$

and

$$A_2 = \frac{0.7}{4} + \frac{1}{5} \tag{32}$$

There is also a mapping

$$y = f(x_1, x_2) = x_1 + x_2 \tag{33}$$

Fuzzy set B induced by the mapping (33) therefore represents numbers close to 8. In this case $B \subseteq Y = \{7, 8, 9\}$. By applying the extension principle, we get:

$$B = f(A_1, A_2) = \frac{min(1, 0.7)}{7} + \frac{max[min(1, 1), (0.8, 0.7)]}{8} + \frac{min(0.8, 1)}{9}$$
$$= \frac{0.7}{7} + \frac{1}{8} + \frac{0.8}{9} \tag{34}$$

2.1.3 Fuzzy Reasoning

Approximate reasoning is based on a generalized fuzzy reasoning rule *modus ponens* [53]. It is an extension of the classic rules of reasoning from bivalent logic. The analogous extension is used for the rule *modus tollens* [53].

The rule *modus ponens*, defined in the way presented in Table 1, is considered the basic scheme of reasoning.

Table 1. Extended (fuzzy) rule *modus ponens*, where $A, A' \subseteq X$ and $B, B' \subseteq Y$ are fuzzy sets, and x, y are linguistic variables.

Premise	x is A'
Implication	**IF** x is A **THEN** y is B
Conclusion	y is B'

The rule uses the concept of linguistic variable whose values are expressed by natural language [116]. Instead of numerical values, the following concepts are used: high, medium, low, etc. It should be noted that these are typical values used by people in communication. Each value of the linguistic variable is identified with a dedicated fuzzy set. Extended rule *modus tollens* is defined in Table 2. Accordingly, in Tables 3 and 4 an example of the application of approximate reasoning is given based on defined rules.

Table 2. Extended (fuzzy) rule *modus tollens*, where $A, A' \subseteq X$ and $B, B' \subseteq Y$ are fuzzy sets and x, y are linguistic variables.

Premise	y is B'
Implication	**IF** x is A **THEN** y is B
Conclusion	x is A'

Table 3. Example of action of the extended rule *modus ponens* for linguistic variable x -power consumption, y - speed of electric train for sets of linguistic variables accordingly $T_1 = \{\text{small, medium, medium-high, high}\}$ and $T_2 = \{\text{small, medium, high, very high}\}$.

Premise	Power consumption is medium
Implication	If power consumption is high then speed is very high
Conclusion	Speed is medium

Fuzzy Rules

These rules are analogous to the classic IF THEN rules. In this case however, both linguistic variables occur in both the predecessor and the consequent, whose values are represented by fuzzy sets. This type of rule takes the form:

IF x_1 is A_1 **AND** x_2 is A_2 **AND** ... **AND** x_n is A_n

THEN y_1 is B_1 **AND** y_2 is B_2 **AND** ... **AND** y_m is B_m

where A_i are fuzzy sets, such that $A_i \subseteq X_i \subset R$ for all $i = 1, \ldots, n$ and B_j are fuzzy sets such that $B_j \subseteq Y_j \subset R$ for all $j = 1, \ldots, m$, $x_1, x_2, \ldots, x_n$ are input linguistic variables, and $y_1, y_2, \ldots, y_m$ are output linguistic variables such that $x = [x_1, x_2, \ldots, x_n]^T \in X_1 \times X_2 \times \ldots \times X_n$ and $y = [y_1, y_2, \ldots, y_m]^T \in Y_1 \times \ldots \times Y_m$. Sets X_i and Y_j mean spaces of input and output variables respectively and R denotes the relation established between elements of these spaces.

These types of rules can be used for describing behaviour of a dynamic control system without the need of going into the analytical details of the problem in question. While implementing systems based on fuzzy rules, databases of rules are created. Individual rules are linked by means of the logical operator OR, which makes it possible to create complex structures.

2.2   Selected Models of Uncertain Information

2.2.1   Rough Sets

Rough sets are one of the approaches to approximation of concepts. They can form the basis of many decision systems in different fields of operation, e.g. data reduction, attribute selection, decision rules generation, knowledge discovery from data (KDD[1]) etc. [107]. Rough sets use the concept of *information system* [39] on the basis of which the context of processed objects, their features and values is presented.

Table 4. Example of action of the extended rule *modus tollens* for linguistic variable x - power consumption, y - speed of electric train for sets of linguistic variables accordingly $T_1 = \{\text{small, medium, medium-high, high}\}$ and $T_2 = \{\text{small, medium, high, very high}\}$.

Premise	Speed id high
Implication	If power consumption is high then speed is very high
Conclusion	Power consumption is medium-high

---

[1] Knowledge discovery in databases.

Definition 8. *Information system is an ordered quadruple* $SI = \langle U, Q, V, f \rangle$, *where U is a set of objects, Q is a set of attributes (features),* $V = \cup_{q \in Q} V_q$ *is a set of all possible values of features and* $f : U \times Q \to V$ *is an information function, such that* $\forall_{x \in U, q \in Q} f(x, q) \in V_q$.

Table 5 presents an example of an information system. This system is a simple and transparent notion, which makes it possible to carry out the necessary analysis of data. For this particular system the sets belongings to the quadruple from the definition are the following:

1. $U = \{1, 2, 3, 4, 5, 6\}$
2. $Q = \{$Salaries, Agreement type, Seniority, Credit decision$\}$
3. $V_{q_1} = \{A, B, C, D\}$, $V_{q_2} = \{UP, UZ, UD\}$, $V_{q_3} = \{2, 3, 5, 7, 10\}$, $V_{q_4} = \{$Yes, No$\}$.

Table 5. Example of an information system where values of the *salaries* attribute represent quantified quota ranges.

Customer	Salaries (q_1)	Agreement type (q_2)	Seniority (q_3)	Credit decision (q_4)
1	A	UP	5	No
2	C	UZ	2	Yes
3	D	UD	3	Yes
4	B	UP	10	Yes
5	A	UZ	3	No
6	A	UD	7	No

Decision tables are a special case of information systems.

Definition 9. *Decision table is an ordered quintuple* $DT = \langle U, C, D, V, f \rangle$, *where: U, V and f have the same meaning as in the Definition (8), C is a set of conditional attributes, D - is a set of decision attributes.*

Information tables can be easily transformed into decision tables by specifying conditional attributes and decision attributes. Table 6 presents the Table 5 transformed into a decision table.

Presentation in the form of decision tables can be interpreted as *IF THEN* rules described in the predecessor by conditional attributes and in the successor by decision attributes and their values. The object representing customer 3 for the example presented in Table 6 may take the form of the following rule:

$$R_3 : \textbf{IF } c_1 = D \textbf{ AND } c_2 = UD \textbf{ AND } \quad c_3 = 3 \textbf{ THEN } d_1 = \text{Yes}$$

An *indiscernibility relationship* is used to analyse decision tables [88]. This relationship is the relation of equivalence due to the fact that it is reflexive, symmetric and transitive. This relation gives the possibility of reducing the set of data recorded as a decision table and as a consequence, it translates into the amount of data required for processing by the entire decision system.

Table 6. Decision table transformed from the information system presented in Table 5

Rule	Salaries (c_1)	Agreement type (c_2)	Seniority (c_3)	Credit decision (d_1)
1	A	UP	5	No
2	C	UZ	2	Yes
3	D	UD	3	Yes
4	B	UP	10	Yes
5	A	UZ	3	No
6	A	UD	7	No

Definition 10. *Objects $x, y \in U$ are **discernible** by $A \subset C$ if and only if:*

$$\exists a \in A : a(x) \neq a(y) \tag{35}$$

for any objects x and y from space U.

Definition 11. *Objects $x, y \in U$ are **indiscernible** by $A \subset C$ f and only if when:*

$$\forall a \in A : a(x) = a(y) \tag{36}$$

Thus, the relation of indiscernibility is defined in the following way:

Definition 12. ***Indiscernibility relation** IND defined in space $U \times U$ is the relation fulfilling the following condition:*

$$xINDy \Leftrightarrow \forall q \in P : f_x(q) = f_y(q), \tag{37}$$

where $x, y \in U$ and $P \subseteq Q$.

This relation divides the set on which it is defined into the family of separable subsets called abstraction classes [25]. These classes consist of objects that are in relation to a given object. The formal definition is as follows:

Definition 13. ***Abstraction class** of relation IND in space U is a set of all objects $x \in U$ which are in relation IND. For all $x_a \in U$ there is exactly one set marked with the symbol $[x_a]_{IND}$:*

$$[x_a]_{IND} = \{x \in U : x_a INDx\} \tag{38}$$

Presented definitions are tools for working with rough sets. They do not define the rough sets themselves. Rough sets approximate classic sets. There are the lower and upper approximations which are defined as follows:

Definition 14. ***Lower approximation** of the set $X \subseteq U$ is the set $\underline{A}X$ described as follows:*

$$\underline{A}X = \{x \in U : [x]_{IND} \subseteq X\} \tag{39}$$

This set consists of objects $x \in U$, which certainly are a part of the set X based on attributes P.

Definition 15. *Upper approximation of the set* $X \subseteq U$ *is the set* $\bar{A}X$ *described as follows:*

$$\bar{A}X = \{x \in U : [x]_{IND} \cap X \neq \emptyset\} \tag{40}$$

This set consists of objects $x \in U$. It cannot be explicitly stated, basing on attributes P, whether they are elements of the set X.

Definition 16. *Boundary region of the set* X *is defined as:*

$$B_{IND}(X) = \bar{A}X \setminus \underline{A}X \tag{41}$$

Taking into account the above mentioned definitions, the rough set is:

Definition 17. *The set* X *is called a* **rough set** *if its boundary region is non-empty.*

Figure 4 presents a schematic representation of individual regions constituting a rough set. There are many applications of this theory. One is to reduce the number of attributes processed by the system to make a decision. The other is to minimize the number of instances needed to accomplish the goal. The first problem is known in literature under the concept of features selection and the second one as instances selection. Both issues form a broad topic and are not the direct subject of this dissertation. Publications [18, 48, 73] contain detailed information on this topic.

Decision systems do not need all attributes to make a decision. The process of selecting the subsets of attributes needed in the decision-making process leads to the designation of reducts. Reducts are a minimal set of attributes which retain the characteristics of the whole set of attributes. Rough sets apply two concepts of reducts: information and decision reducts [85].

Definition 18. *A set of attributes* $B \subset Q$ *is called an* **information reduct** *of table* Q *if and only if:*

– *B retains the indiscernibility of set* Q
– *B is irreducible.*

In practice, this means that if x and y are indiscernible by Q and B is an information reduct, then x and y are also indiscernible by B. In addition, no proper subset of B meets this property, i.e. does not retain the indiscernibility of Q.

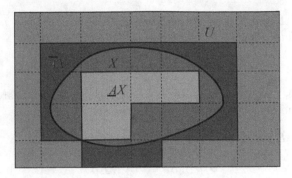

Fig. 4. Graphical representation of the rough set. X - classical set included in U, $\bar{A}X$ - upper approximation of the set X, $\underline{A}X$ - lower approximation of the set X

Definition 19. *A set of attributes $B \subset C$ is called a **decision reduct** of table C if and only if:*

- *B retains the indiscernibility of set C in relation to decision $d \in D$*
- *B is irreducible*

These conditions mean that for all $x, y \in U$, if $d(x) \neq d(y)$ for discernible objects x and y by C and B is a decision reduct, then x and y are discernible by B. In addition, no proper subset of B exists which would retain the condition of discernibility.

Reducts are a very important tool in decision systems. They specify the minimum number of attributes which must be processed in order to achieve the correct results. In decision tables DT there is a set denoted as $RED(DT)$, which means the set of all reducts of the decision table DT. If a set of attributes C of the decision table DT has a cardinality $card(C) = n$, then the number of all reducts is limited by the inequality:

$$card(RED(DT)) \leq \binom{n}{\frac{n}{2}} \tag{42}$$

2.2.2   Extension of Classical Fuzzy Sets

While the popularity of fuzzy sets and their practical applications increased, certain imperfections were noticed. There are different ways of modifying the existing theory or creating new models. All of them are more or less related to fuzzy sets, but all approaches definitely took into account the modelling of uncertainty or incompleteness of information.

Interval valued fuzzy sets - interval fuzzy sets (IV) were proposed by Zadeh in the 1970's as natural extending classic sets [111]. This theory was developed by many researchers, who independently developed many mathematical descriptions.

Interval valued fuzzy set A in a certain space U is denoted as a set in the following form:

$$A = \{(F^A(x), x) : x \in U, F^A(x) : X \to Int([0,1])\} \tag{43}$$

where $F^A(x) = [F_-(x), F^-(x)]$ is an interval valued membership function, and $Int([0,1])$ is a set off all subintervals $Int([0,1]) = \{[a,b] : a, b \in [0,1]\}$. $F_-(x)$ is called a lower membership function and $F^-(x)$ is called an upper membership function. If we take $F^A(x)$ and

$$G^B(x) = [G_-(x), G^-(x)] : x \in U \tag{44}$$

then all basic operations on sets are directly analogous to classic fuzzy sets. They are presented the following way:

– sum of sets $A \cup B(x)$

$$A \cup B = [S(F_-(x), G_-(x)), S(F^-(x), G^-(x))] \tag{45}$$

– product of sets $A \cap B(x)$

$$A \cap B = [T(F_-(x), G_-(x)), T(F^-(x), G^-(x))] \tag{46}$$

– negations (complement) of set $\neg A$

$$\neg A = [1 - F^-(x), 1 - F_-(x)], \tag{47}$$

where T and S mean T-norm and S-norm.

The main idea behind these kinds of sets is to gain a better modelling uncertainty. If the exact membership degree is hard to designate and it is possible to designate an interval, these kinds of sets should be applied, maintaining the standard fuzzy sets methods.

Intuitionistic Fuzzy Sets

The concept of two membership functions was also applied to intuitionistic fuzzy sets (IFS). However, this case involves the membership function of an element (designated as $F_*^+(x)$) and the function of non-membership (designated as $F_*^-(x)$), which additionally are subject to the following condition:

$$\forall x \in U : (F_*^+(x) + F_*^-(x) \leq 1) \tag{48}$$

Basic operations on IFS are defined as follows for two sets described with membership functions (F_*^+, F_*^-), (G_*^+, G_*^-):

– sum $A \cup B$

$$(S(F_*^+(x), G_*^+(x)), S(F_*^-(x), G_*^-(x))) \tag{49}$$

– product $A \cap B$

$$(T(F_*^+(x), G_*^+(x)), T(F_*^-(x), G_*^-(x))) \tag{50}$$

– negation (complement) $\neg A$

$$(F_*^-(x), F_*^+(x)) \tag{51}$$

The idea of considering relationships between positive and negative examples describing the state of objects is present in numerous domains. One can find it in psychology in the form of a mechanism to describe similarities and differences as well as in the processing of compound objects by means of similarity. There are scientific papers describing the approach, according to which the description of similarity has two components [112].

There are clear relationships between IFS and interval valued fuzzy sets described earlier, noticeable especially in the form of negation. Based on the limitation (48) it is easy to transform IFS into IV. It can be done the following way:

$$F^A(x) = [F_-(x), F^-(x)] = [F_*^+(x), 1 - F_*^-(x)] \tag{52}$$

Shadowed Sets

Shadowed sets (SS) are not as strongly connected with fuzzy sets as sets described above. However, it is possible to construct the SS based on fuzzy sets. From the point of view of a definition a shadowed set is a mapping in the form:

$$f_{Ci} : U \to \{0, 1, (0, 1)\}, \tag{53}$$

where U is the universe of objects. Simultaneously, a set of all shadowed sets defined on the space U takes the following form:

$$\{0, 1, [0, 1]\}^U \tag{54}$$

SS can be constructed based on a given fuzzy set A with a known membership function f_A. For discrete values of memberships $\mu_1, \mu_2, \ldots \mu_n$ for set A, $SSA^\frown$ induced by set A is constructed by the selection of a certain value $\alpha \in [\mu_{min}, \frac{\mu_{min} + \mu_{max}}{2})$ by means of which the fuzzy set A is divided into three areas. The areas are defined by means of the function of α-approximation of the fuzzy set A in the following form:

$$f_{Ci}(\mu) = \begin{cases} 0 : f_A(\mu) \leq \alpha, \\ 1 : f_A(\mu) \geq (1 - \alpha) \\ [0, 1] : w.p.p. \end{cases} \tag{55}$$

The optimum α parameter should be selected in a way for the function value that follows to be minimum:

$$V(\alpha) = |\psi_1 + \psi_2 - \psi_3|, \tag{56}$$

where $\psi_1 = \sum_{\mu_i < \alpha}$, $\psi_2 = \sum_{\mu_i \geq (\mu_{max} - \alpha)}$, $\psi_3 = card(I) : I = \{i : \alpha < \mu_i < \mu_{max} - \alpha\}$ The optimal α parameter is designated in the following way [122]:

$$\alpha_{opt} = \arg \min_\alpha V(\alpha) \tag{57}$$

Table 7. Table presenting the sum operation of shadowed sets. S means *shadow* defined as an interval $[0, 1]$

$A^{\frown}(\mu) \cup B^{\frown}(\mu)$	0	S	1
0	0	S	1
S	S	S	1
1	1	1	1

Table 8. Table presenting the product of shadowed sets operation. S means *shadow* defined as an interval $[0, 1]$

$A^{\frown}(\mu) \cap B^{\frown}(\mu)$	0	S	1
0	0	0	0
S	0	S	C
1	0	S	1

It can be noted that the areas designated are analogous to rough sets as well as positive, negative and boundary regions designated accordingly. From this point of view, SS is the conventional bridge between fuzzy sets and rough sets.

Shadowed sets are isomorphic with three-valued logic [9]. Operations executed on these sets correspond to the respective operations within this logic. The basic operations are defined as follows:

- sum of $A^{\frown} \cup B^{\frown}$ is presented in Table 7.
- product of $A^{\frown} \cap B^{\frown}$ is presented in Table 8.
- negation (complement) $\neg A^{\frown}$ is presented in Table 9.

Table 9. Table presenting the negation (complement) of shadowed sets. S means *shadow* defined as an interval $[0, 1]$.

$A^{\frown}(\mu)$	$\neg A^{\frown}(\mu)$
0	1
S	S
1	0

Soft Sets

This theory was initiated by a Russian scientist Molodtsov in [57]. It is an alternative approach to expressing uncertainty, fuzziness or incompleteness of information about objects. It has its own mathematical apparatus, which distinguishes soft sets (SF) from different set theories, e.g. fuzzy sets, rough sets, etc. [115].

A pair $<F, E>$ is called soft set if and only if F is a mapping of E in the form of sets, which are all subsets of the universe U. If P(U) is denoted as a set of all subsets of U, mapping is obtained in the following form:

$$F : E \to P(U) \tag{58}$$

In other words, SF is a parameterized family of subsets of set U. Any set $F(\alpha), \alpha \in E$ belonging to the family of subsets, may be understood as a set of α-approximations of elements of SF.

Example 22. *An example of SF may be a set (F, E) determining the attractiveness of a given thing, which a given person is going to buy, e.g. attractiveness of a car, which is to be bought by a certain John Smith. In this case U is a set of cars taken into account when buying, E is a set of parameters, which are single definitions or whole sentences, defining e.g. $E = (e1 = sports, e2 = fast, e3 = expensive, e4 = red)$. In such a case, in order to define SF one has to enumerate pairs in the following form: sports car, expensive car, fast car, etc.*

Different operations can be defined for soft sets. They can take the following form, e.g.:

– *sum*
  If (F, A) and (G, B) are the SF then the sum of these sets is denoted as $(F, A) \cup (G, B)$ and defined as follows:

$$(F, A) \cup (G, B) = (O, A \times B), \tag{59}$$

  where $O(\alpha, \beta) = F(\alpha) \cup G(\beta)$ for all $(\alpha, \beta) \in A \times B$
– *product*
  If (F, A) and (G, B) are the SF then the product of these sets is denoted as $(F, A) \cap (G, B)$ and defined as follows:

$$(F, A) \cap (G, B) = (H, A \times B), \tag{60}$$

  where $H(\alpha, \beta) = F(\alpha) \cap G(\beta)$ for all $(\alpha, \beta) \in A \times B$
– *complement*
  Complement of SF (F, A) is denoted as $(F, A)^c$ and defined in the following way:

$$(F, A)^c = (F^c, \neg A), \tag{61}$$

where $F^c : \neg A \to P(U)$ is a mapping defined as:

$$F^c(\alpha) = U - F(\neg \alpha), \tag{62}$$

while the $\neg A = \{\neg a_i\}, \forall_{a_i} \in A\}$.

Soft sets are used in various fields, i.e. in game theory. They can be used to define possible states of player behaviour by defining a set of strategies. The case is different for the classic approach, which is based on modelling the cost function [71].

2.3 Similarity Issues

2.3.1 Introductory Information

Ontology

The name comes from philosophy, but now it is also frequently found in the field of artificial intelligence (AI). The formal definition (one of many) was introduced in 2001, described in [103]. Its meaning is as follows: ontology is a system marked as $O = \{C, R, H_c, rel, A, L\}$, which specifies the structure of concepts, relationships between them as well as theory defined on a model, where: C is the set of all concepts of the model and the concept is called the idea of representing a group of objects with common characteristics. R is a set of non-taxonomic relations defined as named connections between concepts [4], H_c - a collection of taxonomic relationships between concepts, rel - defined non-taxonomic relationships between the concepts, A - a set of axioms, L - exicon defining the meaning of concepts (including relations). L is a set of form $\{L_c, L_r, F, G\}$, where L_c - lexicon definitions for concepts, L_r - lexicon of definitions of a set of relationships, F - references to concepts, G - references to relationship.

Literature has described many interesting applications of ontology in pattern recognition, image analysis or modelling situational awareness by AI systems [44]. In all cases ontology is a tool for modelling the structure of concepts and relationships describing a selected part of the local context in which the system is described.

In the context of this work, ontology is used as a set of concepts describing objects with its structure and relations. It is used for designating reducts of features as well as describing features to which they are compared. It is a necessary tool for the recognition and identification process [4].

In the simplest sense, ontology is a set of concepts connected with one another through named relationships. Ontological concepts can create hierarchies by grouping more specific concepts into more general entities. This form is used, for example, to model mereologic relations, which describe dependencies between parts of objects [69].

Compound Objects

Objects, in general, can be divided into two groups: compound objects (X_c) and simple objects (X_s). A simple object is any element of the real world that has its representation capable of being expressed by the adopted ontology (O). In addition, the following properties arise from their ontological representation:

1. Objects always belong to a certain class or a fixed number of classes in ontology. A single object may belong to several classes.
2. An object has a property within a class. Features may vary by class.
3. An object may be in relation to other objects in the same ontology.

A compound object is composed of other objects defined by means of ontology (connects them) and creates a new entity. A compound object has its specification, which describes the structure, relations and connections between sub-objects. Compound objects satisfy the following additional properties:

1. We can extract from them a minimum of two objects that can be independent entities.
2. Component objects are interrelated by relationship with ontology.

2.3.2 Similarity of Compound Objects

Similarity can be seen as a relationship that results in some sense from identity. Identity is an intuitive equality of objects. In other words, it is the equality of attributes of entities compared. Identity is the supreme form of similarity. Rules for determining the identity of entities were created and named identity of indiscernible. Its origin is attributed to Gottfried Wilhelm Leibniz and its meaning is as follows:

$$\forall x \forall y [\forall P(Px \leftrightarrow Py) \rightarrow x = y] : x, y \in U \tag{63}$$

and

$$\forall x \forall y [x \neq y \rightarrow \neg \forall P(Px \leftrightarrow Py)] : x, y \in U, \tag{64}$$

where x and y are objects and P is a property. In other words, formula (63) means that for any objects x and y from the universe U, if they have exactly the same values of all properties, these objects are identical in the space in question. Similarly, formula (64) means that for any object x and y, if x is not identical to y, then there must be at least one property from the space U which differs.

Intuitively, similarity is a certain kind of incomplete identity. Two similar objects are those that are primarily comparable and for which a degree of similarity can be obtained. Objects can be compared, if they have common or distinguishing features. The feature, however, is the attribute describing the object. Attributes have values. Similar objects to some extent have common attributes, and comparison of these attributes gives the possibility of determining the degree of similarity. It is commonly understood that the statement *object is similar* means that one object resembles the other or is *almost* the same. These statements are, of course, very imprecise, but certainly possible to be mapped using the appropriate modelling techniques (e.g. fuzzy sets described in Sect. 2.1). By following this intuition, one can determine when two objects fail to fulfil the definition of identity, but there is very little left to be fulfilled. The first case is a quantitative approach. We are dealing with a set of attributes describing both objects, where the most attributes of these objects are equal, although there is at least one attribute for which equality is not fulfilled. These objects are almost certainly identical (in colloquial speech), but from the strict point of view these objects are similar to a certain degree. The second approach is not limited to examining attributes that characterize identities. Its attention is focused on the remaining attributes. These attributes do not meet the condition of identity, but one can try to determine the degree of similarity for them. This is called a qualitative approach. It may involve a situation in which no identities are found on any attribute, and yet these objects will be similar to a certain degree.

The scale of similarity is most often interval $[0, 1]$, where 0 means a total lack of similarity and 1 is interpreted as discernibility to given attributes, and thus, according to the principle of Leibniz, as an *absolute identity*. Similarity and the very comparison operation are indispensable elements of the world around us. In many cases, these elements are necessary to determine the state of the object. In practice, weight, size, capacity, duration or other characteristics of objects are determined. Each of these elements requires knowledge of a certain reference concept, by which one can specify a given object parameter, e.g. kilogram, litre, second, etc. In spite of the introduction of reference values, the feature of the object can be expressed in a countable way. At the same time, objects have common reference points for all.

One can distinguish several types of approaches to defining similarities. Selected approaches will be discussed in the further part of this paper.

2.3.3 Selected Methods of Expressing Similarities

In literature, the problem of similarity is quite widespread, but it is usually not the main research point, but merely a means to achieve other goals. In most cases similarity is equated with distance in a certain space of features [17]. In this case, the metric is considered in the form:

$$d : X \times X \rightarrow [0, +\infty], \tag{65}$$

which satisfies the following properties $\forall x, y, z \in X$:

1. $d(x, y) = 0 \Leftrightarrow x = y$
2. $d(x, y) = d(y, x)$
3. $d(x, y) \leq d(x, z) + d(z, y)$.

There are various metrics that suit the type of space and the problem that is to be solved. This solution allows one to convert the problem of determining similarity between objects to the problem of distance measurement in a coordinate system determined by features. This is a relatively common approach, but not always sufficient to solve complex problems. It should be noted that there are very strong constraints associated with the metric. In the case of a generally understood similarity, the condition of symmetry is often not possible to be met, not to mention the condition of transitivity. That is why there is a need for other approaches as well. Undoubtedly, the common element of many solutions is the feature vector. However, the essence of the problem lies in how the vector is constructed and how it can adapt to new situations.

The next step in evolution related to methods of implementing similarity involves approaches based on ontological relationships between objects and concepts [19,50,100,120]. In this context, individual ontological concepts are treated as features that contribute to comparing objects. The set of these features constitutes entry into the process of determining the minimum set of essential features. This process comes down to the designation of a kind of reduction of features similar to information reductions encountered in data mining [121], i.e. a minimum set of attributes that uniquely identify or classify a given object. There are many reducts that consist of different features. The process of selecting the best reduct is based on domain knowledge about the problem, information about implementation and many other factors [99]. Ontology and reduct ensure the proper design of a feature vector. However, they do not directly support the method of calculating similarity. Therefore, these methods, after the selection of features, use other methods described earlier, or dedicated methods based on comparison of ontology are constructed [49]. These methods are very complicated and depend on the construction of a particular ontology.

Another approach that replaced distance thinking was the *contrast model* created by Amos Tversky. This model was created on the basis of the study of human perception of similarity [112]. In this model, not only the common features, but also distinguishing features of objects play an important role. Consequently, the model also examines aspects reducing similarity between objects and determines their impact on the value of the degree of similarity [35]. The common formula of the similarity function in the proposed contrast model is as follows:

$$sim(x, y) = \theta f(X \cap Y) - \alpha f(X - Y) - \beta f(Y - X) : \ \theta, \alpha, \beta \geq 0, \qquad (66)$$

where X and Y are sets of features describing object x and y respectively, $X \cap Y$ determines common features for x and y, $X - Y$ determines feature existing in x and not existing in y, $Y - X$ determines features not existing in x, and existing in y. Function f is a scale factor, while θ, α and β are parameters of the model. It is easy to see that for $\alpha = 0$ and $\beta = 0$ the model is limited to common features of objects. On the other hand for parameters $\theta = 0$ and $\alpha = 1, \beta = 1$ we get:

$$- sim(x, y) = f(X - Y) + f(Y - X), \qquad (67)$$

which is a dissimilarity [66].

From the point of view of modelling similarity, it is important to be able to deal with imprecision of the description and its effect on the result. Another method of representing object similarities involves fuzzy sets described in Sect. 2.1 [52]. In particular, the fuzzy relation is an ideal tool for such purposes. According to formula (27) it is defined on the Cartesian product of two not fuzzy sets. In this case the sets include elements for which similarity is determined. There are many similarity measures based on fuzzy sets in literature. However, the common approach is based on the analysis of common features of objects, i.e. those at the intersection of sets $A \cap B$ or complement, in the form:

$$sim(x, y) = 1 - \mu(x, y), \qquad (68)$$

where $\mu(x, y)$ is a membership function of a given relation which designates the degree of difference between objects. The same approach can be used in building similarity functions, which will be used for the purposes of calculation of degrees for individual features or the full feature vector. An important element of the method is the ability to make fuzzy sets of the results obtained [114].

Slightly different methods can be used when comparing object structures or their topological relationships. In cases like these, apart from attributes and their values, constraints related to the location of the object in space or the internal structure of the object are imposed. This kind of similarity can also be expressed by means of methods described above, but only on a case-by-case basis. This is why certain standardized methods to deal with such problems have been sought. The fields dealing with them include e.g. rough mereology and near sets [67,68].

The main idea behind rough mereology is the examination of the extent to which the object is a part of the other object using a properly selected function of rough inclusion [27]. This function can be treated as an example of an asymmetrical measure of similarity [69]. A typical example of the inclusion function, and at the same time a similarity based on the multiplicity of common components, is the following formula [83]:

$$sim(X,Y) = \mu(X,Y) = \frac{card(X \cap Y)}{card(X)}, card(X) \neq 0, \qquad (69)$$

where X means a set of sub-objects included in the object x, and Y is a set of ingredients of object y.

The rough inclusion function is a method comparing parts of objects, their quantities, types or other relationships in the ontological hierarchy. Therefore, it can be interpreted as a measure of similarity that takes into account structural dependencies of objects.

In this paper, structural similarity will be calculated on the basis of the sum of similarities between the sub-components with a fixed structure of object. Sub-objects are extracted by means of decomposition. Similarity values of sub-objects are treated as additive, multiplied by the respective weighting factors. Consequently, the similarity function is based on knowledge of composition of a given object and its significance for the whole object. Certain functions are used to define the relationship between an object and its parts (defining how to make an object from an underlying object). These functions and modelled dependencies are applied to similarities, so the result is also the value of similarity (referring to the main object). An example of similarity function of this kind can be as follows:

$$sim(x,y) = \frac{w_1 sim(x_1,y_1) + w_2 sim(x_2,y_2) + \ldots + w_n sim(x_n,y_n)}{(w_1 + w_2 + \ldots + w_n)} \qquad (70)$$

where x_i are sub-objects of x, and y_i are sub-objects of y for $i = 1, \ldots, n$.

To summarize, there are many different methods of processing and defining similarities. Many of them are related to specific cases of use, where use is subject to special considerations. It is worth pointing out that methods listed here were chosen from among many other equally useful methods, such as the use of rough sets [62]. At the same time, a universal approach will be proposed in this paper to combine the majority of methods described in this section. Consequently, it will be easier to compare similarity results. In addition, methodologies for different cases and different types of facilities will be established.

2.4 Object Identification Issues

2.4.1 The Foundations of Decision Systems

Decision System Support (DSS) deals specifically with decision-making in systems that can be assigned to other categories of systems, i.e. data processing systems, risk management systems, etc. Their construction is based on three main elements: knowledge base, reasoning model and user communication interface. The knowledge base gathers facts on the basis of which the system can learn some characteristics. These data are the source of training sets and test sets. The reasoning model represents a certain mathematical description of reasoning, but can be implement by means of a variety of knowledge representation methods, such as fuzzy sets, rough sets, network models, etc. The communication interface fulfils the role of an element informing about the results of the operation of the model as well as a certain parameterization of input data. It should be borne in mind that DSS systems are not autonomous decision-making systems. They only support the decision-making process. This means that decision is made by an intelligent individual, i.e. a human being. The system provides certain hypotheses or suggests a decision based on accumulated, yet limited knowledge. However, the final responsibility for the decision lies on the person. The suggested decision of the system can be ignored or modified in specific cases.

The operation of the DSS reasoning model largely comes down to performing the classification process. The model is equipped with learning mechanisms, so that it can adapt to a given problem. Based on the training data derived from the knowledge base, it is possible to learn the rules of decision-making in a given context. Such a model is called a *classifier*.

For the purposes of formal description, training data are saved as a set $T = \{x_1, \ldots, x_n\}$. The respective elements x_i of set T include pairs consisting of the information vector and the decision label in the following form:

$$x_i \rightarrow (\langle x_{i1}, \ldots, x_{in} \rangle, d) \tag{71}$$

The information vector consists of conditional attributes of the space of features describing the object, while d represents values of the decision attribute. A certain mapping of the input data to output values is saved in pairs of the training set T. The analytic form of mapping is usually unknown. By denoting this mapping as:

$$f : T \rightarrow D, \tag{72}$$

where D means a set of labels of the decision space, a function of *supervised classification* [81] is created. This function is an approximation of the analytical function representing a given phenomenon through available samples of the training set.

From among objects in the training set one can specify decision classes in the form of sets of objects with the same value of the decision attribute. Classifiers are used for the purposes of the final classification. Their task is to predict the value of the decision attribute based on the set of training data. Classifiers are

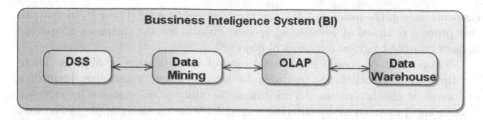

Fig. 5. Visual block diagram of the structure of a business intelligence system.

usually products of pre-classification because they are not generally known at the beginning. The classifier can be expressed in the form of the following function:

$$f(x_{i1}, \ldots, x_{in}) = d \tag{73}$$

or, in the case of multi-decisions, in the following form:

$$f(x_{i1}, \ldots, x_{in}) = \langle d_1, \ldots, d_m \rangle \tag{74}$$

Classification is a two-phase process. The first step is to create an approximate model (classifier) based on input data. This stage is divided into two sub-stages. The first sub-stage consists in learning based on the T training set (see Sect. 2.5 for details). The second is testing and verifying classifier qualities using test data (details in the Sect. 2.4.3).

As a result of evolution, business intelligence systems (IB) are now considered. These are AI systems divided into further subcategories of the BI system. The structure of such systems considers DSS as one of the subsystems existing concurrently with data mining [10], and data analysis (OLAP) subsystems as well as the data warehouse [11]. A flowchart of the BI system is presented in Fig. 5.

Selected Methods Applied in Decision Systems
Decision-making systems process large amounts of information at different levels of generality. In order to determine the state of an object or phenomena, they base on both low-level sensory data and more aggregated concepts. In many cases, determining the exact value is very expensive, so the approximation of terms is used to allow reducing the computational cost while maintaining good quality of results. There are several techniques based on approximation that are used by decision systems. Artificial neural networks [12], which are an approximation of a function solving a given problem, is one example. For more information, see Sect. 2.5.3 of this chapter.

2.4.2 Identification Models as Decision Systems
Identification is a process of recognizing the instance of object. It occurs in many aspects of everyday life, from computer systems, as one of the authentication

elements, to e.g. the process of opening the doors using a RFID card[2]. In general, this process is aimed at identifying specific objects for the purposes of specific actions relating to given instances of objects.

Identification of people is a special case of identification of objects. There are three main identification methods: by means of user knowledge, identifiers or biometric characteristics. Identification in information systems by entering a login and password or by entering a card PIN forms an example of the first group of methods. These methods use knowledge possessed or entrusted theoretically only to a given user. Of course, in practice, these methods may turn out to be ineffective due to faulty security support. Various keys, cards or access codes form examples of applying identification in the second group of methods. They represent additional equipment transferred only to a specific person, to identify that person. The last group of methods is the biometrics that have been intensively developed in recent years. It concerns individual characteristics of a human being, which are constant, unique and permanent. These features include e.g. fingerprints, hand geometry, face image, iris or retina of the eye. All these methods use unambiguous features of a person. By recognizing a unique feature, one can determine with high probability the identity of the person.

The identification model can thus be treated as a special case of the classification model. The difference is that in this case decision classes are one-element classes. This is due to uniqueness of a given feature or uniqueness of the set of features. Identification methods are analogous to classification methods, the difference being that they process features directly connected with a given instance, unique to it, or as generic features used for classification, but with a specific combination. This process is therefore a kind of a more detailed (less generalized) classification.

Therefore, identification can be seen as a function, whose arguments are, similarly to classification, features of an object, and the instance of the object is the output. This function is analogous to a classifier and can be called an *identifier* in the sense of a method or tool for carrying out identification. An identifier

$$f_{id} : U \to U, \tag{75}$$

may be expressed in the following form:

$$f_{id}(x_{i1}, \ldots, x_{in}) = inst, \tag{76}$$

where *inst* is an instance of an object from space U.

Typically, the number of features on the input is small, because they are unique features of the instance. However, identification by a unique combination of more general features is possible. In this case, the feature's input vector will be multi-attribute. A learning process analogous to classification can be considered in respect of the problem of identification. However, this process is reduced to assigning a given combination of values to the instance of the object. In the case of identification, the model (identifier) is usually given.

---

[2] pl.wikipedia.org/wiki/RFID.

It is worth noting that objects which have been identified can automatically be classified. This is due to the fact that by identifying specific instances, knowledge of the membership of a given instance to a given class is also discovered, e.g. if one identifies John Smith, they can automatically classify it into classes: humans, mammals, etc.

2.4.3   Testing the Quality of Classification and Identification

One can use a variety of available methods to perform a quality test of a classifier or identifier. However, each method is closely related to a phase of learning. The first one is re-sampling. The method consists in random division of the dataset into two subsets: the learning and the test set. Usually, it is divided in the following way: $\frac{1}{3}$ for the size of the learning set and $\frac{2}{3}$ for the testing set. The first set is used for learning the model (described in the next section). The second one is used for testing the classifier or identifier obtained. The procedure is repeated k times, where usually k is 30. Each iteration of k attempts gives results, which are saved. At the end, they are averaged in order to calculate the final result of the model. The advantage of this method is the randomness of learning and test sets. The disadvantage, however, is the strong interdependence of both sets due to repetitive elements.

Another method is the k-*fold* cross validation. There are several variants. The basic variant assumes random division of the set into k equal parts. It usually uses $k = 10$. Similarly to the previous method, this one covers both: the learning part and the testing part. The procedure consists in k times learning the model on $k-1$ sets and testing the model on the set k, not used in the learning process. For each iteration, results are stored in the contingency table. They are taken into account in the final result. Variety at each step of the set that is being tested is the essence of this testing methodology. Consequently, each element of the data set is tested exactly once in the entire process. Figure 6 presents the general diagram of workflow for $k = 4$.

Another variation of cross validation is k-times 2-fold cross-validation. The difference lies in the number of tests for each element. In this case, each item is tested twice. After the first run, the split is re-drawn. As in the case of the first variant, results are saved and aggregated into the final form.

The *leave one out* method is specific variation of cross-validation. The learning phase consists of $k-1$ elements, while testing is carried out on a single item remaining. The procedure is performed k-times, so that each item is included in the test set.

According to general characteristics of cross-validation methods, this is a method that takes into account the randomness of training and test sets, and (very importantly) the independence of test sets. However, it should be noted that there is a strong correlation between learning sets due to repetitive elements, which may have a negative impact on the quality of the learned model.

In order to evaluate classifiers or identifiers for individual samples as well as global results, the so-called confusion matrix is used. It has a form of two-way tables, which record the results achieved by the tested model. The result is

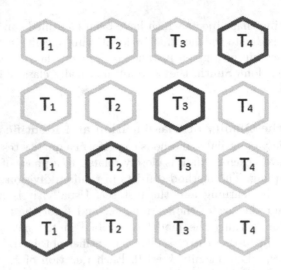

Fig. 6. Diagram of cross-validation for $k = 4$. Light cubes present training sets, dark cubes - test sets. The consecutive lines symbolise the respective iterations of processing.

presented in the form of four values which denote the tendencies in classification or identification between the predicted classes with respect to real object class labels. Table 10 presents the confusion matrix table.

Table 10. Confusion matrix of efficiency classes of classification or identification. Acronyms used: TP (true positive) - number of correctly classified (or identified) cases; FN (false negative) - number of wrongly unclassified cases, while they were positive cases; FP (false positive) - number of wrongly classified cases, while they were negative cases (should not be assigned to the class); TN (true negative) - number of cases correctly excluded from the class, because they are negative.

	Predicted condition positive	Predicted condition negative
Real condition positive	TP (correct answer)	FN (type II error)
Real condition negative	FP (type I error)	TN (correct answer)

Many measures are based on efficiency classes of classification or identification, which specify the models. Selected measures will be presented in the further part of this section.

Accuracy - this is the quotient of correct scores to all cases considered.

$$ACC = \frac{TP + TN}{TP + TN + FP + FN} \tag{77}$$

Fig. 7. Visualization of results with high precision and low accuracy.

Precision - is responsible for the spread of results, denoted with symbol PR. The higher the precision, the smaller the spread. Figure 7 is a symbolic presentation of high precision, but low accuracy results; Fig. 8 - results with high spread and high accuracy; Fig. 9 - results with high precision and high accuracy. Results of the precision measure are calculated on the basis of quantity of efficiency classes in the following way:

$$PR = \frac{TP}{TP + FP} \qquad (78)$$

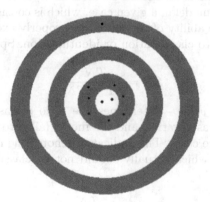

Fig. 8. Visualization of results with low precision and high accuracy.

As the above examples show, precision does not take accuracy into account. It is therefore possible to get a precise results, but with low accuracy.

Recall - is denoted in literature by means of different symbols. In the case of this publication, the *RE* symbol is used, which comes from the term *RECALL*. It defines the ratio of correctly classified results (*TP*) to all classified results

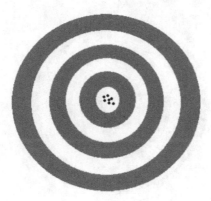

Fig. 9. Visualization of results with high precision and high accuracy.

$(TP + FN)$. In other words, this is a measure that determines the ability of the model to correctly classify or identify objects. It only concerns information about the number of real labels properly classified or identified. This measure is determined by the following formula:

$$RE = \frac{TP}{TP + FN} \tag{79}$$

Specificity - it determines the ratio of the number of correct cases, assuming that no classification took place and the case was negative, to the number of all cases in which no classification should occur. This measure deals with elements not given labels by the model in a given case, which is consistent with the actual class. It determines the ability of the model to properly exclude possession of a feature that is subject to classification or identification. Specificity is expressed by the formula:

$$SP = \frac{TN}{FP + TN} \tag{80}$$

Negative predictive value - determines the ratio of cases where the actual value of a given label was negative, and the model failed to return a given label, to all predicted negative cases. This indicator denotes the number of cases with negative classification, which actually should not be classified. The measure is expressed by the formula:

$$NPV = \frac{TN}{TN + FN} \tag{81}$$

False positive rate - determines the ratio of results to be falsely classified by assigning the label to actual results without this label. The dependence is expressed by the formula:

$$FPR = \frac{FP}{FP + TN} \tag{82}$$

False negative rate - expresses the ratio of results for which the model has failed to return the label, to the real condition positive, i.e. those which should be given a label. The factor is expressed by the formula:

$$FNR = \frac{FN}{FN + TP} \tag{83}$$

F1-score - is a harmonic mean of precision and recall values. It is a very popular measure of effectiveness of classifiers and identifiers. It is expressed by the formula:

$$F1score = \frac{2 \cdot PR \cdot RE}{PR + RE} \tag{84}$$

This measure returns average information about the ratio of the correct classification of results and the precision of their results.

Matthews correlation coefficient (MCC) - is a ratio of the real and predicted classes returned by the model. The values of this measure are between -1 and 1. The maximum value denotes the excellent prediction of the model, 0 indicates the randomness of prediction, while -1 represents the total disapproval between prediction and the real label. The coefficient is expressed by the formula:

$$MCC = \frac{TP \cdot TN - FP \cdot FN}{\sqrt{(TP + FP)(TP + FN)(TN + FP)(TN + FN)}} \tag{85}$$

2.5 Selected Aspects of Learning in Respect of Decision Models

There are many ways of learning and acquiring knowledge in decision systems. Their division is based on numerous criteria. The main criterion is the learning strategy and output derived from the algorithm [6]. The following types of learning algorithms can be distinguished:

- *Supervised learning* - implemented by creating the mapping function of inputs and given outputs. The classification problem is a typical example.
- *Unsupervised learning* - implemented only on the basis of input data by means of methods capable of detecting certain regularities or object similarities.
- *Semi supervised learning* - combination of both methods mentioned above in order to generate a classifier function.
- *Reinforcement learning* - aiming at automatic acquisition of procedural knowledge based on interactions with the environment. Each action causes a certain environmental reaction, which is a hint for the algorithm.
- *Transduction* - similar approach to supervised learning, although the learning outcome is not a function, but prediction of possible output values, which are based on input training data.
- *Meta-learning* - algorithm that uses different learning methods to verify a set of hypotheses. Its purpose is to select the best learning method in respect of a given problem.

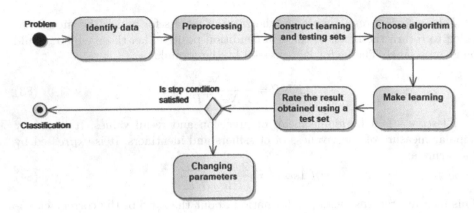

Fig. 10. UML diagram of a general supervised learning algorithm in the form of activity diagram.

2.5.1   Selected Machine Learning Techniques

Supervised Learning

One of the most popular approaches to machine learning. However, the boundary condition is to have proper data in order to apply this procedure. The method can only be used, if it is possible to construct learning and testing sets. In practice, this assumption sometimes turns out to be too strong, which prevents the use of this method. However, assuming the possession of relevant data, preprocessing is performed to improve data quality, remove duplicates, etc. Next, the learning set is constructed, which has a valid output for the selected input elements. Thus, n pairs of input and output examples are obtained, by which the algorithm approximates the function modelling a given phenomenon. The number of elements of the learning set is 33% of the total quantity. Its elements are selected randomly, so as to ensure independence of the selection of objects. The remaining part of the initial set becomes the test set, which is necessary to evaluate the learned function. The next step is to choose the supervised learning technique (e.g. back propagation [75]).

The learning process starts next. It consists in providing an object from the training set as input of a function and calculating the function values for the current coefficients. Results obtained are evaluated. The evaluation function indicates quality of the solution obtained. Depending on the outcome, decisions are made to correct coefficients of the function learned.

A generalized supervised learning algorithm is shown in Fig. 10 in the form of a UML activity diagram. The procedure described is repeated many times for individual objects from the training set until the stop condition is met.

Unsupervised Learning

Method used for problems, for which test sets either do not exist or are difficult to construct. This method uses knowledge derived directly from data. It involves detection, analysis and modelling. It comes down to the analysis of clustering based on various criteria, including object similarity criteria. Discovering knowledge from data allows to identify features, with respect to which the grouping should be performed. The problem of automatic grouping is very complex. There is no universal grouping algorithm for each data type. One should use knowledge domain of the problem and select an algorithm on a case-by-case basis. The additional difficulty is that the number of groups to be created as a result of the algorithm is unknown in the majority of cases. However, there are certain conditions that a data group should fulfil, i.e.:

1. Homogeneity within each groups
2. Heterogeneity between groups.

The first point means that the elements within a group should be similar, while the second point indicates that the elements of different groups should be as distinct as possible. The learning procedure is based on data availability, but without a decision-making label. The structure of clusters changes during processing. The construction of clusters depends on the order of processing objects from data, but once the whole set is processed, results should be the same. Parameters for evaluation of the result, acceptable similarity, etc. are other important elements. These parameters affect the number of object groups at the output. This number is an important factor that affects the quality of grouping. As a criterion for assessing the correctness of grouping, the so-called *grouping quality indicators* [77] are used. There are many indicators of this type. The basic one relies on the assessment of distance of all objects from the centres of groups.

2.5.2 Evolutionary Methods

The popular learning method used in decision systems is the use of evolutionary methods. In general, these methods are algorithms modelled on the surrounding nature and natural evolution. The theory of evolution is one of the basic theories in biology. It allows to describe and understand the processes occurring in the world of living organisms and plants. Within this theory the concept of natural selection, presented by Darwin in the nineteenth century, is discussed.

Based on these experiences, many algorithms have been constructed that imitate mechanisms taking place in nature. Initially, these methods had different names, because they were developed independently by different scientific centres. After some time, however, it was noted that these methods have a common background and can be generalized and included in one standard procedure of dealing with numerous variants.

The subject of the algorithm is the population of solutions called individuals. It is denoted as P^t. In a special case P^0 is called the initial population. Algorithm 1 presents a general evolutionary algorithm pattern with typical components [5], which are characterized by functionality, but can be implemented in a variety of ways. In the further part of this section, the respective elements of the algorithm and examples of implementation will be discussed.

Genetic algorithm
t=0
Initialization P^0
Evaluation P^0
while *stop condition* **do**
| $T^t = \text{Reproduction}(P^t)$
| $O^t = \text{Genetic operations}(T^t)$
| $\text{Evaluation}(O^t)$
| $P^{t+1} = \text{Succession}(P^t, O^t)$
| t=t+1
end

Algorithm 1. Pseudo-code of the evolutionary algorithm

Population is a set of *individuals* representing example solutions to given problems, e.g. values of unknown parameters of a given method. Quantity of the population is a given parameter. Subsequent individuals are described by *genotype*, which consists of one or many chromosomes. Chromosome is a string of ordered *genes*, which code the individual. Gene, in turn, is defined as a feature called attribute of the chromosome. Subsequent genes have values representing variants of a given feature. These values are called *alleles*. The respective alleles cumulatively form a structure called a phenotype, which is a decoded structure providing a specific solution.

There are several methods of representing individuals in the form of chromosomes. Binary representation is one of the methods. It is when subsequent individuals are described as chromosomes coded in the binary system. Different structures of chromosomes are used, both single and complex. It depends on the problem to be solved. The respective genes take the value 0 or 1, which is equated with the presence or inexistence of a given feature. Subsequent stages depend on the selection of the method of representation and of evolutionary methods. The necessity of coding and decoding the solution is a certain disadvantage, yet the big spectrum of available methods used at further stages of processing is an advantage.

The initial population may be formed on a random basis or on the basis of domain knowledge (if available). However, in general, randomized methods are used, which take into account different distributions and parameters. In certain cases not all combinations of values of genotypes are valid. It may lead to the creation of invalid individuals. Therefore, validation of correctness of an individual and the procedure of handling a case of this type has to be taken into account.

The initial population created is subject to evaluation. It consists in the calculation of values obtained by means of the optimized method, assuming that parameters represented by a given individual were used. These values are remembered throughout the life cycle of the population. At this point of the algorithm the very evolutionary cycle starts. The stop condition is considered in the first place. If it is not fulfilled, *reproduction* is performed.

This process is based on the selection of individuals to participate in *genetic operations* generating new solutions. Considering the evolutionary terminology, this process consists in the identification of parents most entitled to the offspring. According to the general rule coinciding with natural selection, they are parents with the best matching factor. Different reproduction variants can be considered, depending on the number of individuals in temporary population, i.e. resulting from reproduction. If the number of individuals is similar to the number of individuals of the output population, a repeatable method of selection may be considered. This allows multiple copies of the same individual. There are several methods of selecting individuals [55]. The most popular is the roulette method, whose name is derived from the game. It consists in symbolic designation of fields on the roulette wheel of size proportional to the adjusted value of the fitting function of a given individual. Taking into account the fitting values of the whole population for i'th individual, probability of selection can be expressed in the form of the following formula:

$$P_s(ch_i) = \frac{f_{fit}(ch_i)}{\sum_{j=1}^{K} f_{fit}(ch_j)} \tag{86}$$

where f_{fit} is the fitting function, and K - is the size of the population. According to the roulette principle, draw is carried out and the roulette wheel division (wheel segment) is implemented with probabilities of the formula (86). It is expressed in the following way:

$$v(ch_i) = P_s(ch_i) * 100\% \tag{87}$$

Population T^t obtained in the reproduction process is then subject to genetic operations, which consist essentially in *crossing-over* and *mutation*. The first operation is much more important because of the probability of its execution, which is assumed to be the value of the range $[0.5, 1]$. Crossing-over is the exchange of genes between individuals, resulting in a new individual (descendant). Mutation, on the other hand, is an operation on a single gene, which is much less likely. Usually, it is defined as the value of the range $[0, 0.1]$, which is consistent with intuition derived from biology.

The crossing-over begins with the selection of pairs of individuals subject to exchange of genes. There are many types of the abovementioned operation. One of the easiest and most widely used methods is single-point crossing-over. It consists in random selection of points for individual chromosomes, in which points alleles change. One new individual is created as a result of the operation, with genes identical to the first parent, and from the division point identical with the second parent. It should be noted that the cross-over operator strictly relates to selected representation. The example described is easy to apply to binary representation, but not always to real representation. However, many examples of operators suitable for different representations have been described in literature [5, 30].

Mutation causes the chromosome to change its value to another. This operation can take various forms, depending on representation. As mentioned earlier, this takes place in the case of relatively low probability. There are several variants of mutations that come down to the choice of genes for mutation. Random selection for each gene, to determine if a given gene is to be mutated, is one of the solutions.

Once genetic operations are carried out, new individuals are evaluated and succession is performed. It involves creating a new population P^{t+1} from P^t and O^t, i.e. from the previous population and the population formed after genetic operations. Selection is carried out similar to that in reproduction. It should be noted that this does not have to be the same selection method as in other places of the algorithm. There are also several approaches to constructing a set, from which selection is carried out. One can combine both sets and select individuals from the whole set or keep the division in assumed proportions, e.g. 30% of the first set and 70% of the second.

The single cycle of evolution ends at this point of the algorithm. The cycle is performed repeatedly, until the stop condition is reached. This condition can be defined in absolute terms, in the form of a determined number of iterations (epochs), or in relative terms, in the form of the number of the final iterations, where solution failed to improve. Other methods to complete processing described in literature are also possible [20]. When iterative processing is complete, the final result is returned in the form of the best individual, i.e. one that has the highest fitting function value. A detailed diagram of the evolutionary algorithm is shown in Fig. 11.

2.5.3   Neural Networks

Introductory Information
Neural networks are inspired by the biological model of human nerve cells. A single cell, called a neuron, forms a network with axons and synapses connecting dendrites of other cells. This makes it possible to transmit an electric impulse equated with a signal.

An artificial neuron model was developed in the 1940's. However, the model considered originally in the form of a single neuron turned out to be too weak a mathematical tool. Based on the idea of the operation of a human brain, an attempt was made at combining artificial neurons to form a network. There are over a hundred billion cells in the human brain. Creating such huge networks of artificial neurons is not possible due to computational and time complexity. However, the start of research in this field in the 1960's allowed the return to the neuron concept as an effective mathematical tool for modelling complex computational and optimization problems. Only the form of a network proved sufficiently large and flexible to apply the approximation of complex phenomena and problems.

Over many years, the use of neuronal networks has led to the development of many artificial neuron models. Selected models are presented below.

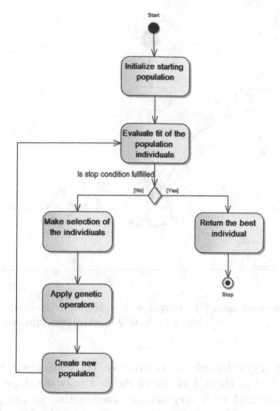

Fig. 11. UML diagram of a general evolutionary algorithm in the form of activity diagram.

Model of a Neuron and Its Selected Types

The generalized model of an artificial neuron has n inputs of the neuron, denoted by $x_0, x_1, \ldots, x_n$. Inputs are represented by a vector called input signal (denoted by $x = [x_0, \ldots, x_n]^T$). In addition, each entry is associated with a synaptic weight w_i for $i > 0$, which comes from the weights vector (denoted by $w = [w_0, \ldots, w_n]^T$). At the same time, w_0 is the highlighted weight, which is a threshold value. There is only one output from a single neuron, which gives the result as a function modelled by the neuron in the form:

$$y = f(s), \tag{88}$$

where the input argument s is described as the following relation:

$$s = \sum_{i=0}^{n} x_i w_i \tag{89}$$

The value calculated by the neuron is additionally subject to the activation function, which aims to filter signals passing through the neuron. Figure 12 presents

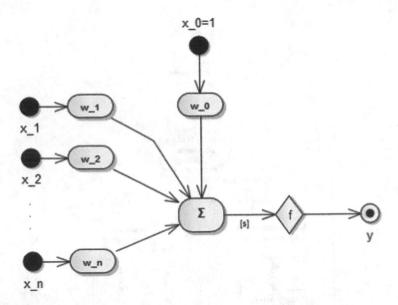

Fig. 12. General neuron model presented as a UML activity diagram. $x_0, \dots, x_n$ - neuron inputs, $w_0, \dots, w_n$ - weights, s - signal, f - activation function, y - output

the neuron model in graphical form. In practice, there are many types of neuron models. The simplest of them, but one of the first ones developed, is *perceptron* [109]. It is characterized by a very similar construction to the general neuron model described above. Its action can be described by the following relation:

$$y = f\left(\sum_{i=1}^{n} w_i x_i + \theta\right), \tag{90}$$

where function f may be a unipolar or bipolar step function assuming values $\{0,1\}$ or $\{-1,1\}$ respectively. Because of the activation functions, the perceptron can classify signals in one of the two classes. For n inputs of the perceptron the $n-1$ dimensional hyperplane called the decision boundary is formed, which divides the space of input vectors into two subspaces. The decision boundary is expressed by the following formula:

$$\sum_{i=1}^{n} w_i x_i + \theta = 0 \tag{91}$$

The diagram of structure of the perceptron is presented in Fig. 13.

A different type of the neuron model is *adaline* (Adaptive Linear Neuron). This type of neuron has the structure almost identical to that of the perceptron. However, they differ in learning methods. The algorithm for determining the output signal is identical to that of the corresponding algorithm in the perceptron. However, in this model, the signal s is compared to the pattern signal d at

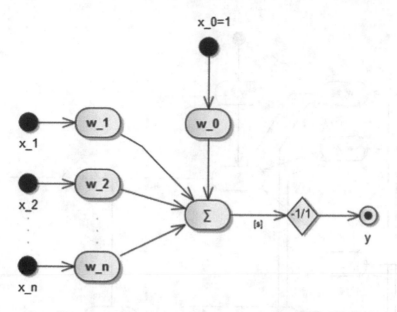

Fig. 13. Perceptron model presented as UML activity diagram, $x_0, \ldots, x_n$ - inputs of neuron, $w_0, \ldots, w_n$ - weights, s - signal, $-1/1$ - bipolar or unipolar activation function, y - output

the end of the linear part of the neuron. This generates an error that is expressed by the following formula:

$$\epsilon = d - s \tag{92}$$

This neuron's learning involves choosing weights in such a way as to minimize the function expressed by the following formula:

$$Q(\boldsymbol{w}) = \frac{1}{2}\epsilon^2 = \frac{1}{2}[d - (\sum_{i=0}^{n} w_i x_i)]^2 \tag{93}$$

This kind of measure of error is called the mean square error. The outline of adaline neuron is presented in Fig. 14.

Another type of the neuron model is a *sigmoidal neuron* [77]. The internal outline of its structure is analogous to that of the *adaline* model, and therefore the perceptron. The difference lies in the use of the sigmoidal activation function (unipolar or bipolar). The unipolar function takes the following form:

$$f(x) = \frac{1}{1 + e^{-\beta x}} \tag{94}$$

while the bipolar function is expressed by the following formula:

$$f(x) = tanh(\beta x) = \frac{1 - e^{\beta x}}{1 + e^{-\beta x}} \tag{95}$$

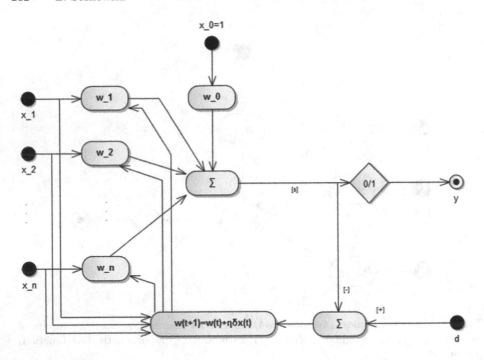

Fig. 14. Adaline neuron model presented as an UML activity diagram, $x_0, \ldots, x_n$ - inputs of neuron, $w_0, \ldots, w_n$ - weights, s - signal, $0/1$ - unipolar activation function, d - pattern signal, η - learning factor, δ - difference between pattern signal and signal calculated from neuron, y - output

The sigmoidal neuron diagram is shown in Fig. 15. The differentiation of the activation function is a significant advantage of the sigmoidal neuron. In addition, derivatives of these functions are calculated in a very simple way. Consequently, one can use gradient methods for their learning. The output signal of this neuron can be formulated as the relation:

$$y(t) = f(\sum_{i=0}^{n} w_i(t)x_i(t)), \qquad (96)$$

while the measure of error Q in the following form:

$$Q(\boldsymbol{w}) = \frac{1}{2}[d - f(\sum_{i=0}^{n} w_i x_i)]^2 \qquad (97)$$

The last neuron model presented in this section is the *Hebb neural model*. The structure of the neuron does not differ from the previously described structures. However, a different method of weight learning is used (the so-called Hebb rule). This rule is available both with and without a supervisor. Its version for unsupervised learning for a single neuron consists in the modification of weights

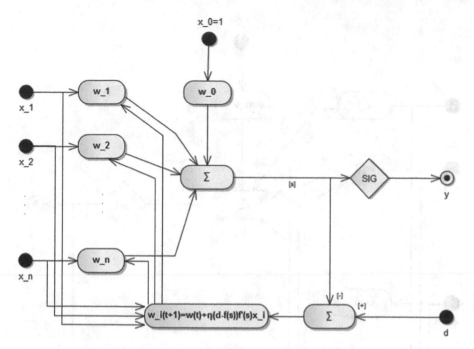

Fig. 15. Sigmoidal neuron model presented as an UML activity diagram, $x_0, \ldots, x_n$ - inputs of neuron, $w_0, \ldots, w_n$ - weights, s - signal, SIG or f - sigmoidal activation function, f' - derivative of the activation function, d - pattern signal, η - learning factor, y - output

in proportion to the value given on the i-th input and the output y. This relation can be expressed in the following form:

$$w_i(t+1) = w_i(t) + \Delta w_i, \tag{98}$$

where Δw_i is expressed by the formula:

$$\Delta w_i = \eta y x_i, \tag{99}$$

and η is the learning factor. It can be noted that there is no pattern signal in the rule given in formula (98). The supervised version consists in a slight modification of formula (99) and takes the form of the following relation:

$$\Delta w_i = \eta x_i d \tag{100}$$

The Hebb neuron is presented in Fig. 16.

Multilayer Unidirectional Networks
Single neurons have fairly limited capabilities of modelling problems. This is due to the fact that one neuron having n inputs divides n-dimensional space into two halves separated by $n-1$ dimensional hyperplanes. With that in mind, structures

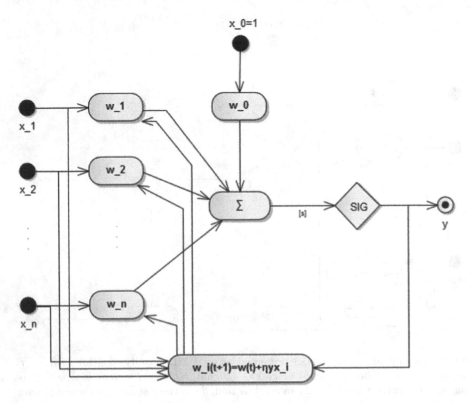

Fig. 16. Hebb neuron model presented as a UML activity diagram, $x_0, \ldots, x_n$ - inputs of neuron, $w_0, \ldots, w_n$ - weights, s - signal, SIG - sigmoidal activation function, d - pattern signal, η - learning factor, y - output

composed of many neurons arranged in layers were invented. These structures are called *neural networks* [41]. There are many types of neural networks characterized by a variety of structures. There are radial networks, self-organizing networks, cellular networks, multilayer networks and more. However, only multilayer networks will be used for further considerations. This type of network assumes the existence of a minimum of two layers - input and hidden. There are neurons located in the layers, which receive data on the input, process them and pass the calculated output value to further neurons as input data in the subsequent layer. Connections between individual neurons have their weights that take part in the processing of signal passing through the network. The flow of signals is unidirectional (from input to output).

The general diagram of a multilayer neural network is illustrated in Fig. 17. Multilayer networks are characterized by connections between individual neurons. Neurons are not connected in a given layer, but have connections with all neurons of the subsequent layer. Consequently, the output signal is received at the input of all neurons in the subsequent layer. Neural networks can be subject to learning. Similarly to the case of a single neuron, the learning element is the

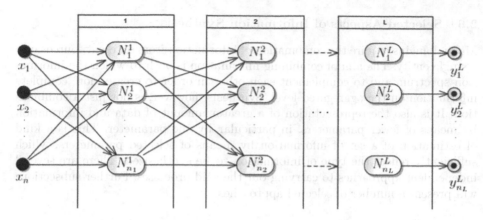

Fig. 17. General model of unidirectional multilayer network. Designations: $x_1, \ldots, x_n$ - inputs of net, N_k^i - k'th neuron in layer number i, y_k^L - k-th output of network from layer number L.

vector of weights. However, in the case of networks, the number of weights is appropriately greater. They occur at each connection between neurons arranged between layers. The goal is to calculate all weights in such a way that, for a given sample, the network approximates set points d_i with sufficient precision. There are several algorithms for learning multilayer networks, but one of the most effective is *backward propagation of errors* [75]. This method is based on the use of gradient optimization algorithms. The main feature of this method is that the target function is described as the sum of squares of differences between the current values of the output signals and the values derived from the training sample. For a single training sample (x, d), the objective function assumes the form:

$$Q(\boldsymbol{w}) = \frac{1}{2} \sum_{i=1}^{m} (y_i - d_i)^2 \qquad (101)$$

where $\boldsymbol{w}$ is the vector of weights, and m is the number of net outputs.

The learning of the network by means of backward error propagation consists of several stages. The first stage is the presentation of the training sample x and the calculation of the individual signal output values of the hidden layer (v_i) and the output layer (y_i) neurons. The second step is to minimize the objective function expressed by formula (101). Details of the algorithm are discussed in a variety of publications, e.g. in [77].

One of the fundamental dilemmas faced when creating networks is structure. The number of network layers is chosen on a case-by-case basis. The network has at least two layers: input and output layer. However, it may also contain hidden layers. As part of the network design, neuron types and quantity are selected. The number of neurons in the respective layers of the network may vary.

2.6 Selected Aspects of Information Synthesis

Many domains require the combination of information from different sources into a single entity. The aim of combining information is both to widen the information spectrum and to complement each other in order to create more complete information on the aggregated level. Synthesis, however, is not just a combination. It is also the representation of a greater amount of data and information by means of fewer parameters, in particular by one parameter. This is a kind of estimation of a set of information by means of selected parameters, which sufficiently reflect the type of information to be synthesized. There are several independent approaches to carrying out the said processes. Further subsections will present a number of selected approaches.

2.6.1 Basic Methods

One of the ways to synthesize information is to use selected central tendency measures [40]. This facilitates carrying out numerical characterization of distribution of a given variable or feature. Measures of the central tendency consist of three basic subgroups: mode, median and average. The first two subgroups are clearly defined.

Mode is the most frequent value of a given set. This is a very simple characteristic that reflects the intuitions of popularity of a given value among all values tested. In other words, it is a frequency-based method.

The Median, on the other hand, is the middle value, i.e. the one that is exactly in the middle of the ordered set of values. The median formula depends on whether the number of elements in the set is even or odd. For an odd number of elements, it takes the following form:

$$x_{med} = x_{(n+1)/2},\tag{102}$$

and for an even number of elements:

$$x_{med} = \frac{1}{2}(x_{n/2} + x_{(n/2+1)})\tag{103}$$

where n is the number of elements of a set of values.

Average value may appear in different forms. There are several types of the average value. The selected types will be described below.

Arithmetic Mean

The most widespread of the mean values. This value is expressed by the following formula:

$$avg_{mean} = \frac{1}{n}\sum_{i=1}^{n} x_i : n > 0,\tag{104}$$

where n is the number of elements of a set of values. The average value is a consistent estimator, sensitive to outlier elements [40]. However, the simplicity of its calculation is the advantage.

Geometric Mean

This mean can be used in the case of the average rate of change of phenomena presented in the form of dynamic series. It is calculated according to the following formula:

$$avg_g = \sqrt[n]{\prod_{i=1}^{n} x_i} : x_i > 0, \tag{105}$$

where n is the number of elements of a set of given values. This mean makes sense for positive values only.

Harmonic Mean

Method of expressing the mean value used for items of equal signs (all positive or negative). It is the inverse of the arithmetic mean calculated from the inverse of the feature value. This mean is expressed by the following formula:

$$avg_h = \frac{n}{\frac{1}{x_1} + \frac{1}{x_2} + \dots + \frac{1}{x_n}} : x_i > 0, \tag{106}$$

where n is the number of elements of a set of values [40].

Root Mean Square

A specific type of mean that maintains good properties for values with different signs. This mean is used for the purposes of estimation of the order of magnitude of the series of numerical data. The mean is expressed by the formula:

$$avg_q = \sqrt{\frac{x_1^2 + x_2^2 + \dots + x_n^2}{n}} : n > 0, \tag{107}$$

where n is the number of elements of a set of values [40].

Truncated Mean

A mean value, which is not much sensitive to outlier values. Its calculation consists in ordering the value and rejecting a fixed number of outlier values (the greatest and the smallest). The number of discarded values is the k parameter of the mean. An arithmetic mean is calculated using the rest of the values. The truncated mean is an estimator not much sensitive to outlier observations [40]. The mean is expressed by the following formula:

$$avg_c^k = \frac{1}{n - 2k} \sum_{i=k+1}^{n-k} x_i : k < \frac{n}{2} \tag{108}$$

Weighted Average

This kind of mean takes into account the weight of a given element. It is based on arithmetic mean, but weights cause specific multiplications of selected elements, so that their significance and impact on the calculated value increases. This mean can be very useful. However, it should be borne in mind that weight is, on the one hand, a flexible mechanism for modelling the preferences of a given element and, on the other, enforces additional obligation to designate weight [40]. The mean is described by the following formula:

$$avg_w = \frac{\sum_{i=1}^{n} w_i x_i}{\sum_{i=1}^{n} w_i} : w_i \geq 0, \sum_{i=1}^{n} w_i > 0 \tag{109}$$

Winsorized Mean

The last mean presented is very similar to the truncated mean. The difference, however, is that there is no removal of the outliers, but substitution with minimum and maximum elements takes place respectively. The number of overridden elements is controlled by parameter k [40]. The described mean is defined as:

$$avg_{wns}^k = \frac{1}{n}[(k+1)x_{k+1} + \sum_{i=k+2}^{n-k-1} x_i + (k+1)x_{n-k}] \tag{110}$$

2.6.2 Fuzzy Methods

Other approaches to the synthesis of information may be presented by means of fuzzy sets described in Sect. 2.1. To this end, synthesis input data must be presented in the form of fuzzy sets. This is not always possible, which is why this method can be used only in selected cases.

However, if data can be presented as fuzzy sets, the synthesis can be interpreted as searching for the intersection of each of these sets. In that case elements (result of synthesis) must belong to each of these sets simultaneously.

If one assumes that the respective fuzzy sets are denoted by A_i, where i is an index, aggregation takes the following form:

$$A_1 \cap A_2 \cap \ldots \cap A_m \tag{111}$$

Each set of this kind has a membership function $\mu_{A_i}(x)$, and thus the intersection of fuzzy sets defined by formula (111), is denoted as:

$$\mu_{A_1 \cap A_2 \cap \ldots \cap A_m}(x) = T(\mu_{A_1}(x), \mu_{A_2}(x), \ldots, \mu_{A_m}(x)), \tag{112}$$

where T means T-norm [77].

There is also an opposite interpretation of synthesis in the form of a sum of sets. It consists in completing information collected in individual sets. In that case it takes the following form:

$$A_1 \cup A_2 \cup \ldots \cup A_m, \tag{113}$$

and in the case of fuzzy sets it assumes the form:

$$\mu_{A_1 \cup A_2 \cup \ldots \cup A_m}(x) = S(\mu_{A_1}(x), \mu_{A_2}(x), \ldots, \mu_{A_m}(x)), \tag{114}$$

where S means S-norm (or alternatively T-conorm) [77].

2.6.3 Method Based on Voting

Election algorithms are a field of voting and selection of candidates as a result of the election process. It also deals with the issue of securing election procedures so that unauthorized actions are difficult to take. The purpose of this field is also to create algorithms taking into account preferences of voters in the best and fairest way. These methods are divided according to the number of candidates selected, the method of voting (the ability to cast only one vote or many), etc. However, regardless of the type of method, they allow the optimization of the selection function of one or more candidates. Therefore, they are also suitable for solving problems of artificial intelligence. These methods are used both in the field of optimization of functions and of classifiers [46].

Election algorithms use the term *candidate* and *voter*. The first type represents the candidate object, i.e. the one becoming one of the possible solutions. As a result of election the candidate can become a winner. The voter is a person or object with the right to vote in elections for defined candidates. The very electoral process is equated with a certain function, which assumes a set of voters and a set of candidates at the input. Its calculation gives a subset of candidates constituting winners of the election process. This function can be expressed as:

$$f^E : V \times C \to W, \tag{115}$$

where V is a set of voters, C is a set of candidates, and $W \subseteq C$ is a set of winners [23].

One of the criteria for the classification of election algorithms is the so-called *Condorcet criterion*. This criterion is fulfilled if the winner is more preferred by voters than any other candidate. If all the possible pairs of candidates are generated, the *Condorcet candidate* is the one who wins all his pairs. A candidate like this does not always exist [54].

Selected voting methods, which can be easily applied in AI, will be discussed in the further part of this section.

Plurality Voting

Popular type of voting in real elections in many countries. This is a single-winner voting system. Each voter casts only one vote. The winner is the candidate with the highest number of votes.

The main advantages of this electoral system include simplicity of the voting process, in which one vote is cast for one candidate. This necessitates choosing the most preferred candidate. Another advantage of the system is the limitation of parties and constituencies, which take part in voting. It is assumed that this system leads to a two-party political system in the longer term.

However, this system has certain disadvantages. First of all, it does not reflect electoral preferences of all voters, but only the majority who voted for the winner. This can create situations, when a winner is completely rejected by voters who did not vote for him. This creates divisions within the voting community. Another problem is the desire to avoid wasting votes, i.e. voting for a candidate who has lesser chance of winning than others. Typically, this results in a vote for one or two candidates who have the greatest chance of victory. This voting system thus eliminates the importance of the individual, which in many cases is not a desirable phenomenon. This voting system does not meet the Condorcet criteria. This approach is easy to implement in AI systems [46].

Example 23. $V = \{v_1, v_2.v_3, v_4, v_5\}$, $C = \{c_1, c_2, c_3\}$.

The vote consists in casting one vote for exactly one candidate. Example voting is shown in Table 11.

Table 11. Table presenting example majority voting of voters from set V for candidates from set C. The winner is candidate c_2, who was cast the highest number of three votes.

Voter/candidate	c_1	c_2	c_3
v_1	1	0	0
v_2	0	1	0
v_3	0	0	1
v_4	0	1	0
v_5	0	1	0
Sum	1	**3**	1

Borda Count

Single winner voting method. However, voices are given to every candidate. The vote is based on the ranking of candidates who would reflect the preferences of the voter [7]. Then, individual ranking entries are given points. There are several different scoring methods. The simplest is linear allocation of points, inversely proportional to the position occupied, e.g. if the ranking consists of n elements, the first place gets n points, the second $n-1$, and so on to the last position receiving 1 point. A modified version of this method consists in giving a selected element as many points as the element exceeds, e.g. first position gets $n-1$ points, second $n-2$ points, and so on until the last one gets 0 points. There is also a non-linear version of the point assignment function, which differentiates the meaning of each element to a greater extent [78].

Rankings of individual voters, created by means of a given method, are then summed up. Points of individual candidates are aggregated. The winner is the candidate with the highest number of points.

This method, although not fulfilling the Condorcet criterion for selected cases, is considered to be a compromise-based approach. It takes into account the preferences of individual voters in relation to all candidates and allows for the selection of the candidate with the highest level of support. The disadvantage of this method is its complexity. With a large number of voters and candidates classic elections are very difficult to conduct. As far as application is concerned, this method is used for the purposes of conducting competitions in many sports disciplines. It is also widely used in optimization and AI applications.

Example 24. $V = \{v_1, v_2.v_3\}$, $C = \{c_1, c_2, c_3, c_4, c_5\}$.

Voting is done by creating as many rankings as there are voters. Rankings show voting preferences of voters in a way that the candidate ranked higher is more valued by a given voter. Table 12 presents example rankings of voters from set V. Votes are summed up and winner is chosen next. The summary ranking is presented in Table 13.

Table 12. Table presenting example rankings for Borda count method by voters from set V for candidates from set C. The position of local winners is marked in bold.

Id	Candidate	Points	Id	Candidate	Points	Id	Candidate	Points
1	c_3	4	1	c_4	4	1	c_5	4
2	c_1	3	2	c_1	3	2	c_1	3
3	c_2	2	3	c_5	2	3	c_4	2
4	c_5	1	4	c_2	1	4	c_3	1
5	c_4	0	5	c_3	0	5	c_2	0

Table 13. Aggregate ranking of candidates with the winner of the Borda count voting marked in bold.

Id	Candidate	Points
1	c_1	9
2	c_5	7
3	c_4	6
4	c_3	5
5	c_2	3

Approval Voting

Approval voting is another type of single-winner voting. It consists in voting for each candidate reflecting support or disapproval. In this respect, this method is similar to the Borda count, but in this case there is no support value, only the declaration of acceptance. Consequently, each voter who participates in voting designates a set of supported candidates. The winner is the candidate who is accepted by the largest number of voters [14]. It can be implemented by assigning binary values to a given candidate from the set $\{0,1\}$, where 0 denotes disapproval and 1 means acceptance of a given candidate. The function modelling the voting for a candidate may take the following form:

$$f^c : C \to \{0,1\}, \tag{116}$$

where C is a set of candidates.

Example 25. $V = \{v_1, v_2.v_3, v_4, v_5\}$, $C = \{c_1, c_2, c_3\}$.

Voting consists in expressing support or disapproval in respect of each candidate. Example voting is shown in Table 14.

Table 14. Table presenting example approval voting of voters from set V for candidates from set C. Designations: 1 - acceptance, 0 - disapproval. The winner is a candidate c_1, who reached the highest number of four acceptance votes.

Voter/candidate	c_1	c_2	c_3
v_1	1	1	1
v_2	1	1	0
v_3	1	0	1
v_4	0	0	0
v_5	1	1	0
Sum	4	3	2

Copeland's Method

This is one of the methods fulfilling the Condorcet criteria. It consists in pairing and conducting a *tournament* with individual duels between the respective candidates. The winner of the duel is the candidate who gathers the largest number of votes. Once the winner and the loser are announced, points 1 and -1 are allocated respectively. In the case of a tie, both candidates are assigned a fixed value of *alpha*, which is a parameter of the method. However, it is generally assumed that $\alpha = 0$. The winner remains the candidate with the highest number of points reached as a result of all duels [79]. The result is returned in the form of a ranking of all candidates. This method is considered to be a tournament method and is used very often in various sports disciplines. Its advantage is transparency and comprehensibility for the majority of people. It is also possible to implement the method in computer systems, although computational complexity is $O(n^2/2)$. However, good qualities associated with the very idea of selection, its precision and social support make the method attractive for application also to AI problems.

Example 26. $V = \{v_1, v_2, v_3, v_4, v_5\}$, $C = \{c_1, c_2, c_3\}$.

Voting takes place by creating all possible pairs between candidates (not taking order into account) and carrying out individual duels. Table 15 presents results of particular duels and Table 16 contains the final result.

Table 15. Table presenting example Copeland's method voting of voters from set V for candidates from set C. Designations: Candidate 1/2 - pair of candidates for which the duel takes place; Support 1/2 - support of particular voters for candidate 1 and 2 respectively; Result 1/2 - points awarded to candidates 1 and 2 respectively.

Candidate 1	Candidate 2	Support 1	Support 2	Result 1	Result 2
c_1	c_2	v_1, v_2, v_4	$v_3, v_5,$	1	-1
c_1	c_3	v_1, v_4	$v_2, v_3, v_5,$	-1	1
c_2	c_3	v_5	$v_1, v_2, v_3, v_4,$	-1	1

Table 16. Table presenting the results of voting by means of the Copeland's method in respect of voters from set V for candidates from set C.

Id	Candidate	Wins	Defeats	Sum
1	c_3	2	0	2
2	c_1	1	-1	0
3	c_2	0	-2	-2

Range Voting

This type of voting procedure provides for assessment of each candidate from a specific range. The range can be any. Commonly used ranges are in the form of $[0, 1]$, $[0, 100]$, but also in the form of sets of discrete values, e.g. $\{1, 2, 3, 4, 5\}$. Votes cast for each candidate are either aggregate or average (depending on the method applied) [89]. One candidate with the highest number of points wins. This method does not meet the Condorcet criteria.

Example 27. $V = \{v_1, v_2, v_3, v_4, v_5\}$, $C = \{c_1, c_2, c_3\}$.

Voting takes place by awarding each candidate an assessment of a set of values. In this example, the possible set of rating values is defined as the interval $[0, 100]$. The method of combining the points of individual voters was established as an arithmetic mean. Table 17 presents the results of voting and in summary the results of the election.

Table 17. Table presenting an example range voting related to voters from set V for candidates from set C. The winner is candidate c_1, who reached the highest average score calculated from votes of individual voters

Voter/candidate	c_1	c_2	c_3
v_1	23	45	75
v_2	56	50	34
v_3	70	90	5
v_4	95	5	20
v_5	60	40	30
Mean	**60.80**	46.00	32.80

Weighted Voting

This type of voting makes it possible to favour selected voters. This means that different voters have different voting powers. In practice, this is often the case in voting of shareholders of trading companies [45]. Individual shareholders have as many votes as shares. In this way, the vote is multiplied, so that its strength grows or decreases, depending on the weight of the vote. A special case is when all weights are equal. Then weighted voting comes down to the above-mentioned range voting. This electoral system does not meet the Condorcet criteria. However, it is distinguished by the previously described scaling of power of a given vote. The function calculating the number of votes for a given candidate may take the form:

$$f^{c_i} = \sum_{i=1}^{n}(w_i \cdot rank_i^c(v_i)) \tag{117}$$

or

$$f^{c_i} = \frac{\sum_{i=1}^{n}(w_i \cdot rank_i^c(v_i))}{\sum_{i=1}^{n} w_i} : \sum_{i=1}^{n} w_i > 0, \tag{118}$$

depending on the method used to combine results from individual voters.

Example 28. $V = \{v_1, v_2, v_3, v_4, v_5\}$, $C = \{c_1, c_2, c_3\}$.
   Voting comes down to awarding each candidate a rating from a set of values. In this example, a possible set of rating values is defined as the range $[0, 100]$. In addition, weight examples were given for individual voters: $60, 20, 10, 6, 4$. The voting method for individual voters was determined on the basis of formula (118). Table 18 presents the results of voting and the summary of results of the election. Comparison of Tables 17 and 18 shows that the same votes cast by voters produced completely different results.

Table 18. Table presenting example weighted voting related to the voters set V and the set of candidates C. The winner is candidate c_3, who received the highest average rate calculated on the basis of ranks of particular voters in relation to particular strength of the vote defined by weights.

Voter/candidate	c_1	c_2	c_3
v_1	23	45	75
v_2	56	50	34
v_3	70	90	5
v_4	95	5	20
v_5	60	40	30
Weighted average	40.10	47.90	**54.7**

3   Comparators

3.1   Introductory Information

So far, a comparator has been said to be as a specialized logic element used for the purposes of comparing two numbers received on input. It has specified the value of the majority relationship between those numbers. The most common comparators have been the binary ones used for comparing binary coded numbers. The concept of a comparator has gained popularity in the 1950's and was quickly applied in electronics. Used until today as an analog comparator [51] and digital comparator [110], it specializes in comparing analog and digital signals. Comparators of this kind are atomic components forming parts of larger components.

A major step in the evolution of comparators was the development of programming languages. More and more computer programs started to emerge. These programs operated on data stored in structures and objects. The comparison operation was the basic operation performed during calculation. It made it possible to determine the relationship between inputs. In this case, comparators began to receive not only numbers on the input, but also simple data types corresponding to attributes of the object. The comparator's response, however, remained unchanged and took the form of a single value indicating the state of relation of the majority.

A *compound object comparator* is analogous, but dedicated to processing complex objects represented as data entities. It is designated by com^{ref}. It can be interpreted as the following function:

$$\mu_{com}^{ref} : X \times 2^{ref} \rightarrow [0,1]^{ref}, \tag{119}$$

where $X \subseteq U$ is a set of input objects to be compared and ref is a set of reference objects from which similarity is inferred. $[0,1]^{ref}$ denotes a space of vectors v of dimension $|ref|$, where each i-th coordinate in $v[i] \in [0,1]$ corresponds to an element $y_i \in ref$, $ref = \{y_1, \ldots, y_{|ref|}\}$. In the following parts of the paper ref shall be called the *reference set*, while each $Y \subseteq ref$ will be referred to as the *reference subset*. Additionally, $a(x)$ is the function representing the object $x \in X$ for a given attribute a. This representation is then used by the comparator for processing x. Similarly, each reference object $y \in Y$ is processed by means of its representation $a(y)$ for a given attribute a. With ordered elements of the reference set ref, i.e. $ref = \{y_1, \ldots, y_{|ref|}\}$, the comparator may be presented in the form of the following function:

$$\mu_{com}^{ref}(x, Y) = Sh(F(v)), \tag{120}$$

where Sh is a (result) *sharpening function*, F is a function responsible for filtering the result before sharpening and v is a vector of dimension equal to cardinality of ref, composed of proximity (similarity, closeness) values between the object x and each of the reference objects in ref. Typically, F is based on the combination of certain standard, idempotent functions, such as min, max, top, etc. [70] When Y is a proper subset of ref, the positions in v corresponding to $y_i \notin Y$ are filled with zeros. Non-zero elements of v determine the degree of similarity (proximity) between the object x in question and each element of the reference subset Y. With that in mind, the *proximity vector* can be defined as:

$$v[i] = \begin{cases} 0 & : & y_i \notin Y \\ sim(x, y_i) & : & y_i \in Y \end{cases} \tag{121}$$

The value of similarities $sim(x, y)$ used above is calculated by means of a fuzzy relation [36] combined with additional mechanisms, described in subsequent section.

The structure of the compound object comparator, the way it operates, and the input and output are considerably different as compared to earlier approaches. Previous approaches were based on atomic values passed to the comparator. The result was a simple fixed answer, e.g. $\{0, 1\}$ or $\{1, -1, 0\}$, where each element expressed the value of relationship between input elements (majority, minority or equality) [59]. In the case of compound object comparators, *similarity* is taken into account, rather than the value of a single relation. This is one of the fundamental differences between these approaches, and it makes a radical difference. This fact facilitates the analysis of the problem in a much broader context. The approach outlined in this paper contains several innovative solutions, inexistent so far in the field of comparators. One of them is the aforementioned *reference set*. This set contains instances of objects representing domain knowledge of a given problem. It is a kind of knowledge used for the purposes of comparing and designing the proximity vector. It is a collection of elements by which similarity is expressed. Issues of structure of this kind of set are described in the publication [99].

Another new solution for comparators is the exception rules mechanism. They allow the exclusion of parts of the solution from deliberation, without changing the fundamental membership function of the relation (base similarity function). These rules implement the exception mechanism by means of defined similarity functions in an unchanged form [84]. This simplifies the comparator design and improves clarity and ease of managing changes.

Another new solution introduced in the field of comparators is the local mechanism for the quality control of solutions [96]. It is based on the fact that a solution of low value of similarity is not taken into account. The threshold value for which the solution ceases to be interesting is determined on a case-by-case basis. This is one of the comparator parameters that can be learned.

Comparators of this kind are characterized by two other significant differences. The first one is a mechanism for separating results, called *sharpening mechanism*. Its task is to strengthen the differences between the best and the medium and weak solutions. The second difference is the occurrence of a solution filter block, which determines the suitability of a solution based on comparison with results of other pairs of objects. From this point of view, this mechanism is based on competitiveness between reference objects and not on a certain independent rule, as in the case of the previously mentioned mechanism for controlling the quality of individual solutions. Competition is determined by comparing the results of similar measurements of individual pairs of objects.

The undisputed advantage of comparators is their universality, which is associated with the method of dealing with any type of object processed. This technique can be used in solving problems with decisions related to text objects, graphic objects, video or audio material, etc. [92]. Methods used for measuring similarities and reference sets are different, depending on the problem in question. However, the basic operating principle is the same.

3.2   Definitions of Comparator Components

3.2.1   Reference Set

Ordered non-empty set $ref = \{y_1, \ldots, y_{|ref|}\}$, which is a subset of universe U, selected for input objects $x \in X$, is called the reference set. This set consists of objects for which comparisons are made in the comparator. These objects are grouped into a set because of certain relationships and attributes, or certain representations of knowledge. Individual elements are separate object instances with combinations of attribute values that span the space of possible object states that are relevant to a given comparator or a group of comparators. The reference set is a kind of analogy to the notion of a generalized decision that occurs in rough sets [63]. This term denotes a set of decision classes (see Sect. 2.4.1), possible for a given object. However, comparators do not deal with decision classes, but with subsets of reference objects that are possible for an input object, which is to a certain extent analogous to the notion in question. The reference set is necessary for the comparator for comparison purposes and the possibility to return the value of similarity. However, for a particular comparator on the input, a reference subset is passed $Y \subseteq ref$, the elements of which are used for comparison in a specific comparator.

3.2.2   Base Similarity

The membership function of a fuzzy relation [36], defined as follows:

$$\mu : X \times ref \rightarrow [0,1], \tag{122}$$

where ref is a reference set and X is a set of input objects. According to the definition, the function reaches the value of the range $[0, 1]$, where 0 means total lack of similarity, and 1 total affiliation to the relation, which in the case of comparators is interpreted as indiscernibility in terms of the examined feature (according to the Definition 11 in Sect. 2.2). In a given comparator, exactly one membership function is defined. It is used to calculate values of particular coordinates of the proximity vector (formula (121)). Different comparators may have different membership functions, although this is not required. It is therefore possible to use the same base similarity functions in many comparators, but its choice is dictated mainly by the type of object processed and the problem solved. Papers [93, 97, 102] present the examples of membership functions possible to apply in relation to different types of compound objects.

3.2.3   Threshold Function

Function defined as:

$$t_h(z) = \begin{cases} 0 & : & z < p \\ z & : & z \geq p \end{cases}, \quad p \in [0,1], \tag{123}$$

where z is an argument representing a function of the base similarity function for a given pair of objects (x, y) in a given comparator [117]. The main purpose of this function is to limit too weak solutions, e.g. those for which similarity is too low. If the value of p is not reached, the value is set to 0. The p parameter is

228   L. Sosnowski

defined individually for each comparator. It can be set by an expert or learned
by means of the machine learning procedure.

This mechanism relates to the concept of α−cut of a fuzzy set (Definition 7).
Each compound object comparator forms a fuzzy set. The threshold function
transforms the resulting set into its p−cut by applying the function of the formula
(123).

3.2.4 Exception Rules

A family of sets indexed by a variable i is defined, such that for each reference
object y_i there is a set of rules $Rules_i$. This set contains rules for the reference
object y_i, but their argument is an input object $x \in X$. A single rule can take the
form of predicates bound by logical operators, such as the conjugate, alternate
or negation [28]. If a given logical sentence is true, then the rule is fulfilled. The
general form of the rule is:

$$r_j : X \to \{0,1\},   (124)$$

where j is an index of a rule in the set $Rules_i$, which denotes the id number of
the defined rule. The whole mechanism can be modelled as a function:

$$Exc^{ref}_{Rules_i}(x) = max_{j=1}^{|Rules_i|}\{r_j(x)\}, \; x \in X   (125)$$

On the functional side, these rules allow one to exclude parts of solutions from
the discussion, without changing the base similarity function. This simplifies the
design of comparators and increases their clarity and ease of change management.

3.2.5 Similarity Function

A function defining similarity between a single pair of objects (x, y), fulfilling:

$$sim : X \times ref \to [0,1]   (126)$$

It specifies values of the respective coordinates in the proximity vector v (formula (121)). It is a combination of three predefined elements: the base similarity
function, the threshold function and exception rules. The general form of this
function is expressed by the formula:

$$sim(x, y_i) = \begin{cases} 0 : & Exc^{ref}_{Rules_i}(x) = 1 \vee y_i \notin Y \\ t_h(\mu(x, y_i)) : & w\,p.p. \end{cases} , \; x \in X, \, y_i \in ref,$$

$$(127)$$

where t_h means threshold function, μ is the base similarity function, $Exc^{ref}_{Rules_i}$
is the exception rules function and i is the index of proximity vector for which
the similarity is calculated, whereas $Y \subseteq ref$.

3.2.6   Filtering Function

Function designated as F filtering the values of the vector $\boldsymbol{v}$. The selection of the function significantly affects the properties of the comparator. The result of the function is always the vector in the form:

$$F(\boldsymbol{v}) = \langle h(\boldsymbol{v}[1]), \ldots, h(\boldsymbol{v}[|ref|]) \rangle, \tag{128}$$

where $v[i] = h(\boldsymbol{v}[i])$ and h is one of the idempotent functions. An example of a simple filtering function is the max function, which selects the highest values.

The essence of the filtering function is to filter results so as to minimize the number of non-zero results returned by the comparator. In a sense, this is a function that transforms the proximity vector into a sparse vector (but not in every case). Filtration takes place in conjunction with values of all coordinates. It implements the assumption of competition of individual reference objects between one another. This mechanism operates on a full set of results generated through the calculation of similarity for each pair of input objects to particular reference objects. Consequently, filtration takes into account all combinations of results and is able to compare the said results. It is therefore based on the selection of the best results in the sense in question, regardless of the level of individual similarities (controlled by another, previously described mechanism).

3.2.7   Sharpening Function

The aim of this feature is to increase the distance between the best, medium and the worst results. It is focused on the analysis of individual values of the proximity vector. The domain and codomain of this function is a vector with the dimension $|ref|$. This function has the form:

$$Sh : [0,1]^{ref} \rightarrow [0,1]^{ref}, \tag{129}$$

Sharpening can take place through different transformations, but in this dissertation the following shall be used:

$$Sh(\boldsymbol{v})[i] = \begin{cases} max_v \cdot e^{\alpha(v[i]-max_v)}, & \alpha > 0, \\ 0, & otherwise. \end{cases} \tag{130}$$

where $\boldsymbol{v}$ is a proximity vector and max_v is a maximum value of the coordinate of the vector $\boldsymbol{v}$ in the form:

$$max_v = max_{i=1}^{|ref|}\{\boldsymbol{v}[i]\} \tag{131}$$

Function (130) in non-linear and has an interesting property as regards factor $\alpha > 0$ [106]. It is characterized by three basic features:

$$\forall i \in \{1, \ldots, |ref|\} : (v[i] = 0) \Rightarrow (Sh(\boldsymbol{v})[i] = 0), \tag{132}$$

where formula (132) means keeping the unchanged value of zero equal, which prevents the artificially high result;

$$\forall i \in \{1, \ldots, |ref|\} : (v[i] = max_{j=1}^{|ref|}(v[j])) \Rightarrow (Sh(\boldsymbol{v})[i] = v[i]), \tag{133}$$

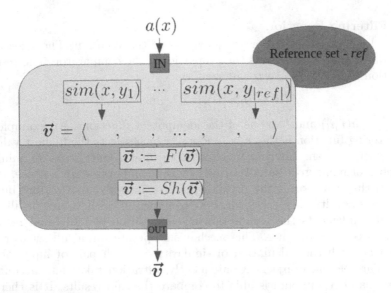

Fig. 18. Block diagram of the comparator's internal structure. The structure consists of two main layers: concurrent - responsible for the calculation of values for the initial proximity vector; sequential - responsible for the processing of values from the previous stage. Blocks marked with *sim* are responsible for the calculation of unit similarity values. The internal structure of such a similarity block is described in Fig. 19 by means of the UML.

where formula (133) means keeping the maximum values unchanged, so that the best result retains its original properties;

$$\forall i, j \in \{1, \ldots, |ref|\} : (v[i] < v[j]) \Rightarrow (Sh(\boldsymbol{v})[i] < Sh(\boldsymbol{v})[j]), \tag{134}$$

where formula (134) means strong monotonousness, which is a prerequisite for fulfilling the assumed role.

3.3 Comparator Architecture

A comparator consists of two main layers (stages) - concurrent and sequential. The first of those layers is responsible for the calculation of values of the proximity vector coordinates for $x \in X$, $y_i \in ref$, with the use of a similarity function given by the formula (127). Similarities for different y_i can be calculated concurrently, because they do not depend on one another. The internal, layered structure of a comparator is shown in Fig. 18. While each coordinate of the vector $\boldsymbol{v}$ is calculated independently, the final calculation of the value of the function (127) has to be performed in a sequence for a given pair of objects (x, y_i). This sequence of operations is illustrated in Fig. 19 in the form of an UML activity diagram [74]. The processing in the first layer ceases when all coordinates of the proximity vector are derived. Only then can the comparator activate the next layer.

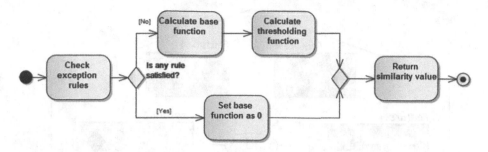

Fig. 19. An UML activity diagram of the *sim* block for a comparator calculating similarity between the input and reference object.

The processing in the comparator's second layer is performed in sequence. Hence, operations, such as filtering by means of $F(v)$ (described in Sect. 3.2.6) and sharpening of the proximity vector with $Sh(v)$ (described in Sect. 3.2.7), are performed one-by-one. The final result of this sequence is the vector given by (121).

Taking a wider look at the comparator for compound objects, one may notice that if all the notions introduced above are combined, it can be expressed as:

$$\mu_{com}^{ref}(x, Y) = Sh(F(< sim(x, y_1), \ldots, sim(x, y_{|ref|}) >)) \tag{135}$$

3.4   Granular Structures

A signal granule is a specific implementation of the information granule defined by Lotfi Zadeh in the 1970's. His definition is as follows: *"An information granule is a clump of objects of some sort, drawn together on the basis of indistinguishability, similarity, or functionality"* [65].

In this case the granule is used to represent an input object and its closest surrounding built from reference objects. This is a representation of a signal moving through the network from layer to layer. The content is different at each stage, depending on comparisons obtained and layers of the network visited. An outline of a signal granule is presented in Fig. 20 [92].

3.5   Required Comparator Properties

Individual comparators can fulfil many properties resulting from properties of the underlying base similarity function $\mu(x, y)$ (Sect. 3.2.2). However, there are several properties that have a very important effect on the correctness and effectiveness of comparator-based methods. The most important ones are presented below. They are to be met by comparators applied in cases described in this paper. The following property assumes that $ref \neq \emptyset$ and $(ref \subseteq U)^3$:

$$\mu_{com}^{ref}(x, Y)[i] = 0, \forall x \in X, Y \subseteq ref, y_i \notin Y \tag{136}$$

---

^{3} $[0,1]^{ref}$ this is a designation of the vector space v with the length $|ref|$, where each i'th coordinate $v[i] \in [0,1]$ refers to the element $y_i \in ref$, $ref = \{y_1, \ldots, y_{|ref|}\}$.

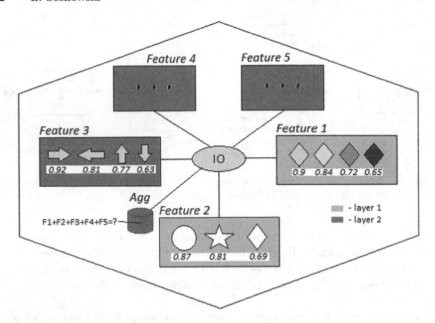

Fig. 20. General scheme of a signal granule used for the transfer of information between comparators, aggregators, layers, in the network of comparators. Feature n - processed feature, IO - input object, Agg - aggregated data, $C2 \ldots Cn$ - features processed by comparators.

$$\mu_{com}^{ref}(y_i, Y)[i] = 1, \forall y_i \in Y, Y \subseteq ref \tag{137}$$

$$\mu_{com}^{ref}(x, Y_1 \cup Y_2) = Sh(F(\mu_{com}^{ref}(x, Y_1)^*) + F(\mu_{com}^{ref}(x, Y_2)^*)),$$
$$\forall Y_1, Y_2 \subseteq ref, Y_1 \cap Y_2 = \emptyset \tag{138}$$

$$\mu_{com}^{ref}(y_i, Y)[j] = 1 \land \mu_{com}^{ref}(y_j, Y)[i] = 1 \Rightarrow \mu_{com}^{ref}(x, Y)[i] = \mu_{com}^{ref}(x, Y)[j],$$
$$\forall x \in X, y_i, y_j \in Y, Y \subseteq ref, \tag{139}$$

where X is a set of input objects, ref is a reference set and F is a filtering function, which determines the final result returned by the comparator (Sect. 3.2.6).

The first property regulates a special case in which the reference subset Y does not contain the element $y_i \in ref$. In particular, this may be the case when $Y = \emptyset$. The comparator should return 0 for this coordinate. The second property concerns the case when one of the reference objects fulfils the role of the input object. Then the corresponding coordinate should reach the value of 1.

Another very important property is the one allowing the division of the reference set into disjoint subsets. This allows parallel processing of this set, which has a very significant effect on performance of the solution. At the same time, this property guarantees the same results as during sequential processing of the entire set. The comparator function marked as μ_{com}^* means that the sharpening function is disabled, which is very important for the property to be fulfilled. When implementing the property (138), it should be noted that for each subset, proximity vectors will amount to ref. Based on the property (136), proper coordinates corresponding to objects $y_i \notin Y_1$ will have the value of zero. Therefore, the sum $F(\mu^*(x, Y_1)) + F(\mu^*_{com}(x, Y_2))$ is realized as a simple sum of vectors $v_1 + v_2$. After summing up the corresponding coordinates, one needs to perform sharpening operations for the resulting vector to obtain the same proximity vector as the one for the comparator operating in the full set $Y_1 \cup Y_2$.

In order to describe the last property required, the relationships of the indistinguishability of features for comparators will be defined. Let us have $x \in X$ and $y \in Y \subseteq ref$. Two objects are indistinguishable in relation to a given feature examined by comparator functions $\mu_{com}(x, Y)$, if:

$$\mu^{ref}_{com}(x, \{y\}) = 1, \forall_{y \in Y, x \in X} \tag{140}$$

The last property means that if two objects are indistinguishable by a particular feature examined by the same comparator, knowing the value of similarity to the third object of one of them automatically gives information about the similarity of the second pair. The indiscernibility of comparators is directly attributable to the indiscernibility function given in the Definition 12.

3.6   Comparator Fuzzy Interpretation

The model of the compound object comparator presented is a functional description. However, it is not the only possible description and in certain cases it is better to use different ones. It is therefore important to move between models in an easy and precise way. This situation occurs when the comparator is described by fuzzy sets. There is a simple method for converting a proximity vector into a type I fuzzy set [77]. According to the formula (121), the result of the compound object comparator is a proximity vector. It should be noted that individual vector coordinates define similarity of a particular pair of objects (x, y), where $x \in X$ and $y \in Y \subseteq ref$. These values reflect the degree of memberships of the fuzzy set. In addition, there are pairs of objects on each coordinate, i.e. it is a fuzzy relation, which is also a fuzzy set. Consequently, the result described in functional terms can be converted to a fuzzy set notation in the following way:

$$R(x, y) = \{((x, y_i), v[i]) : i = 1 \dots |ref|\}, \tag{141}$$

where $v[i]$ is i'th coordinate of the proximity vector, which simultaneously fulfils the condition of the membership function of the fuzzy relation in the form:

$$\mu : X \times ref \to [0, 1] \tag{142}$$

This method is also consistent with the definition of the similarity function of the comparator (Definition 126). The form of the formula (141) is equivalent to Zadeh's notation:

$$R = \frac{v[1]}{(x, y_1)} + \frac{v[2]}{(x, y_2)} + \dots + \frac{v[|ref|]}{(x, y_{|ref|})} \tag{143}$$

In addition, the structure of a compound object comparator demonstrates that this is the α–cut of the fuzzy set, for $\alpha = p$.

3.7 Example

A very simple example will be given in this section, which aims to illustrate definitions and formulas previously presented. It is assumed that the input object is the real number amounting to 2.40, which is in the set of input objects $X = \{2.40\}$. The reference set takes the form $ref = \{1, 2, 3, 4, 5, 6\}$. The compound object comparator is given for the following components:

- base similarity function: $\mu(x, y) = 1 - \frac{|x-y|}{max(x,y)}$,
- parameter of the threshold function: $p = 0.8$,
- filtering function : $F(v) = \langle h(v, 1), \ldots, h(v, |ref|)\rangle$,

$$h(v, i) = \begin{cases} v[i] : & rank(v, i) \geq k \\ 0 & w\,p.p. \end{cases}$$, where $k = 2$ i $rank(v, i)$ returns an index

  from vector's ranking of values,
- sharpening function: $Sh(v)$ according to formulas 130, 131.

Similarity between the input object (Sect. 2.4) and elements of the reference set ref is sought. The first layer of the comparator will return the following vector:

$$v = \langle 0, 0.833, 0.8, 0, 0, 0\rangle$$

The zero values result from the threshold function. In the second layer, the vector is processed by the filtering function, which in this case does not change the vector's form and therefore it still has the form:

$$v = \langle 0, 0.833, 0.8, 0, 0, 0\rangle$$

Then the vector is sharpened, resulting in the form:

$$v = \langle 0, 0.833, 0.780, 0, 0, 0\rangle$$

It is the final proximity vector of the comparator with the filtering function, using the $rank(v, i) \geq 2$. Transition to Zadeh's fuzzy notation results in the form:

$$R = \frac{0.833}{(2.4, 2)} + \frac{0.78}{(2.4, 3)}$$

3.8 Automatic Correction of Results

In the case of processing multiple input objects with certain additional assumptions, the model can improve results on its own by means of a certain competition between objects [87]. This solution produces results with dependence between the respective objects. This procedure works if the following assumptions are met:

- reference objects are unique in terms of the indiscernibility relation,
- input objects are unique in terms of the indiscernibility relation,
- searched mapping of set A to B is injective.

Once the full cycle of model computation for each input object is complete, a matrix of comparison is formed for each comparator. The matrix compares the results for the respective pairs of objects (x, y). If assumptions are met, it is possible to search through the matrix to identify possible wrong results. The point is that all input objects should be associated with different reference objects. The association in this context means that the indication of a comparator is the best local result of comparison. A simple example is shown in Fig. 21. If a reference object is associated more than once, the cases must be checked in detail. The case can take the following form:

$$\exists_{i,k:i\neq k} : (\arg\max_{ref} sim(x_i, y_j)) \cap (\arg\max_{ref} sim(x_k, y_j)) \neq \emptyset \qquad (144)$$

In such cases the pair with the highest resemblance remains unchanged. Other pairs, in which the input object points to the same reference object, are reset. A new maximum value is designated. The procedure is repeated until non-contradictory assumptions are met. This method can, however, lead to the resetting of the result matrix.

	y_1	y_2	y_3	y_4	y_5			y_1	y_2	y_3	y_4	y_5
u_1	**0,91**	0,65	0,23	0,45	0,49		u_1	**0,91**	0,65	0,23	0,45	0,49
u_2	0,82	0,81	0,34	0,42	0,12		u_2		**0,81**	0,34	0,42	0,12
u_3	0,23	0,12	**0,88**	0,23	0,31		u_3	0,23	0,12	**0,88**	0,23	0,31
u_4	0,51	0,43	0,32	**0,94**	0,12		u_4	0,51	0,43	0,32	**0,94**	0,12
u_5	0,49	0,21	0,14	0,39	**0,89**		u_5	0,49	0,21	0,14	0,39	**0,89**

Fig. 21. Method of improving the efficiency of the model by means of competition between objects.

3.9   Selected Issues of Comparators' Learning

3.9.1   Learning of the Reference Set

One of the problems encountered during the formation of the comparator and development of the solution based on this methodology is the proper construction of the reference set. The reference set contains the selected objects.

In practice, specific instances of this object type are processed. In this paper, however, the terms *object instance* and *object* will be used interchangeably. The problem in question is reduced to the selection of objects characterized by attributes and attribute values which make the set of objects representative of the problem to be solved, taking into account the properties of the comparator (examined feature) as well as the specificity of the input object. Therefore, learning the reference set will be an automatic and *intelligent* method of designating a set, making it possible to represent the majority of objects processed.

Designation of a reference set can be done with a set of object instances and a fixed comparator. This method is based on clustering an instance set by means of a similarity function. The *leave-one-out* method (Sect. 2.4.3) is used to perform the learning procedure. This type of validation implies that a given input set is divided n-times into two subsets, while the first one is a one-element set. The learning process consists in the self-organization of objects in clusters. The clustering method is based on similarity calculated by a comparator. One object is selected from each cluster to represent a given cluster within the reference set. There are several known methods for selecting cluster delegates. Certain methods are based on centroids, others on maximizing or minimizing distances between individual clusters. Any of these methods can be used with comparators.

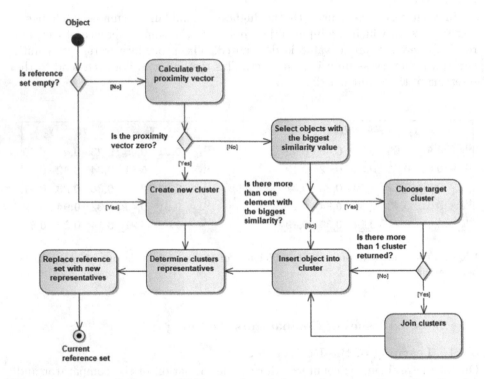

Fig. 22. Activity diagram in UML notation of the performance of one iteration in unsupervised learning of a set of reference objects

A single iteration of learning is presented in Fig. 22. The algorithm starts with an empty reference set. In this case, the first input object becomes automatically the one-element cluster and a representative of this cluster in the reference set. Each iteration of the algorithm determines the proximity vector relative to current representatives of clusters. If the returned vector is nonzero, with correspondingly high similarity values, then the input object is assigned to the cluster represented by the reference object for which the greatest similarity was obtained. If more than one reference object has the same and correspondingly high similarity value, one of the cluster selection methods described in literature is used [15]. The selected method can indicate more than one cluster simultaneously, which should result in merging clusters into one and selecting a new representative of the newly created cluster.

If the zero proximity vector is returned, the object creates a new cluster and automatically supplies it. The size of the cluster corresponds to parameters of comparators p, defining the minimum acceptable resemblance at the output of the respective comparators. Therefore, if fewer clusters are sought, with automatic increase of the tolerance for differences between objects within one cluster, the value of p should be decreased.

An analogous procedure is used for each successive element from the cross validation set, until objects are exhausted. It should be noted that too stringent quality criteria for similarity may result in the creation of separate clusters for each object. Conversely, if parameter values are set too low, all the objects might appear in one cluster.

3.9.2 Learning of the Threshold Function Parameter

Another example of learning by the compound object comparator is the optimization of the threshold parameter p. As defined in [102], the parameter p is the minimum acceptable value of similarity in the proximity vector of the comparator. In order to automatically designate it, a supervised method should be used, where learning is based on the training set. One can base this procedure on the re-sampling method (with k of at least 10). The data set is divided into two subsets. The first is called the learning set which consists of $\frac{1}{3}$ of all elements, and the second is a testing set with the quantity of $\frac{2}{3}$ of all elements. In the case of a single comparator, it is necessary to determine the limit value of similarity under which no correct solution is found. One of the available methods is the *simple local search*. The starting point of the algorithm is 1. The quality of the solution is determined by the function in the form:

$$f_{eval}(p) = \sum_{u \in U} (f_{recall}(x) - (1 - p)), \tag{145}$$

where

$$f_{recall} : X \rightarrow \{0, 1\} \tag{146}$$

is the function returning the value 1 in the case of obtaining a correct decision label from the training set for the pair of objects (x, y) or the value of 0.

The higher the value of function (145), the better the quality of the solution. Neighbouring solutions are generated by modifying the parameter values by a fixed constant, such as 0.01 (addition or subtraction). The stop condition can take the form of monitoring the repetitive solutions. When the limit value is exceeded, the quality of the solution will start to swing (increase then decrease repeatedly). Once this is identified, processing should be discontinued.

3.10   Selected Extensions of Comparators

3.10.1   Composite Comparator

A comparator made from other comparators, logically linked into one coherent element with compatible input and output, marked as com_∞^{ref}. From a logical point of view, it performs comparison operations divided into smaller steps. This type of comparator is very important because of the ability to model problems of any level of complexity. This comparator is implemented by means of the network described in the next chapter. Its task is to connect other comparators to make more general blocks. It facilitates the presentation of functionality in diagrams and understanding of the operation. Each composite comparator can be replaced with a set of simple comparators, which are additionally linked by certain relationships resulting from the structure of the network.

3.10.2   Classifying Comparator

A comparator with a defined class of objects, where, in the case of returning a non-zero proximity vector, the classification occurs. It is designated as com_{cls}^{ref}. It fulfils the role of a classifier that allows to build complex solutions on the basis of previous conclusions. In practice, this comparator is used mainly with the *max* filtering function. The interpretation of results is reduced to classifying the object with maximum similarity to the class represented by this comparator. It is therefore very important for the threshold function to be able to limit all incorrect solutions to avoid misclassification. Classifying comparators can be combined into compound comparators.

3.11   Summary

The theory outlined above systematizes the concept of a comparator as the smallest decision-making unit in such systems. This unit is responsible for local comparisons related to a feature represented by functions used in the comparator. The comparator in question results in the fact that the input object matches the reference objects to the maximum extent. It accomplishes this through the prism of the feature for which it had been created. It expresses the degree of similarity of the input object to the reference elements specified, however, in the context described by its component parts corresponding to a given aspect.

The comparator, along with the other elements described, may be an inherent decision-making system, but for solving relatively simple problems. An essential feature of a comparator is that it represents the function of similarity. Therefore, these solutions can be categorized into a group of systems called similarity based reasoning. Their main feature is the ability to model similarities between objects and their proper interpretation [84].

The comparator is to some extent analogous to a neuron. In neural networks it is also the smallest decision element, made up of smaller subassemblies. On the basis of a single neuron one can also create certain limited decision systems. Both elements involve machine learning. In the case of a neuron weights are always involved as far as learning is concerned. In the case of a comparator, this applies to the reference set as well as to the control parameter of the minimum acceptable value of similarity.

The comparator approach, however, has important features distinguishing it from others. First and foremost, it is the clarity and intuitiveness of the solutions built. This gives one the opportunity for a relatively easy modification of the solution.

From the point of view of practical implementation, it is also a concurrent approach guaranteeing adequate scalability [98]. Comparator-based methods are not optimal in many cases, but they are good enough for practical use and often better than other suboptimal methods. The comparator research has shown that a single comparator has considerable potential. However, even more can be obtained, if the power of many comparators is combined into a well-ordered, systematized and thoughtful way. This approach has developed into research on comparator networks [96].

4 Comparator Networks

4.1 Introductory Information

Comparator networks are logically related elements modelling the similarity of compound objects representing phenomena, processes or other entities. The structure of the network allows one to break down the task of analysing a compound object into pieces. It also allows its local processing, aggregating partial results and presenting results in the form of a proximity vector pointing to similarity of an input object and a kind of knowledge base, such as a reference set. In addition, the process of network construction can take into account the task of minimizing the number of reference sets and also utilize knowledge of the essence of the structure of objects examined. The first case models homogeneous networks and the second one - heterogeneous networks [100].

These networks are used in the modelling of recognition processes and identification of objects using similarity. These two processes are the main areas of application of artificial intelligence (AI). Comparator networks thus broaden the collection of available methods for modelling these complex issues, while offering a concise methodology for dealing with various types of problems.

4.2 Definitions of Components of Comparator Networks

The operation of a comparator network can be interpreted as a calculation of the function:

$$\mu_{net}^{ref_{out}} : X \to [0,1]^{ref_{out}}, \tag{147}$$

where the input object $x \in X$ is an argument and ref_{out} is a reference set for the network's output layer. The target set (codomain) of $\mu_{net}^{ref_{out}}$ is the space of proximity vectors. The proximity vector from the target space is v. Such a vector has values that are similarities of specific reference objects to input objects x.

By ordering the reference set, i.e. if $ref = \{y_1, \ldots, y_{|ref|}\}$ is used, the value of the network function is:

$$\mu_{net}^{ref}(x) = \langle SIM(x, y_1), \ldots, SIM(x, y_{|ref|}) \rangle, \tag{148}$$

where $SIM(x, y_i)$ is the value of *global similarity* established by the network for an input object x and a reference object y_i. Global similarity depends on partial (local) similarities calculated by the respective elements of the network (unit comparators). Through application of aggregation and translation procedures at subsequent layers of the network these local similarities are ultimately leading to the global one.

4.2.1   Layers in Comparator Network

Each comparator network is composed of three types of layers: input, intermediate (hidden/internal) and output. A given network may have several internal layers [100]. The said layers group comparators on the basis of a common purpose of processing a particular piece of information (attributes) about the object in question. Each layer contains a set of comparators working in parallel and a specific translating/aggregating mechanism. The translating and aggregating mechanisms are necessary to facilitate the flow of information (similarity vectors) between layers. As sets of comparators in a particular layer correspond to a specific combination of attributes, the output of the preceding layer has to be aggregated and translated to fit the requirements. This is done by elements called translators and aggregators, respectively. The translator converts comparator outputs to information about reference objects that can be used in the next layer. The role of the aggregator is to choose the most likely outputs of the translator, in case there is any non-uniqueness in assigning information about input objects to comparators. The operation of a layer in the comparator network can be presented as the following mapping:

$$\mu_{layer}^{ref} : X \to [0, 1]^{ref_l}, \tag{149}$$

where $x \in X$ is an input object and ref_l is the reference set for the layer.

Within a given layer only local reference sets associated with comparators in that layer are used to establish (local) similarities. However, through aggregation and translation these local similarities become the material for synthesis of output similarity and a reference set for the layer. This synthesis is based on a translation matrix, as described in [101]. Function (149) is created as a superposition of the comparator function (119), local (layer) aggregation function and translation. The local translation operation is responsible for filtering the locally aggregated results.

The input and internal (hidden) layers in the comparator network contain comparators with function (119) as well as translators and local aggregators. The output layer contains the global aggregator responsible for returning the final result. The components of the comparator network are briefly described below. Details are presented in [95, 101].

4.2.2 Local Aggregator

The local aggregator processes partial results of the network at the level of a given layer. The aggregator's operation depends on the type of reference objects and the output of comparators. It can be represented as:

$$f_{agg}^{ref_l} : [0,1]^{ref_1} \times \ldots \times [0,1]^{ref_k} \to [0,1]^{ref_l}, \tag{150}$$

where k is the number of comparators in a given layer l, i.e. the number of inputs in the aggregating unit (local aggregator). ref_l is the output (resulting) reference set for layer l composed by means of the *composition rules* from reference sets ref_i ($i = 1, \ldots, k$) used by comparators in layer l.

4.3 Translator

The translator is a network component associated with the adaptation of results of one layer to the context of another layer (the one to be fed with). In other words, this element expresses the results of the previous layer (their reference objects) in reference objects of the current one. It uses reference objects of the next layer, taking into account the relationships between objects of both layers [92]. The translator is defined by means of the translation matrix:

$$M_{ref_l}^{ref_k} = [m_{ij}], \tag{151}$$

where $i \in \{1, \ldots, m\}$, $j \in \{1, \ldots, n\}$ for m and n denotes the cardinality of ref_k and ref_l, respectively. The matrix $M_{ref_l}^{ref_k}$ defines the mapping of objects in the set ref_k onto objects in the set ref_l. In practice, ref_k is the total of reference sets for all comparators in a given layer and ref_l is the target reference set. Values in the matrix are within $[0,1]$.

Depending on the type of mutual relationship between both sets (generalization or decomposition), the matrix assumes different values. In the case of decomposition, there are many elements of set ref_l for a single element of set ref_k, which is emphasized by multiple 1 values in the line corresponding to ref_k. In the case of generalization, the values of the matrix are usually less than 1.

4.3.1 Projection Module

This network unit appears in selected layers whenever there is a need for selecting a subset of coordinates (project the vector onto subspace) in the proximity vector to be used in further calculations. Selection of a particular coordinate may be based on its value (above/below threshold) and/or on the limitations regarding the number of acceptable coordinates. For the i-th coordinate in the proximity vector the projection can be the following:

$$\mu_{proj}(v[i]) = \begin{cases} v[i] & projection(v[i]) = 1 \\ 0 & projection(v[i]) = 0 \end{cases} \tag{152}$$

where $i \in \{1, \ldots, |ref|\}$ and $projection(a)$ for $a \in [0,1]$ is a function in the following form:

$$projection : [0,1] \rightarrow \{0,1\}, \tag{153}$$

The function $projection$ is the actual selection mechanism. It determines whether or not a given coordinate is to be set at 0. This function can be defined as a threshold, maximum, ranking function, etc.

4.3.2 Global Aggregator

The global aggregator is a compulsory element of the output layer. Unlike local aggregators, which process results within a single layer, the global one may process values resulting from all layers at the same time. In the simplified, homogeneous case, when all layers use exactly the same reference set, the global aggregator may be expressed by:

$$\mu_{agg}^{ref_{out}} : \left([0,1]^{ref}\right)^m \rightarrow [0,1]^{ref_{out}}, \tag{154}$$

where m is the number of **all** comparators in the networks, i.e. the number of inputs to the global aggregator.

In the more complicated, heterogeneous case, the sets in subsequent layers and comparators may differ. In this case the aggregator constructs the resulting (global) reference set ref_{out} in a way that every element $y \in ref_{out}$ is decomposed into y_1 in the reference set ref_1, y_2 in the reference set ref_2 and so on, up to y_m in the reference set ref_m. For a given input object $x \in X$ the value of similarity between x and each element of $ref_1, \ldots, ref_m$ is known, as this is the output of the corresponding comparator. To obtain the aggregated result we use:

$$\mu_{agg}^{ref_{out}} : [0,1]^{|ref_1|} \times \ldots \times [0,1]^{|ref_m|} \rightarrow [0,1]^{ref_{out}} \tag{155}$$

It can be noted that formula (155) is similar to that of the local aggregator (150). The essential difference is that the local aggregator is limited to a subset of comparators in a given layer, while the global one takes into account all comparators of the network.

4.3.3 Summary

Taking into account all the definitions above, comparator networks can be expressed as a composition of mappings in subsequent layers:

$$\mu_{net}^{ref_{out}}(x) = \mu_{layer-out}^{ref_{out}}(\mu_{layer-int}^{ref_{k-1}} \cdots (\mu_{layer-in}^{ref_1}(x)) \ldots), \tag{156}$$

where ref_i means the reference set corresponding to layer i and ref_{out} is the reference set for the network as a whole. The general scheme of the comparator network is shown in Fig. 23.

4.4 Designing Comparator Networks

A comparator network consists of layers communicating with one another through the translator. The central element of the layer is the aggregator with which other elements are connected. Its task in each layer is to choose the best solution for further processing. It is a cohesive element, because only this element facilitates one-way communication between individual elements. Comparators (different types), projection modules and the translator are elements of the layer, too (see Fig. 23). Each comparator passes results to the aggregator. Comparators are not directly connected. Individual comparators examine the defined features independently. They carry out their calculations concurrently and return independent results. The purpose of the translator is to convert the result of the layer to a useful and acceptable form for the next layer or for the output of the network.

The obligatory elements of the layer are the aggregator and the translator. However, the comparator is the main computing unit, so the lack of this element in the layer causes the lack of decision. The output layer, which is used for global aggregation of network results, is the exception.

Graphical network modelling uses diagrams. Due to the wide availability of UML modelling tools, comparator networks use symbols of that tool, but with different meanings. Comparator networks are modelled, in particular, with symbols available in activity diagrams. Detailed descriptions of the comparator network elements and their meanings are presented below.

1. ● *Input* - the place where the input object x is converted into its representation created for each comparator in the input layer and also a starting point for processing. Information granule created around the input object [118] is the result of this process. It will be called a *signal granule* in this paper [92], due to its specific form. The input object on comparator network diagrams is symbolized by a large black dot. In modelling tools, this symbol represents the initiation of processing within the UML activity diagrams.

2. *Layer* - part of the network which groups processing, bound by a common context referring to an object or a pattern sought. Context is defined by domain knowledge related to the object as well as relationships with other objects and sub-objects, such as decomposition or generalization. The layer as a whole forms part of sequential processing of the network (individual layers are processed in specific order), whose constituent elements can be processed concurrently. This allows to explore contextually relevant features using different comparators. There are three types of layers: input, intermediate and output. Layers of the first and the last type are mandatory, while intermediate layers are optional. The number of intermediate layers may be greater than zero or equal to zero without the upper limit. Their number depends on the nature and complexity of the problem in question. On the diagrams, the layer is labelled as a rectangle with a label representing its name. In modelling tools, this symbol corresponds to a partition in UML activity diagrams.

3.   ⬭ *Comparator* - unit for comparing compound objects, described in detail in Sect. 3, mentioned here only for information purposes. There are several comparator extensions. Different types can occur together in one layer. The extension characteristics are shown in Sect. 3.10. Comparators are represented on the diagram by the oval, with the name inside. The name signifies the feature compared by a given comparator. In modelling tools, this symbol signifies a single activity in the UML activity diagram.

4.   *Aggregator* - mandatory element of each network layer responsible for the synthesis of results obtained by comparators. It is the main decision-making element of the network, both at the local level (in the input layer or the intermediate level) and globally (in the output layer as a global aggregator). There are many methods available to perform this synthesis. Selected methods will be described later in the paper. It is important, however, that this element can be learned. Network performance depends on the aggregation methods used in the majority of cases [95]. In the diagram, the aggregator is represented by a black wide horizontal or a vertical line. In modelling tools, this symbol is available for activity diagrams and represents the processing symbol fork or join.

5.   ◇ *Translator* - network component responsible for transforming the result of a given layer (proximity vector) into a reference set of the next layer or a final result. It is not just about changing the representation of an object or changing the vector into a set, but also changing the level of generality or detail of objects. Translator can convert the resulting object into more objects at a higher decomposition level or also assign a more general object based on the object passed. Classic examples are hierarchical relationships, e.g. in the administrative division of the country, with hierarchies, such as voivodeship, county, commune. A voivodeship consists of many counties, a county consists of many communes. There are additional assumptions here that the element of hierarchy is entirely contained in the object at its higher level. Consequently, it is easy to imagine expressing the object of voivodeship, e.g. Mazowieckie, as a collection of counties belonging to this voivodeship or as a collection of communes. If there are no assumptions about inclusions of the object, coefficients defining the level of object inclusion must be considered. The value of coefficients depends on the feature to which they relate, e.g., surface, capacity, etc. Every type of translation is modelled through the so-called *transition matrix* described in Sect. 4.2 and given in template (151). In the diagram, the translator is represented by a diamond, which in modelling tools signifies a decision in notation of an activity diagram.

6.   ⊗ *Projection module* - module selecting attributes of a proximity vector (its coordinates) which are taken into account at the subsequent stage of network processing. In the final projection, this module determines attributes returned as a result of processing. Projection is performed due to the adopted functions working on attribute values. These functions may be: a subset of

k best results, only the best results, a full set of results, etc. On network diagrams, the projection module is represented by a circle with the x sign in the middle. In modelling tools, this symbol means *end of flow* in UML activity diagrams.

7. ◉ *Output* - the point at which the processing result is obtained and the operation of the network ends. The final result is represented by a proximity vector which uniquely identifies the degree of similarity between individual reference objects and the input object. For each object of the reference set, any value in the range $[0, 1]$ is possible. The response of the network may be a zero vector, in particular. On network diagrams, output is represented by a double dot. In modelling tools, this symbol signifies the *end of processing* in UML diagrams.

4.5   Types of Networks

A homogeneous network is based on a monolithic nature of an input object. It uses the properties of objects processed without intruding into their internal structure or relationships between other objects. In this paper, this type of network will be called *type I*. It is characterized by a specific structure that processes relatively simple features in initial layers. The purpose of such a structure is to quickly and relatively inexpensively calculate a set of candidates for solutions, from which the final solution will be selected at a later stage. Consequently, the processing of complex features takes place only for selected elements. This type of network processes the signal granule on the input of each layer based on a monolithic input object. For that reason, the passing of results between layers is relatively simple, since the result set is a subset of the set existing in a given layer. The translation operation in this kind of network is limited to identicalness transformation [93].

From the point of view of performance analysis, type I of the network optimizes the reference set's size by processing the subsequent layers. This approach minimizes the number of comparisons, resulting in increased performance. On the other hand, reducing the size of the set increases the risk of eliminating the correct solution. From this perspective, it is a classic problem of balance between speed and efficiency and accuracy of the solution [91].

Due to the limitation of the reference set and the fact that the final solution is the proximity vector, this approach promotes solutions in the form of a sparse vector. In the proximity vector, coordinates responsible for eliminated objects have zero values.

Typical uses for this type of network include problems solved by means of the *hash method* described in [105]. This type of network is a specific implementation of this approach by means of compound object comparators.

4.6   Heterogeneous Network

The structure of this type of network is based on the internal structure of the object processed. Internal relationships, their values and properties are consid-

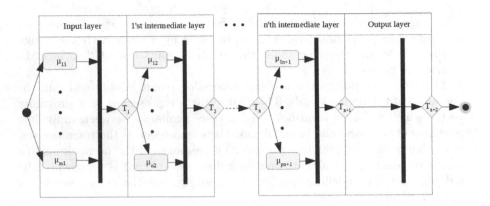

Fig. 23. General scheme of a comparator network in UML-like representation. Notation: com_{ji} - comparators, T_j - translators. Symbols: oval - comparator, thick vertical line - aggregator, rhombus - translator, encircled cross - projection module.

ered. In the further part of this paper, such networks will be called *type II networks* or *heterogeneous networks*.

Each object is described by means of ontology defined by concepts and relationships between them [120].

The object is an instance of the class it represents. For the purpose of this kind of network, attention is paid to hierarchies of concepts related to object description, which is defined by two specific relations: generalizations and decomposition. Generalization is the relation of being a sub-object of another object, while the decomposition relation is the relation of being a parent (superobject) of a set of sub-objects [103]. Depending on the position of concepts in the ontological hierarchy, different relationships may be used. Taking into account the hierarchy, with the concept represented by the input object on top, decomposition is the only possible operation. Otherwise, where the object under consideration is a representative of concepts within the hierarchy or at the bottom of hierarchy, it is possible to use both relationships or just generalization relationships, respectively.

When designing a network, the purpose of processing to be achieved is the important element. It affects the number of hierarchy levels related to the object processed that will be of interest. Each layer of the projected network corresponds to hierarchy levels in which the object is well-described and in which there is reference to the nature of the problem solved. In addition, comparators of a given layer refer to features defined at the same level of hierarchy, to which the entire layer refers. By using these relationships, the network penetrates the structure of the object or makes a generalization and processes objects at the higher level of hierarchy [99].

As described earlier, individual hierarchy levels of a compound object are treated as a context in the comparison process. The decomposition operation causes the appearance of different sub-objects in the selected context and the

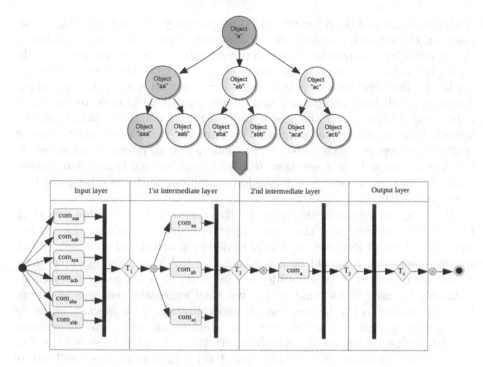

Fig. 24. Creation of a heterogeneous network implementing decomposition relationships based on the structure of a compound object - a general flowchart

generalization operation causes the appearance of superobjects, which by definition are more general and contain the processed object in the set of their sub-objects. These two types of relationships will be described in greater detail.

Generalization can be used, if the input object has a parent object, of which it is a part and which is described in the ontology used to describe the object processed. Considering the structure of the object shown in Fig. 24 and assuming that the input object is defined at *aa*, one can use the generalization relation to the object *a*. The purpose of such an operation may be, similarly to type I networks, to restrict the reference set only to objects whose parent is a superobject identified. Another reason for implementing this relationship is the lack of possibility to define similarity at the level defined by the input object and to move to a higher level in hierarchy to determine the approximate solution. This is precisely the case where there is no possibility of determining the proximity vector for a given input object (zero vector is returned). Then, by implementing the generalization relationships, one can determine similarity between the superobject containing the input object and the corresponding reference set. With a well-defined relationship between the input object and its superobject, this solution is a certain approximation for a given input object.

Decomposition is the implementation of the inverse generalization operation. In this case, one takes into account sub-objects forming part of the structure of the input object and the reference objects, respectively. By implementing decomposition, the network processes the decomposition tree bottom-up. This means that the first layers represent the lowest decomposition levels of the object. At the input, the network receives a compound object, which is processed by individual layers. However, each layer is dedicated to a certain level of decomposition. Therefore, each layer comparator decomposes the signal granule to the designated level. This results in the need to process a large number of object pairs for a single input object (signal granule). There are also many proximity vectors. A single vector represents the sub-object's similarity to all reference subobjects at the decomposition level in question. In the aggregation and translation process, proximity vectors are processed to the form compliant with the subsequent level of hierarchy [93].

By using an example analogous to the generalization relation based on Fig. 24, it is assumed that the input object is a. By determining the greatest similarity at the object level aa, partial similarity is calculated on a higher level. At the same time, it should be noted that the a object of the example above is also composed of other sub-objects. Therefore, it is necessary to calculate similarity for each of its components, and then aggregate and convert (translate) the results to the highest level of generalization. This mechanism works by calculating local similarities, and then synthesizing them into increasingly general results. By successive generalizations, there is a global resemblance to a given object. The main idea of this kind of network is to express similarity of the input object through similarity of its sub-objects. This is the direct analogy to behaviour in rough mereology [68], whose theoretical basis may be useful in constructing aggregators for this type of network.

4.7   Granular Processing of Information in Networks

The signal granule is defined in Sect. 3.4. It is responsible for the centralized storage of data processed by the network and its components, such as comparators, aggregators, translators and projection modules. The form and complexity of the granule depends on the type of network for which it is used and the reference set used. The simplest form of the granule relates to the type-I comparator network, because the reference set used in this case contains monolithic objects, for which neither the subobject nor the superobject relations are considered. Individual results of layers or comparators represent a subset of the main reference set (encoded as a proximity vector). In the case of type-II networks, the granule contains references to decomposed objects at different levels of ontological hierarchy as well as to aggregated objects. It also contains links of objects and values of similarities on individual levels of generalization of an object. Upon completion of processing in the case of both types of nets, the granule contains complete information generated during processing. By centralizing information, global aggregation can be done simply by using granular data. Figure 25 presents a diagram showing a model of interaction of the network with a signal granule during processing when the sample input object is x.

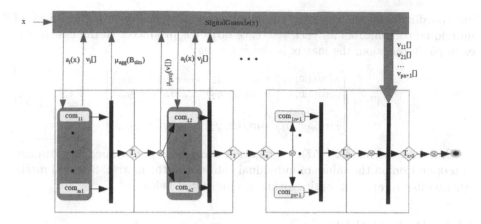

Fig. 25. Diagram of the flow of signals through a network of comparators, indicating the interaction and type of data exchanged. $A_i(x)$ - representation of a feature for an object x, $v_i[]$ - proximity vectors returned by a particular comparator, B_{sim} - matrix of decomposed objects constructed from proximity vectors of comparators, T_i - translators, $com_{ij}^{ref_i}$ - comparators

4.8 Aggregation of Partial Results

According to the definition of comparator networks and formula (147) a proximity vector appears at the output. The vector dimension is the size of the output reference set specified for the network. The problem of aggregation of partial results is the transformation of a matrix made from local proximity vectors into a vector representing the global values of similarity. This vector is generated by the synthesis of local vectors using the aggregation function according to formulas (154) or (155). This process optimizes the selection of elements of a reference set for the ranking based on values of similarity between objects.

A matrix consisting of local vectors has the dimension $m \times n$, where m is the number of comparators and n is the cardinality of the reference set considered to be the sum of local subsets. The set is built by adding elements of further local reference sets not included in the previously added sets. In the case of a local aggregator, values for all comparators in a given layer are taken into account. In the case of a global aggregator, values of all comparators of the entire network are available, but may be taken into account for any of them, depending on the intention of the network designer. It is assumed that there are n comparators in a given layer. For a fixed reference set ref for a layer or the whole net, each comparator returns the proximity vector of the size m. Then the matrix from each of the proximity vectors derived from comparators $com_1^{ref_1}, com_2^{ref_2}, \ldots, com_n^{ref_n}$ is constructed in a way that each subsequent vector is a row of the matrix A_m^n. The values of a given vector are placed in positions corresponding to reference objects. The remaining positions in the matrix line are added with zero values. Individual elements of the matrix correspond to the values of the similarity function of a given comparator and the reference object specified, i.e. individual coordinates of the calculated proximity vector. Consequently, an image of individual comparators for each reference object is produced. In the case of the example in question, the matrix is the following:

$$A_m^n = \begin{bmatrix} sim_1(x,y_1) & sim_1(x,y_2) & \dots & sim_1(x,y_m) \\ sim_2(x,y_1) & sim_2(x,y_2) & \dots & sim_2(x,y_m) \\ \dots & \dots & \dots & \dots \\ sim_n(x,y_1) & sim_n(x,y_2) & \dots & sim_n(x,y_m) \end{bmatrix} \qquad (157)$$

Based on the matrix (157), aggregation consists in performing an optimization operation on the values of individual columns of the matrix. Selected methods possible to apply in optimization are presented below.

4.8.1 Basic Methods

The use of averaging methods is an intuitive approach to aggregation of local results. The respective matrix columns (157) represent the similarity of the input object to a particular reference object. Individual values in lines represent different features. For the feature in question, one value is sought, which represents the general similarity, taking into account each feature processed. One of the approaches is to estimate the value of similarity by the mean value. Certain aggregation functions that can be used in both Type-I (homogeneous) and Type-II (heterogeneous) networks are presented below.

The default aggregation method for comparator networks is the arithmetic mean. It is based on calculating the average mean value of similarities of individual reference objects, taking into account values derived from different comparators (different features). These are the operations of averaging the values from columns of the matrix, which can take the following form:

$$\mu_{agg-mean}^{ref}(A_m^n) = \langle \frac{\sum_{i=1}^{n} sim_i(x,y_1)}{n}, \dots, \frac{\sum_{i=1}^{n} sim_i(x,y_m)}{n} \rangle, \qquad (158)$$

where $\langle \dots, \dots, \dots \rangle$ is a vector and sim_i is the similarity function of the i'th comparator. A slightly modified method is weight-based aggregation indicating the features (comparators) that are more important than others. In this case, the average value from individual columns of the matrix is also calculated, but this time the values are multiplied by weight. The general formula of this kind of aggregation is as follows:

$$\mu_{agg-wa}^{ref}(A_m^n) = \langle \frac{\sum_{i=1}^{n} w_i sim_i(x,y_1)}{\sum_{i=1}^{n} w_i}, \dots, \frac{\sum_{i=1}^{n} w_i sim_i(x,y_m)}{\sum_{i=1}^{n} w_i} \rangle \qquad (159)$$

This type of aggregation is very effective, provided one has a well-balanced weight. It makes it possible to learn weights, which increases the attractiveness of this method [95].

Another variant is the average harmonic. Similarly, it is used to count values from elements of the matrix column (157). The formula describing this type of aggregation is as follows:

$$\mu^{ref}_{agg-hm}(A^n_m) = \langle \frac{n}{\sum_{i=1}^{n} \frac{1}{sim_i(x,y_1)}}, \ldots, \frac{n}{\sum_{i=1}^{n} \frac{1}{sim_i(x,y_m)}} \rangle \tag{160}$$

By adding weights to individual components of the sample, a variant is obtained by means of the weighted harmonic mean. It may be presented in the following form:

$$\mu^{ref}_{agg-hw}(A^n_m) = \langle \frac{\sum_{i=1}^{n} w_i}{\sum_{i=1}^{n} \frac{w_i}{sim_i(x,y_1)}}, \ldots, \frac{\sum_{i=1}^{n} w_i}{\sum_{i=1}^{n} \frac{w_i}{sim_i(x,y_m)}} \rangle \tag{161}$$

In this case, the problem of determining weights w_i is also present, which at the same time is an additional advantage of this method due to the possibility of learning by means of known methods of other fields, e.g. using genetic algorithms [26].

The last examples include the geometric mean and the weighted variant of this function. Formulas for these aggregate function cases are as follows:

$$\mu^{ref}_{agg-gm}(A^n_m) = \langle \sqrt[n]{\prod_{i=1}^{n} sim_i(x,y_1)}, \ldots, \sqrt[n]{\prod_{i=1}^{n} sim_i(x,y_m)} \rangle \tag{162}$$

for geometric mean and:

$$\mu^{ref}_{agg-gw}(A^n_m) = \langle (\prod_{i=1}^{n} sim_i(x,y_1)^{w_i})^p, \ldots, (\prod_{i=1}^{n} sim_i(x,y_m)^{w_i})^p \rangle \tag{163}$$

for weighted geometric mean, where $p = \frac{1}{\sum_{i=1}^{n} w_i}$.

There are many other examples of aggregation functions. As shown in the above examples, the method of operation is in every case analogous. The starting point is the partial result matrix (157), on the columns of which the averaging operations are performed. Although these methods are relatively simple, results obtained by means of these methods are often satisfactory. Their undisputed advantage is the low computational complexity and the lack of sophistication of methods of calculation. This makes it possible to quickly and efficiently prepare the results for the purpose of further calculations.

4.8.2   Method Based on Voting

Aggregation of local results involves the synthesis of data and does not necessarily rely on typical aggregation functions. The optimization method derived from a completely different field, i.e. the electoral algorithm, is an example. The method is based on voting algorithms that lead to election rankings. These methods aim at correct mapping of voters' preferences, e.g. the selection of candidates supported by as many voters as possible. Some of these methods fulfil the Condorcet criterion [22], which guarantees that the chosen candidate has more support than any other. Consequently, these methods, although used for a completely different field, are all about optimizing a function representing voters' opinions. This method involves two basic concepts: the candidate and the voter. The first concept represents the object subject to voting and included in the potential solution. The voter is the object voting for a candidate. The voting methods may vary, depending on how one votes. In certain cases the voter casts only one vote, in other cases more votes. There are also methods of counting votes. However, each case results in the ranking of the candidates and the place in the ranking, according to the adopted voting rules (electoral law). The electoral law also determines the method of nominating winners, e.g. how votes are translated into decisions about the choice of a candidate.

In the case of application of these methods to optimization problems the basic concepts used by these methods will be applied as well. Roles will be assigned to candidates and voters. In comparator networks the problem of optimization relates to aggregators. In the case of these network components it is possible to use election algorithms as a method of selecting solutions. As in the previous types of aggregators, once objects are processed by comparators, proximity vectors are obtained. Selection has to result in receiving one proximity vector. Candidates are represented by reference objects. Solutions are based on them and provide basis for proximity vectors. In addition, each vector has values defining similarity. In the case of electoral algorithms this value is interpreted as the power of votes for candidates. Comparators are equated with voters. It's the comparators that cast votes for candidates, allocating a corresponding level to each referenced object. This results in defining two basic concepts necessary for the election. The next section will describe the adaptation of the selected election algorithms used in the context of comparator networks.

Majority Voting
Majority voting is a very simple method. It is easily implemented in comparator networks. Voting for a candidate is accomplished through interpretations of similarities in the proximity vector. For the proximity vector in the form:

$$\mu_{com}^{ref}(x, Y) = \langle sim(x, y_1), \ldots, sim(x, y_{|ref|}) \rangle, \tag{164}$$

which originates from a given comparator, there are similarity values connected with the reference objects on the respective coordinates. There are as many vectors as comparators at the input. Each proximity vector is treated as a vote cast. In this type of voting, each voter can vote only once. It is assumed that the vote of a given comparator is the one with the highest similarity value in the proximity vector. The function of voting for j'th comparator can be presented as follows:

$$vote^j(y_k) = \begin{cases} 1 & : k \in argmax_{i=1}^{|ref|} v[i], i, k \in I, \\ 0 & w\,p.p. \end{cases} \tag{165}$$

where I is the set of indexes of coordinates of proximity vector v, and j is the number of a comparator. However, the problem of a vector with many identical values, and in particular the value of 1 for each coordinate, remains. This is the first exception to the original method of voting. However, such solutions are allowed because the value of 1 is the value of insensitivity to a feature, so it highly desirable. Each comparator votes analogically. Finally, the ranking of reference objects is based on the total number of votes. The more votes, the higher the ranking. The vote rating function can be presented in the following form:

$$score(y_i) = \sum_{j=1}^{n} vote^j(y_i),$$
(166)

where n is the number of comparators. The result is an object with the highest number of votes calculated according to the formula (166). In the case of a comparator network, a result consisting of more than one winner is allowed. This is the second exception to the original voting method, but in this case this is a desired property. The winners of voting receive 1 as a value of coordinate of a proximity vector, and the rest receive 0 value.

This method has a linear computational complexity. From this point of view it is very useful. Its optimization advantages are limited because preferences of voters not voting for the winner are not taken into account at all.

Borda Count

In the Borda Count method, a ranking for reference objects is constructed from each proximity vector. The ranking is based on points awarded to each candidate and its value of similarity. Points are awarded according to the following rules:

$$vote^j(y_i) = |ref| + 1 - RankPos(y_i),$$
(167)

where $|ref|$ is the number of all reference objects, j is the index of a comparator, $RankPos$ is the function returning the position in ranking resulting from sorting objects in descending order in relation to similarity. Objects with the same value of similarity receive the same position (ex-aequo).

Each comparator gives a point-based ranking. For each reference object, point values from the respective rankings are added according to the formula:

$$score(y_i) = \sum_{j=1}^{n} vote^j(y_i),$$
(168)

where $vote^j(y_i)$ means the point value given to the object y_i by the n-th comparator. Consequently, a final ranking is obtained. It provides basis for the determination of the final proximity vector. Vector values can be determined by means of the following functions:

$$v[i] = \frac{score(y_i)}{n * |ref|} : |ref| > 0, n > 0$$
(169)

This method is characterized by very good computational properties. It has linear computational complexity. The method of awarding points for the position in the ranking is an important element. In this case, a linear model was suggested, although non-linear models favouring the higher places in the ranking are used more frequently.

Copeland's Method

This algorithm is a tournament method that deals with the individual preferences of voters. It can be used only when there is more than one candidate. Candidates are reference objects, so attention will be paid only to cases when the reference set fulfills the condition: $|ref| > 1$. Comparators of the layer or the entire network are voters (depending on the type of aggregator). For each comparator, a scoreboard is created showing votes for each candidate. In order to calculate points for individual candidates, the Cartesian product of the set of reference objects $ref \times ref$ is generated. Each of the resulting pairs is assigned the following points: $+1$ for victory, -1 for defeat, 0 for tie. The result for a single pair is determined by the value of the object's similarity to the input object x, according to the following rule:

$$
result(y_i, y_j) = \begin{cases} 1 & : sim(x, y_i) > sim(x, y_j) \\ 0 & : sim(x, y_i) = sim(x, y_j) \\ -1 & : sim(x, y_i) < sim(x, y_j) \end{cases} \tag{170}
$$

wherein points are awarded to both objects respectively in the position of victory or defeat. Consequently, a scoreboard is created with the ranking of candidates. The global value of points for the object y_i to determine the rank of the k'th comparator is made by adding the points for victory and the points for defeat (which are negative) and can be expressed as:

$$
vote^k(y_i) = \sum_{j=1}^{|ref|} result(y_i, y_j) \tag{171}
$$

Result tables for each comparator are generated analogously. In order to create a global ranking that constitutes the result of the election, points of individual candidates are summed up. The score of a single candidate can be calculated on the basis of the following formula:

$$
score_{global}(y_i) = \sum_{j=1}^{n} vote^j(y_i) \tag{172}
$$

The proximity vector values returned by the aggregator are determined in the following way:

$$
\langle \frac{n(|ref| - 1) + score_{global}(y_1)}{2n(|ref| - 1)}, \dots,
$$
$$
\frac{n(|ref| - 1) + score_{global}(y_{|ref|})}{2n(|ref| - 1)} \rangle : |ref| > 1, n > 0, \tag{173}
$$

where n is the number of comparators, and ref is the reference set with at least two elements.

This method has a square computational complexity. It does not require any additional parameters.

Approval Voting

In this method, comparators fulfil the role of voters. They vote for each candidate in the form of reference objects. The method assumes that each voter casts as many votes as there are candidates. The votes are of two kinds: in the form of acceptance or disapproval. This method requires slightly greater adaptation so that it can be successfully used to aggregate partial results in comparator networks. As in other methods, the main determinant of the vote is the value of similarity stored in the proximity vector. However, in this case, each reference object must be accepted or rejected. Therefore, the limit value of the similarity was introduced, below which a vote was treated as disapproval and above - as acceptance. This value, called the acceptance factor, is marked as K_{af}. The initial value of this parameter is 1. Using the proximity vector of a given comparator, the vote for each candidate is performed by means of the following function:

$$vote^j(y_i) = \begin{cases} 1 & : sim(x, y_i) \geq K_{af} \\ 0 & : w\,p.p. \end{cases} \tag{174}$$

where 1 is interpreted as acceptance and 0 as disapproval. Voting is performed for each comparator, using its proximity vector. Then votes cast by individual comparators are summed up. The total number of votes cast for a candidate is measured by means of the following formula:

$$score(y_i) = \sum_{j=1}^{n} vote^j(y_i), \tag{175}$$

where n means the number of comparators. Due to the fact that the initial value of the factor $K_{af} = 1$, it may be that no candidate has *majority*. Majority means that the number of votes of acceptance is greater than half of all votes. Therefore, this method was adopted in an iterative way. The first iteration is executed for the start value of the acceptance factor. If none of the candidates reaches majority, the value of the factor is reduced by 0.01. This operation is repeated until one of the candidates reaches majority or the factor reaches the lowest possible value. The final value of the proximity vector is determined by means of the following dependencies:

$$v = \langle h(y_1), \ldots, h(y_{|ref|}) \rangle, \tag{176}$$

where $h(y)$ is in the form:

$$h(y_i) = \begin{cases} \frac{score(y_i)}{n} & : score(y_i) > \frac{n}{2} \\ 0 & : w\,p.p. \end{cases} \tag{177}$$

The general computational complexity of this algorithm is linear. In the case of a comparator network, however, there are additional operations that have a significant effect on time efficiency of the algorithm.

Weighted Voting

This type of voting operates on weights determining the strength of a given vote. As in previous cases, votes are cast by comparators constituting voters. Candidates are the reference objects and the vote is identified by the value of similarity in the proximity vector. The output vector values can be determined in the following way:

$$v[i] = \frac{\sum_{j=1}^{n} w_j sim^j(x, y_i)}{\sum_{j=1}^{n} w_j} : \sum_{j=1}^{n} w_j > 0, w_j \geq 0, \tag{178}$$

where n is the number of comparators. This method has been chosen for its extraordinary simplicity and the very important possibility of application of machine-learning for weights. Weights ensure strength and effectiveness of this method.

4.8.3   Structured Methods Inspired by Rough Mereology

In heterogeneous networks, knowledge of the structure of an object and its relationships is needed in the majority of cases in order to properly aggregate results. Relationships in question relate to being part of one object in another, searching for common or similar parts between objects as well as analysing components. Such topics fall within the scope of mereology. Its extension operating on degrees of fulfilment of the relation is called rough mereology [68]. Appropriate inclusion functions can be good sources of similarity or composition function [27]. In this context, the said role is fulfilled by the aggregator. In other words, it seeks answers to the question of how to infer the similarity of the whole object from the collected information about subobjects and their similarity.

For heterogeneous networks, the aggregation operation is strictly related to the translation operation. In practice, one can be done only with the other. This is because the translation matrix expressed by the formula (151) contains the so-called *composition rules*. These rules define the way of expressing an object of a further layer by objects of the previous layer of a given network. In type-II networks, the context of the layer relates to the decomposition level of the object processed. Therefore, the composition rule allows to some extent the calculation of the value of similarity of super-objects using similarity of subobjects. Composition rules are as follows:

$$y_i = w_1 y_{i1} + w_2 y_{i2} + \ldots + w_k y_{ik}, \tag{179}$$

where y_i is an object on the higher level of generality, y_{ij} are objects of the layer currently processed, where $j = 1, \ldots, k$, and $i = 1, \ldots, l$ for certain fixed k and l meaning respectively the number of subobjects and the number of main objects, and wi are the rows of the translation matrix (151). For the purposes of the generalization relation, the sum of weights of a given row in the translation matrix must fulfil the condition:

$$\sum_{i \in I} w_i \leq 1, \tag{180}$$

where I is a set of indexes in the row of the translation matrix, which is the size of k. In the case of the sum equal to 1, the transition to a higher level of aggregation is complete. If the sum is less than 1, similarities of subobjects do not fully determine the similarity at a higher level of generality. Then, the sum of weights determines the degree of completeness of the translation. In the contrary case, when the decomposition relation is implemented, the condition is obviously not fulfilled.

The full process of structural aggregation is quite complex. It depends on three main parameters: structure of the object, comparators from which the results are aggregated and reference sets used by comparators. It is generally accepted that each reference set in a type-II network is a subset of the general reference set ref. In other words, there may be many sets processed, labelled as $ref_1, ref_2, \ldots, ref_m$, but they are all subsets of the general set $ref_i \subseteq ref$.

Proximity vectors derived from comparators constitute input of structural aggregation. Additional difficulty is that they produce as many proximity vectors as there are subobjects decomposed from the input object in a given layer. There are $d * m$ proximity vectors, where d - denotes the number of decomposed subobjects and m is the number of comparators. In addition, individual vectors with different reference subsets may have different dimensions. However, when analysing them from the comparators point of view, one can build a matrix in the following form:

$$B_{sim} = \begin{bmatrix} v_1 \\ \ldots \\ v_d \end{bmatrix} \tag{181}$$

where v_j are the proximity vectors of particular decomposed subobjects x_j, resulting from the calculation of similarities with the relevant reference objects, where $j \in \{1, \ldots, d\}$, in the form:

$$v_j = \langle sim(x_j, y_{11}), \ldots, sim(x_j, y_{1n}) \rangle \tag{182}$$

The aggregation process is divided into two stages. The first one involves physical synthesis, i.e. combination of results on the level of the proximity vector. This process can be performed, if more than one comparator uses the same reference subset, e.g. comparison is based on the same object, but different features. In this case, aggregation methods for type-I networks are used, such as averaging or voting. Proximity vectors $d * k$ are created as a results of this operation, where k means the number of distinctive reference sets used in the layer and d is the same as in the previous case, but $k \leq m$.

The second stage involves the translation of objects and the assembly of similarities using composition rules. Consequently, similarities of the proximity vector are calculated on the higher level. In composition rules, higher-level reference objects are substituted and as many equations as reference objects are created at a higher level of generality. When calculating a single value of a vector, a given reference object is substituted with the rule given in the formula (179) and lower-level results stored in the matrix defined by the formula (181) are used.

Since there are similarities for different subobjects on the lower level, particular similarities connected with it by the S-norm [36] are given. Therefore, a single formula calculating the value of the proximity vector at the higher level will take the form:

$$sim(x, y_i) = S(w_1 sim(x_j, y_{i1}) + \ldots + w_k sim(x_j, y_{ik}), \ldots,$$
$$w_1 sim(x_l, y_{i1}) + \ldots + w_k sim(x_l, y_{ik})), \tag{183}$$

where all combinations of reference and input subobjects composed according to the appropriate composition rule given in the transition matrix - formula (151) are in the S-norm.

4.9 Example

Let us assume that there is an input object x_1 divided into two subobjects x_{11}, x_{12} and four reference objects y_1, y_2, y_3, y_4, which are divided analogically into subobjects: $y_{11}, y_{12}, y_{21}, y_{22}, y_{31}, y_{32}, y_{41}, y_{42}$. Let us also assume that the network architecture is composed of two comparators. There are also two reference subsets $ref_1 = \{y_{11}, y_{21}, y_{31}, y_{41}\}$ and $ref_2 = \{y_{12}, y_{22}, y_{32}, y_{42}\}$ supported by comparators com^{ref_1} and com^{ref_2}. There is also a transition matrix given in the form:

$$M_{ref}^{ref_1 \cup ref_2} = \begin{bmatrix} \frac{1}{2} & \frac{1}{2} & 0 & 0 & 0 & 0 & 0 & 0 \\ 0 & 0 & \frac{1}{2} & \frac{1}{2} & 0 & 0 & 0 & 0 \\ 0 & 0 & 0 & 0 & \frac{1}{2} & \frac{1}{2} & 0 & 0 \\ 0 & 0 & 0 & 0 & 0 & 0 & \frac{1}{2} & \frac{1}{2} \end{bmatrix} \tag{184}$$

This matrix defines the composition rules in the form:

$$y_1 = \frac{1}{2} y_{11} + \frac{1}{2} y_{12} \tag{185}$$

$$y_2 = \frac{1}{2} y_{21} + \frac{1}{2} y_{22} \tag{186}$$

$$y_3 = \frac{1}{2} y_{31} + \frac{1}{2} y_{32} \tag{187}$$

$$y_4 = \frac{1}{2} y_{41} + \frac{1}{2} y_{42} \tag{188}$$

The first layer of the network deals with the processing of subobjects, and therefore its task is reduced to the recognition of objects x_{11}, x_{12}, x_{13} by the subobject of the reference set ref. As a result of processing in this layer, two similar matrices are created, denoted as $B_{sim}^{\mu_{com1}^{ref_1}}, B_{sim}^{\mu_{com2}^{ref_2}}$. Each of sim them is made from two four-dimensional proximity vectors. Each of them represents similarity between the respective input subobjects and the reference subobjects.

$$B_{sim}^{\mu_{com_1}^{ref_1}} = \begin{bmatrix} sim(x_{11}, y_{11}), sim(x_{11}, y_{21}), sim(x_{11}, y_{31}), sim(x_{11}, y_{41}) \\ sim(x_{12}, y_{11}), sim(x_{12}, y_{21}), sim(x_{12}, y_{31}), sim(x_{12}, y_{41}) \end{bmatrix} \quad (189)$$

and

$$B_{sim}^{\mu_{com_2}^{ref_2}} = \begin{bmatrix} sim(x_{11}, y_{12}), sim(x_{11}, y_{22}), sim(x_{11}, y_{32}), sim(x_{11}, y_{42}) \\ sim(x_{12}, y_{12}), sim(x_{12}, y_{22}), sim(x_{12}, y_{32}), sim(x_{12}, y_{42}) \end{bmatrix} \quad (190)$$

Next, the second stage of aggregation begins. The purpose of this action is to calculate the proximity vector in the form:

$$v = \langle sim(x_1, y_1), sim(x_1, y_2), sim(x_1, y_3), sim(x_1, y_4) \rangle \quad (191)$$

To simplify the calculations only the first coordinate calculation will be shown. The other coordinates are calculated analogously. The calculation $sim(x, y_1)$ is based on the composition rule in the form:

$$y_1 = \frac{1}{2} y_{11} + \frac{1}{2} y_{12} \quad (192)$$

By substituting these rules and using the first column of matrices (189) and (190), we get:

$$sim(x_1, y_1) = S(\frac{1}{2} sim(x_{11}, y_{11}) + \frac{1}{2} sim(x_{11}, y_{12}),$$
$$\frac{1}{2} sim(x_{11}, y_{11}) + \frac{1}{2} sim(x_{12}, y_{12}),$$
$$\frac{1}{2} sim(x_{12}, y_{11}) + \frac{1}{2} sim(x_{11}, y_{12}), \quad (193)$$
$$\frac{1}{2} sim(x_{12}, y_{11}) + \frac{1}{2} sim(x_{12}, y_{12})),$$

where S is the S-norm according to the formula (12).

4.10   Fuzzy Interpretation of Heterogeneous Networks

Different comparators can refer to different types and levels of reference (sub)objects in the network, using different attributes and parameters. Thus, the first task is to extract for a given $x \in X$ its structural representation, i.e. all its parts and their corresponding attribute values. Moreover, it is not always obvious which parts of x should be compared to particular reference sets. In such cases, a single x can yield multiple possible combinations of assignments of its parts to particular comparators. All such alternative representations, denoted as x', should be processed through the first layers and, later, the most probable assignments of the x's parts to particular categories of reference objects can be derived. One can think about collections of possible representations of x' as information granules $g(x)$ created around input objects $x \in X$ [119].

Inputs to each layer are determined by values of attributes for $x \in X$ or its subobjects. However, subsets of reference (sub)objects, which x is going to be compared to, are induced dynamically by comparators in the previous layers. In the simplest scenario, comparators in preliminary layers aim at reducing subsets of potentially comparable reference objects using relatively easily-computable attributes, leaving more complex calculations to further layers, where the number of reference items to be compared is already decreased. In other cases, initial layers work with attributes specified for subobjects, producing vectors of similarities that need translation to the level of similarities between more compound objects, whose attributes are analysed later. However, the complexity does not need to increase with consecutive layers. In some applications, the first layers can work with relatively basic attributes of compound objects, whose similarities are then translated to lower structural levels for detailed processing.

Types of reference objects can vary from layer to layer or even within a single layer. Comparators in a given layer usually refer to entities at the same level of ontology-based hierarchy of objects in question. However, a given hierarchy level can include multiple types of entities. Let us denote by $\mu_{net}^k(x)$ an outcome of the k-th layer for input object $x \in X$, after applying the above-mentioned operations of translation and aggregation. Let us also denote by $ref_1^{k+1}, \ldots, ref_{m(k+1)}^{k+1}$ reference sets used by comparators in the $(k+1)$-th layer.

The goal in this section is to specify the function $\mu_{net}^k : X \to [0,1]^{ref_1^{k+1}} \times \ldots \times [0,1]^{ref_{m(k+1)}^{k+1}}$, which takes into account similarity vectors obtained from comparators in the k-th layer. Once we obtain $\mu_{net}^k(x)$, we can forward it as a signal granule and prepare subsets $Y(x)_1^{k+1} \subseteq ref_1^{k+1}, \ldots, Y(x)_{m(k+1)}^{k+1} \subseteq ref_{m(k+1)}^{k+1}$ to be utilized by the subsequent comparators. Those two types of granules - the above signal granule and the previously-mentioned information granule $g(x)$ - illustrate a twofold way of dealing with information about objects throughout networks of comparators.

The central part of μ_{net}^k is matrix M_{net}^k with dimensions $|ref_1^k|+\ldots+|ref_{m(k)}^k|$ and $|ref_1^{k+1}| + \ldots + |ref_{m(k+1)}^{k+1}|$, which links the k-th and the $(k+1)$-th layers of the network. In its simplest implementation, it is a sparse Boolean matrix encoding these combinations of reference (sub)objects in sets $ref_1^k, \ldots, ref_{m(k)}^k$ and $ref_1^{k+1}, \ldots, ref_{m(k+1)}^{k+1}$, which structurally correspond to one another. Matrices are created during the process of defining reference sets, whose elements are decomposed due to their ontology-based specifications. Connections can also be additionally weighted with degrees, expressing e.g. to what extent particular subobjects should influence similarities between their parents.

Translation can be executed as a product of M_{net}^k with concatenated vectors of similarities obtained as outputs of comparators $com_1^k, \ldots, com_{m(k)}^k$ in the k-th layer, for each of the possible representations of x gathered in the information granule $g(x)$. Let us enumerate all such representations as $x'_1, \ldots, x'_{|g(x)|}$ and denote by $G_{net}^k(x)$ the matrix of all possible output combinations, that is:

$$
G_{net}^k(x) = \begin{bmatrix}
\mu_{com_1^k}(x_1') \, [1] & \cdots & \mu_{com_1^k}(x_{|g(x)|}') \, [1] \\
\vdots & \ddots & \vdots \\
\mu_{com_1^k}(x_1') \, [|ref_1^k|] & \cdots & \mu_{com_1^k}(x_{|g(x)|}') \, [|ref_1^k|] \\
\mu_{com_2^k}(x_1') \, [1] & \cdots & \mu_{com_2^k}(x_{|g(x)|}') \, [1] \\
\vdots & \ddots & \vdots \\
\mu_{com_{m(k)}^k}(x_1') \, \left[|ref_{m(k)}^k|\right] & \cdots & \mu_{com_{m(k)}^k}(x_{|g(x)|}') \, \left[|ref_{m(k)}^k|\right]
\end{bmatrix} \tag{194}
$$

The mechanism for computing $\mu_{net}^k(x)$ can be as follows:

$$
\mu_{net}^k(x)[i] = \max_j \min\left((M_{net}^k G_{net}^k(x))[i][j], 1 \right) \tag{195}
$$

where $[i][j]$ denotes coordinates of matrix $M_{net}^k G_{net}^k(x)$. Surely, specification of required operations in terms of matrices and vectors facilitates efficient implementation. On the other hand, it is demonstrated below that these calculations can indeed be interpreted by means of the well-known T-norms and S-norms.

Firstly, for a given $x_j' \in g(x)$, column $(M_{net}^k G_{net}^k(x))[j]$ represents possible similarities of x to reference objects in the $(k+1)$-th layer. Each of these similarities is computed as a sum of similarities between components of x (distributed among comparators according to combination x_j') and reference objects in the k-th layer. If it exceeds 1, then of course we cut it down. Thus, similarities between objects at the $(k+1)$-th layer are computed by means of the Łukasiewicz's T-norm of similarities between the corresponding objects at the k-th layer.

Secondly, in order to finally assess similarity of x to a given reference object in the $(k+1)$-th layer, we look at all combinations in $g(x)$ and choose the maximum possible score. Thus, we follow Zadeh's S-norm. Intuitively, our usage of the T-norm corresponds to taking a conjunction of component similarities in order to judge similarity between compound objects, while our usage of the S-norm reflects a disjunction of all alternative ways of obtaining that similarity. From this perspective, our current implementation reflects one of the possible specifications and other settings of the T-norm and the S-norm could be considered as well.

4.11 Selected Issues of Learning of Comparators Network

Comparator networks are characterized by the ability to learn selected parameters, thereby increasing their efficiency by adapting to a particular problem, while maintaining generality related to data. Two basic groups of learning algorithms are taken into account: supervised and unsupervised. The first group requires learning and test data. By means of this group of methods the network can learn the values of the p parameters for the respective comparators, so that their value is as high as possible, but does not eliminate the correct solution. On the other hand, with the increase of the threshold function parameter (formula 123), the number of comparisons to be calculated is reduced.

Another problem that can be optimized this way is learning the network structure. This procedure can be performed if one has information about the possible uses of comparators and defined parameters in the form of aggregators and a reference set. The selection of aggregation functions is a similar problem. In this case, slightly different parameters are set, and suboptimal aggregation functions appearing in the network layers are sought after.

The last element of learning by means of the supervised methods is the determination of comparator weights in selected aggregators. This issue has been described quite extensively in the previous publication [95].

As for self-organizing methods, the construction of a reference set, with a fixed network structure and a set of input data, is an example in comparator networks.

Two most interesting problems from those mentioned above have been selected for further discussion. The second problem is the extension of comparator learning described in Sect. 3.

4.11.1   Issues of Learning of the Homogeneous Network Structure

This section contains one of the two key issues related to the efficiency of the network of comparators, i.e. selecting features to be utilized by the network and the network structure. This issue relates directly to the concept of the object, its features, relationships and dependencies. Objects are defined by the description in ontology [103], which operates on concepts and relationships between them. It specifies a set of features that can be used for network construction. The optimal selection of features for a given problem is therefore the problem of looking for minimal subsets of attributes, which can uniquely identify the objects or their classes. Such minimal subsets of attributes can be referred to as reducts in the rough set theory [64].

The process of selecting a network structure can be transformed into the problem of determining the significance of features and in particular the possibility of their elimination. It was implemented by means of evolutionary algorithms [20]. The learning procedure assumes the existence of a set of features for the processed objects (stored in ontology). The task of the procedure is to select the minimum number of features and to indicate the location of comparators in layers, ensuring the correct operation of the network with the established types of local aggregators, translators and the number of layers.

Individuals represent various configurations of the network structure, e.g. the comparator and its feature assigned to a particular layer. A single chromosome is in the form: $x = \langle (i, k) \rangle$, where i is an identifier of the comparator and $k \in \{0, 1, \ldots, n-1\}$, where n is the cardinality of layers. The value k represents the number of a layer, where 0 stands for no assignment to any layer. The order of comparators in the layer is irrelevant due to concurrent processing. The comparator can be assigned to one layer only.

Processing begins with a random selection of the population. Individuals are randomized multiple times until a full population of correct individuals is formed. An improper individual is the one that cannot be used to construct a correct network, e.g. $\langle (1, 0), \ldots, (n-1, 0) \rangle$. Population size is the parameter of the algorithm. Reproduction operations, application of genetic methods, evalua-

tion and succession take place in each iteration (called an epoch). The scheme of
the applied genetic algorithms is consistent with evolutionary methods described
in literature. Reproduction was based on tournament selection, which accepts
the l parameter indicating the number of individuals selected for the tournament. The number of individuals created in the reproduction process is also the
parameter of the algorithm. Reproduction creates a local population, which is
subject to cross-over and mutation operations. Two individuals are drawn for
the cross-over operation. The procedure involves random assigning of a cut-off
point where the elements of the two chromosomes are replaced. There are genes
of the first chromosome from the zero position to the drawn position and from
the drawn position to the end - genes of the second chromosome. This operation
can, however, result in an erroneous individual. In this case, an individual with
a higher adaptation ratio is returned at the output. The mutation operation
involves adding a random natural number to the current value and executing
$mod\ n$, where n is the number of layers. Mutation is performed with a certain
probability for each chromosome gene. One of the most important parts of the
algorithm, individually adopted to the specificity of the network of comparators is the evaluation function (population fit assessment). The fit assessment
values take into account both the final results of the recognition and the cost
of performing the calculations by comparators. The last value is calculated on
the basis of the unit cost of the comparator's execution and cardinality of the
reference set in a given layer. The general form of the fit assessment function is
as follows:

$$f_X(ch) = \sum_{x \in X} f_{recall}(x, ch) - \left(\frac{\sum_{i=1}^{n-1} \sum_{j=1}^{m} f_{ij}^{cost} * |ref_i|}{|C| * |ref|} - \alpha \right) \qquad (196)$$

where ch is the chromosome representing a given network structure, X is the set
of input objects from the learning set, C is the set of all considered comparators
in a given problem, f_{ij}^{cost} is the function of the unit cost of execution of the
comparator with index j in the layer number i. Values of this function are within
the range $(0, 1)$, ref_i is the subset of the ref set occurring as a reference set in
the layer with index i, and α is the positive value close to zero. Consequently,
the value of the penalty function (the element in the second parenthesis of the
function (196)) is always less than 1, which results in promoting higher the
elements identified more accurately than the simple structure of the network,
although the last element is important as well. In the above mentioned formula,
f_{recall} expresses performance of recognition for an individual from a learning set.
Succession is performed the same way as reproduction, e.g. using tournament
selection. In this case both populations are combined and individuals are selected
for the tournament. The tournament is played as many times as the size of the
target population is. The population is also a parameter of the method. The
stop condition of the algorithm refers to the number of epochs, for which there
is no improvement in the solution.

   One can use the re-sampling method to implement this procedure. It consists
of a random divide of available data into a learning set and a test set typically

in the proportions $\frac{1}{3}$ and $\frac{2}{3}$ respectively. Learning is performed in the first of the sets mentioned, e.g. there is a suboptimal network structure searched, while the solution is tested on the second set. This procedure is repeated k times where $k \geq 10$.

4.12 Learning Issues of the Reference Set for Homogeneous Networks

Learning the reference set is provided in Sect. 3.9.1 relating to the learning of individual comparators. This method remains constant for the comparator network. Figure 26 shows the activity diagram of this algorithm. In the network version, the only difference is the method of calculating similarity. Previously, the proximity vector originated from a single comparator and thus represented similarity of a single feature. In the present case, similarity is calculated by the complete network, which materially affects the quality of the adaptation that can be obtained.

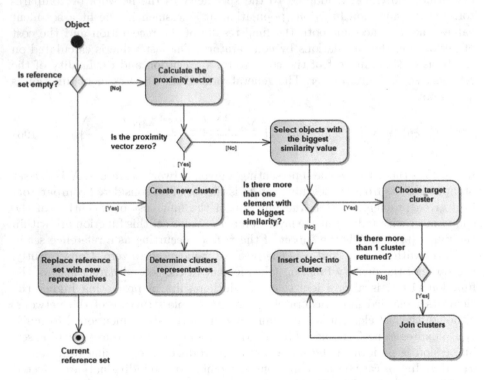

Fig. 26. Activity diagram of one iteration in unsupervised learning of reference objects for comparator networks

4.13   Summary

This chapter contains mathematical and system foundations related to comparator networks. One can look at individual definitions of components that make up the network and trace the mechanisms that work within them. The synthesis methods used in comparator networks, both for homogeneous networks and the more complex for heterogeneous networks, are discussed in detail. In general, comparator networks are designed for the construction of complex decision systems, which aim at classifying or identifying compound objects. However, because these atomic elements have a simple structure and are intuitively intelligible, these solutions are transparent and easy to interpret by humans. These networks reflect local comparisons of objects, their features, values or components and then aggregated information about a higher-level object.

This method resembles to the greatest extent the multi-layer neural networks where similar elements exists. This, however, is quite general. It is easy to see in detail that both networks operate on the basis of other mechanisms. The model of learning is different, though it derives from the same group of algorithms. Interpretation of both networks is also different. Neural networks are not so easily digested and require considerable design experience.

The method of identification and classification based on comparator networks is an innovative approach that can be easily implemented in many fields of operation. It allows modelling simple observations from the designer or expert and combining them into larger sequences, while making it possible to automate the learning of certain elements. These methods have been used in a number of commercial, scientific and research applications. Three selected examples will be described in the next section.

5   Selected Applications of Comparator Networks

5.1   Methodology of Conduct

Methods based on similarity-grounded granularity, especially applied in identification and classification tasks, constitute a noticeable part of Computational Intelligence toolbox. Nevertheless, no unified approach to construction of such systems exists. Comparators and comparator networks are an attempt at providing a framework that is more universally applicable to designing and implementing these kinds of solutions. The process of designing a comparator network for a given problem is, in fact, a *recip'* for granular modelling. It represents a granular system that models a solution to an underlying task. It entails several steps which, depending on a given application, may become very hard or very easy to perform.

– **Domain analysis** - this initial step involves gathering all available domain knowledge of objects that ought to be identified by the system [108]. Ideally, this knowledge is complete enough to build ontology that describes objects, their parts and relationships between them. The most sought-after piece of knowledge at this stage concerns the relation of generalization and specification between the entities in question. In granular terms, this step corresponds to the establishment of hierarchy of infogranules. Once knowledge is collected, it is possible to select the set of features (attributes) of objects to be further used by comparators in the network. Attributes selected at this stage must be sufficient to distinguish between different objects. In this sense, the selection of attribute collections is very similar to calculating information reducts in the theory of rough sets [64]. Moreover, if we combine the process of attribute selection with selecting a set of reference objects, which is sufficiently representative of a given domain, we obtain a framework comparable to the search of information bireducts discussed in [86].

- **Conceptual design** - in this step, on the basis of knowledge gathered during domain analysis, decisions are made regarding the type of network. If the constructed ontology and attribute sets are non-uniform, heterogeneous setting is adopted. If, however, attribute sets and ontology arrangement are highly regular, the simpler homogeneous architecture is used [99]. The decision about the type of network has to be correlated also with the form of a reference set, which has a decisive impact on initial performance of the constructed network. Once the network type and the reference set are established, the actual design stage begins. The number of layers and attribute sets, corresponding to internal and output layers, is determined. Each attribute selected corresponds to a comparator in a selected layer. The actual placement and parameters of each individual comparator are not yet determined - this happens in the next step. To complete the global design step, the type and placement of local aggregators, translators and the global aggregator are determined. Lastly, a general layout of the network, together with a set of (parts of) reference objects, is obtained.

- **Technical design** - at this stage, attention is shifted from general considerations about the network architecture to particular attributes and construction of their corresponding comparators. Design of a comparator responsible for the determination of the level of similarity with respect to a selected attribute can take a form of derivation, usually through iterative approximation, of a membership function in respect of a fuzzy relation of similarity. The procedure determining the configuration of each comparator is analogous to the weight update step in an artificial neural network.

Parameters of each comparator have to be obtained in an efficient way. In order to achieve this, the designer has to provide the comparator with the right format of input. That entails designing adequate aggregators and translators in preceding layers as well as preprocessing objects in order to transform them to the most desired format. Such transformation has to take into account the computational overhead. Modification of object representation is, in essence, equivalent to replacement of the original *crisp* objects by infogranules that later become parts of granules at higher levels of hierarchy [60].

- **Final tuning** - the remaining part is to implement the resulting architecture by means of the dedicated software library and then test it. The work of comparators, local aggregators and translators is finally processed by the global aggregator in order to derive the final output of the network. This final output is a ranking list of objects from the initial reference set that the comparator network considered similar to the one given as the input (query). The network's initial answer may be unsatisfactory. In such a case, further tuning is performed. Such tuning may involve the modification of both the parameters of the global aggregator and the contents of the initial reference set. In case of significant changes in the reference set of objects, a re-run of the previous design steps may be required. Thus, the whole design process becomes both iterative and interactive. Once behaviour of the network meets our requirements, we may use it for the assumed purpose, i.e. identification of similarities between compound objects.

According to the above methodology, network models are created, which can be used to solve specific problems. The overall approach implemented in each network is reduced to the following three phases:

- **representation** - it is used to acquire information about an object, construct a representation of an object and, in certain cases, to reduce its complexity through information granulation techniques [65]. This phase is an important element allowing the adaptation of objects for further processing to increase the efficiency of calculations. This stage prepares data about objects, which are direct inputs for comparators. In the majority of cases, this phase is performed independently for each comparator, since the majority of comparators typically requires different representation of objects.
- **comparison** - the key phase of the process, where individual comparators work directly on pairs of objects: the input object and the reference object. Individual comparators in the network layer are processed concurrently. In general, this phase is the launch of the previously designed network. Implementation of this part of methodology is the basis for final reasoning. Quality of comparisons in this section affects the effectiveness of final reasoning. Execution of this phase provides numerical data on similarity of individual pairs of objects collected in the information granule. At this stage, it is also possible to filter at the level of comparators and the projection module in the network.
- **reasoning** - the final aggregation process (the so-called global aggregation) located in the network layer. It produces the final decision on the ranking of reference objects related to global similarity of objects, constructed from local similarities of individual features. This operation synthesizes information gathered so far, using all the results computed at earlier stages of processing within the network. Based on the returned ranking, a final decision is made to identify or classify the input object to a given object class [95].

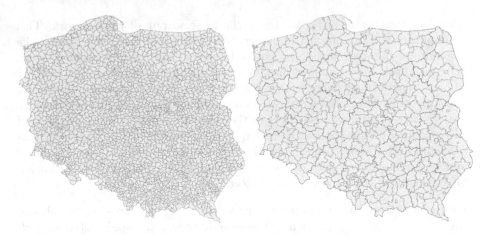

Fig. 27. Contour maps of administrative division of Poland. Left: map of communes; right: map of counties and voivodeships. Individual communes and counties are objects for identification in the experiment in question (Color figure online).

5.2 Overview of Applications

In the following sections, three selected experiments will be presented. They are practical applications of the described methods of treatment. Applications cover completely separate domains. The purpose is to demonstrate that the method is universal because of the problem to which it can be applied. During the research, a greater number of experiments was performer. They are described in publications [84,99]. Therefore, they are not included in this chapter. These three selected examples cover areas related to image processing, risk management based on CBR [1] systems and current applications in the field of text processing. This shows a wide range of applications and, at the same time, one can compare the methods used.

5.3 Identification of Contour Maps

5.3.1 Description of the Problem

In general, this experiment presents the task of identifying compound objects in the form of images, using a finite set of examples, also in the form of images [21]. The task assumes recognition in contour maps of individual administrative areas shown in Fig. 27, based on an independent contour map database. The map database contains objects interrelated by the following hierarchical structure: voivodeship - county - commune. For all reference objects, necessary information is known, including the assigned TERYT[4]. Recognition is based only on visual attributes, such as shape, colour, etc. Solution to the problem is based on comparators discussed in this article and networks built by means of the said

---

[4] National Register of Territorial Divisions.

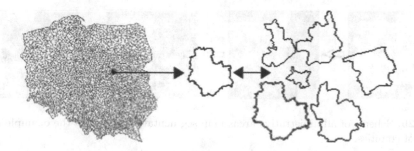

Fig. 28. General diagram which illustrates processing in the presented use. Arrows reflect the respective phases.

comparators. The input object is an image containing a map of a single administrative area (e.g. commune or county), derived on the basis of segmentation of data obtained from one of the maps shown in Fig. 27.

The reference set is a database of individual administrative areas acquired from a different source, in the form of pictures, but of other dimensions, scales, etc. Figure 28 presents a general idea of identification divided into phases.

5.3.2   Suggested Methods and Solutions
Data Segmentation
The first phase of processing involves extraction of individual outline of a commune from the map shown in Fig. 27 to the form of independent maps representing single images. The procedure assumes having a nationwide map, as shown in Fig. 27. The first step is to convert this object into a map of two colours: black and white. However, this should be done in a loss-free manner, e.g. so that no edge element is lost and consequently a given area ceases to be coherent. This can be accomplished by triggering operations, which turn all pixels of different colours from white to black. The experiment will use the RGB^5 colour space. After converting a full map, a two-colour image is obtained with $RGB(255, 255, 255)$ colours reserved for interior of the administrative area and the external part outside the outline of the map of Poland. The second colour is $RGB(0, 0, 0)$, reserved for the contours of administrative maps. Prior to initiation of the extraction procedure, ambiguity of the white colour should be removed, so that the outer region is overcoated with a fixed colour, e.g. $RGB(125, 125, 125)$, using an algorithm called *flood fill*[6]. After this operation, colours uniquely assigned to individual elements of the image are obtained. The next step is to calculate the histogram of the image in order to determine the $RGB(255, 255, 255)$ colour quantity [76], which is indirectly of interest to the extraction algorithm (inside the administrative area on a large map).

The procedure of obtaining single maps is repeated, until the colour histogram shows available white pixels. The histogram counting procedure is

---

[5] https://pl.wikipedia.org/wiki/RGB.
[6] https://en.wikipedia.org/wiki/Flood_fill.

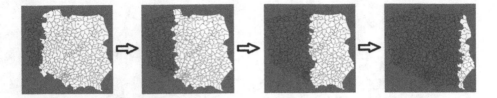

Fig. 29. Scheme of administrative areas map segmentation phase on the example of a map of counties.

repeated after each extraction. In order to accelerate the procedure, extraction is performed from a certain pixel range, starting from the left of the map image until white pixels are exhausted. Then, the interval is shifted to the right by a fixed number of pixels. This is repeated until all white areas are exhausted.

The procedure is to draw one pixel from the white pixels available, memorize its coordinates, and then overlay the area to a given colour, e.g. $RGB(50, 50, 50)$, using the *flood fill*. Then, all $RGB(50, 50, 50)$ pixels are read, and a 1-pixel edge is added with the dedicated algorithm [90], and the entire image is saved as a separate image with coordinates of the starting point in the file name. The new image is saved as the smallest possible rectangle, whose edge is exactly 1 pixel away from the contour of the map in the closest point. The number of pixels in a rectangle is $w \times h$, where w and h denote width and height, respectively. The internal colour of a newly created map is $RGB(250, 250, 250)$ and of the edge colour - $RGB(0, 0, 0)$. The remaining part is of the colour $RGB(255, 255, 255)$. The colour of the interior is necessary to correctly define the contour. Finally, the colour of the interior is replaced with $RGB(255, 255, 255)$.

Once a new map is generated, the area on the big map is repainted to the reserved colour, e.g. $RGB(125, 125, 125)$. This colour indicates areas already processed. The histogram should be recalculated after this operation. The condition for completing the loop is the absence of pixels in the $RGB(255, 255, 255)$ colour. Completion of extraction ends the first processing phase, while preparing the input for subsequent steps. Figure 29 shows the extraction phase diagram.

Image Granulation
Once the areas are isolated, we need a layer to describe them in a way convenient for imprecise comparisons. This reflects the stage of the first model of comparator network operation. In this case, representation uses information granules to represent the respective map objects [65], providing for the possibility of processing data of different noise levels. Resolution of every image is parameterized by integers m and n, where $0 < m < w$ and $0 < n < h$. This means dividing the image into $n \times m$ granules. Parameters m and n can be chosen based on expert knowledge or tuned experimentally. They have a significant impact on the quality of the process. If m and n are too high, the algorithm may not find a sufficiently good solution. If they are too low, many equivalent solutions may be obtained. Resolution may vary from image to image and, actually, it can be recomputed dynamically, if images are stored appropriately (see [84]) (Figs. 30 and 31).

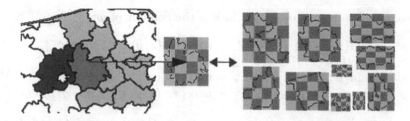

Fig. 30. The Wejherowski County (grey). One of the areas is selected for the purposes of identification (dark grey). Its granulation is compared with granulations of reference objects (the right part). An example of the area (dark) identified as including a smaller area (white) during the first phase of the algorithm is also present.

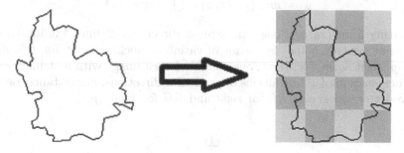

Fig. 31. Example of image granulation from the contour map of an administrative area.

However, when comparing two images, the same m and n should be set up for both of them. $G = \{g_1, \ldots, g_{n \times m}\}$ is obtained for a given image. Each granule corresponds to a rectangular subsurface of the image's polygon. Using granules, various aspects of images can be approximated, such as size, proportions or shape. In this paper, emphasis is put on the following features:

Coverage - the idea is to compute the degree of the granules' overlap with the area represented by a given image. It can be done easily by computing each granule's histogram and reading its score for colour $RGB(255, 255, 255)$. For image x and its granule $g_i \in G$, let us divide this score by the number of pixels in g_i and denote the result as cov_i^a. The value is always less than or equal to 1. These coefficients will be used in the next step in order to define similarities between image pairs.

Contour - the idea is to choose specific extreme points and connect them. For every $g_i \in G$, extreme points look as follows:

$$(x_{\leftarrow}^i, y_{\leftarrow}^i), \quad (x_{\rightarrow}^i, y_{\rightarrow}^i), \quad (x_{\uparrow}^i, y_{\uparrow}^i), \quad (x_{\downarrow}^i, y_{\downarrow}^i) \tag{197}$$

with coordinates defined over C_i, which is the contour of g_i:

$$
\begin{aligned}
x^i_{\leftarrow} &= \min\{x : (x,y) \in C_i\} & y^i_{\leftarrow} &= \max\{y : (x,y) \in C_i, x = x^i_{\leftarrow}\} \\
x^i_{\rightarrow} &= \max\{x : (x,y) \in C_i\} & y^i_{\rightarrow} &= \max\{y : (x,y) \in C_i, x = x^i_{\rightarrow}\} \\
y^i_{\uparrow} &= \min\{y : (x,y) \in C_i\} & x^i_{\uparrow} &= \max\{x : (x,y) \in C_i, y = y^i_{\uparrow}\} \\
y^i_{\downarrow} &= \max\{y : (x,y) \in C_i\} & x^i_{\downarrow} &= \max\{x : (x,y) \in C_i, y = y^i_{\downarrow}\}
\end{aligned}
\tag{198}
$$

The next step is to draw straight lines between the above points and describe them by means of certain linguistic variables. For each image and its related set of granules G, consider (x_0, y_0) in a way that:

$$
\begin{aligned}
x_0 &= \min\{x : (x,y) \in \textstyle\bigcup_i g_i\} \\
y_0 &= \max\{y : (x,y) \in \textstyle\bigcup_i g_i, x = x_0\}
\end{aligned}
\tag{199}
$$

Starting from (x_0, y_0), one can express directions of lines leading to each subsequent extreme point[7] by means of variables, such as *right, up, left, down, right-up, right-down*, etc. Next, one can label each image with a string forming the concatenation of abbreviations of particular directions. For instance, one can use two-letter codes, e.g.: *RR* for *right* and *RU* for *right-up*.

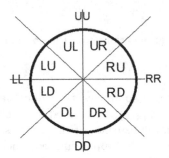

Fig. 32. Method for determining the directional labels for linguistic variables, depending on the angle between the axis X and the section connecting two points of the contour.

The choice of individual linguistic variables determines the inclination of the segment relative to a straight line passing through the point (x_0, y_0), in the form: $y = y_0$. Figure 32 shows how labels are determined, depending on the value of an angle.

In conclusion, the output of this processing phase is a set of approximated contour descriptions, calculated by means of the granulation method and its parameters (m and n). This output is the representation of input and reference objects (Fig. 33).

---

[7] The next extreme point is chosen clockwise, basing on the 8-point neighborhood, remembering the recently visited points in order to backtrack, if necessary.

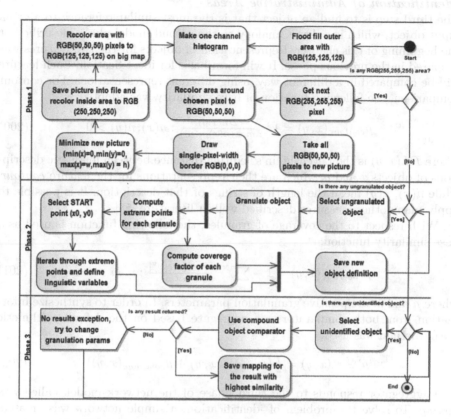

Fig. 33. Activity diagram (UML) of the described algorithm.

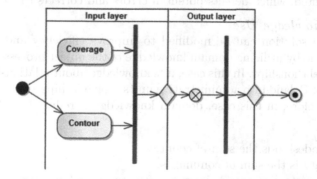

Fig. 34. Network architecture for the recognition of contour maps of administrative areas in Poland.

Identification of Administrative Areas

The third step is to find an object that is the most similar reference to a given input object, which is a direct analogy to the second model phase, described at the beginning of this chapter. Representation of objects for the two features was prepared in the previous phase. It will form basis for identification. Each feature will be compared in a different way, using different representations. For contour comparison, the following function of base similarity was adopted:

$$\mu_{contour}(x,y) = 1 - \frac{DL(x,y)}{max(n(x),n(y))} : n(x), n(y) > 0 \tag{200}$$

where $DL(x,y)$ is the Levenshtein's[8] editing distance between linguistic descriptions of objects x and y, which are their representations for the feature *contour*, while $n(x)$, $n(y)$ mean the length of strings of that description. It is possible to apply many other measures described well in literature [58].

With respect to the coverage of granules, the following function is used as a base similarity function:

$$\mu_{coverage}(x,y) = 1 - \sum_i^{n \times m} \frac{|cov_i^x - cov_i^y|}{n \times m} : n, m > 0 \tag{201}$$

where n and m are positive granulation parameters. In order to synthesize information from both comparators, an aggregate based on the default arithmetic mean method was used:

$$\mu_{agg}^{ref_{out}}(x,y) = \frac{1}{2}\left(\mu_{contour}(x,y) + \mu_{coverage}(x,y)\right) \tag{202}$$

The aggregator responds to the third phase of the network model, called *reasoning*. To solve the problem of identification, a simple network was created, which consists of two comparators described. The network architecture is shown in Fig. 34. No exception rules were applied to its implementation (the set of rules was empty), although this mechanism is often very useful, especially to model expert or domain knowledge. Meanwhile, the mechanism described in Sect. 3.8 was implemented, which detects potential errors and corrects them.

Domain Knowledge Use

The proposed solution can be modified to improve efficiency and shorten the processing time by utilizing domain knowledge of the object processed, its structure and its relationships. In this case, it is knowledge about relationship between the division of Poland into administrative units (voivodeships, counties and communes). Therefore, in this case, domain knowledge is reduced to the following facts:

1. Each voivodeship is the sum of counties;
2. Each county is the sum of communes;
3. Each administrative area belongs exactly to one administrative unit of the higher level (e.g. county belongs exactly to one voivodeship, etc.).

---

[8] https://en.wikipedia.org/wiki/Levenshtein_distance.

Knowledge of this hierarchy allows to limit the number of reference sets by identifying the object at the higher level of hierarchy, e.g. identification of a voivodeship allows to limit counties, and thus communes. Limiting the number of reference sets equals the reduction of the number of comparisons performed, thus reducing time spent on calculations. This also increases efficiency of the method, as there are less similar reference objects.

In order to implement additional domain knowledge, a new network of comparators was developed. The network needs additional input data. First of all, an additional map of Poland is needed with the same dimensions as the initial map - both are shown in Fig. 27. The additional map shows the territorial division of the country into voivodeships and counties. A starting point was recorded during the extraction of communes from the map of Poland. The same point was used to obtain a voivodeship object. This object is relatively easy to identify. The reference set contains only 16 objects. By identifying the voivodeship, we narrow the set of counties. An analogous mechanism works for counties and communes.

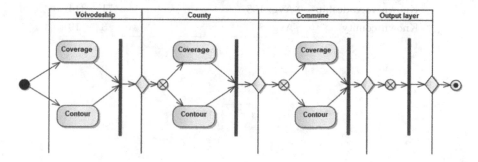

Fig. 35. Multilayer network of comparators, which implements domain knowledge about the hierarchical structure of communes. The purpose of the network is to identify the input object as a contour map of a commune, with minimized comparisons.

The new network has four layers. Its structure is shown in Fig. 35. Individual layers of this network are responsible for identifying elements belonging to hierarchy at a specified level.

5.3.3   Results Description

The experiment consisted in identifying an administrative area based on visual features of a contour map, which was obtained from the map of Poland, shown in Fig. 27. Contour maps of communes, counties and cities with county status with TERYT codes assigned, which were obtained from external sources valid in 2012, were used as reference maps. The first experiment was carried out by means of the prepared comparator network shown in Fig. 34.

It was a two-layer network implementing direct identification, based on two main features: contour shape and surface area limited by contour. For subsequent input objects, the network indicated a proximity vector based on a reference set, whose values were similar to individual contour maps. For each of the 2479 input objects, 5716 were compared, which gives a total of 14169964, i.e. 7084982 performed by each comparator. Of the 2479 input elements 2281 were identified correctly, which gave a 92.01% efficiency. Accuracy is the adopted performance metric, as described in Sect. 2.4.3.

Table 19. Results table of optimization of the number of comparisons using domain knowledge about the object processed. Abbreviations and symbols: Min - Minimum number of comparisons, Max - Maximum number of comparisons.

Variants	The magnitude of the ref. set	Min	Max
No domain knowledge	2874	2874	2874
Only voivodeships	16	16	16
Only counties	379	379	379
Only communes	2479	2479	2479
Known voivodeship	Avg: 155	71	314
Known county	Avg: 8	3	19

Fig. 36. Example of a special case of commune areas containing other areas (marked in darker colour).

The second part of the experiment was based on the application of domain knowledge and development, on its basis, of a network of comparators with successive layers of selection of reference objects, taken into account in the final identification. Consequently, optimizations of the number of comparisons needed to identify the set of input objects have been achieved. The results of this process are shown in Table 19. Depending on the type of additional domain knowledge, the number of comparisons decreased accordingly. This also had an impact on the effectiveness of identification. Using the maximum variant of domain knowledge (voivodeship and county) the number of correctly identified objects increased to 2355, which translates into the increase of ACC to 94.99%. These additionally identified cases were derived from vectors for which the previous method returned some very similar solutions. However, the correct ones were not in the first place. The use of domain knowledge reduced the size of reference elements and thus eliminated competing solutions.

5.3.4 Summary

The results presented prove the method used very effective. The resulting value means that, on average, over 9 out of 10 maps of communes, counties or cities with county rights processed have been identified correctly. It has also been shown that additional domain knowledge can significantly affect computational complexity as well as effectiveness of a solution. With this knowledge, the number of comparisons, despite the use of three times more comparators, fell to 237984 on average, which number is smaller by 2 orders of magnitude. The analysis of misidentified cases showed that many of them were specific cases, which could not be identified directly, without additional support. One such element was the inclusion of one area in the other, which caused data shredding related to the coverage comparator. An example of this is shown in Fig. 36. Resolution of some input objects that was too low was another problem.

In summary, this method achieves a very high performance score, which allows to automate the complex problem of identification in image processing.

5.4   Classification of References in Scientific Publications

5.4.1   Description of the Problem

The practical need for introducing the above scheme arose in the SYNAT project, which was a national R&D program of the Polish government aimed at the establishment of a unified network platform for storing and serving digital information in widely understood areas of science and technology [8]. One of the tasks within the scope of SYNAT was to design and implement a sub-system for searching within repositories of scientific information (articles, biographical notes, etc.), using their semantic content (SONCA). Documents processed were complex by nature and relations discovered between their parts and other entities in data were crucial for proper indexing. One of the complex parts is the reference part, which takes a form of unstructured texts. The purpose of the experiment is to determine semantic patterns of bibliographic references represented as unstructured texts. The result was to be expressed by objects classified in already known classes, such as authors, titles, journals, publication date and so on. As an example, the following text: *Sosnowski, L., Slezak, D.: Networks of Compound Object Comparators. In: Proc. of FUZZ-IEEE 2013 (2013)* should be assigned to the ATRY pattern, where: A stands for authors, T - title, R - proceedings and Y - publication date. Searching for structure in bibliographic references of processed documents was to improve the understanding of the content of a document and thus increase search efficiency of the entire system.

278　Ł. Sosnowski

5.4.2　Suggested Methods and Solutions

The whole process is divided into several stages: preprocessing, parsing and classification. The first stage is responsible for cleaning the data, removing unnecessary data from the text structure and clarifying it (e.g. lowercase, trimmed, etc.). It contains the prefix cleaning procedure, which eliminates unwanted prefixes, e.g. *"[12] D., E., Willard, New trie data structures which support very fast search operations, Journal of Computer and System Sciences, v.28 n.3, p.379-394, June 1984 u00A0[doi>10.1016/0022-0000(84) 90020-5]"* is the input text. The procedure in question cuts the part: *"[12]"*.

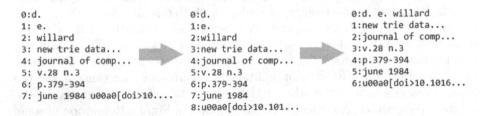

```
0:d.                          0:d.                        0:d. e. willard
1: e.                         1:e.                        1:new trie data...
2: willard                    2:willard                   2:journal of comp...
3: new trie data...           3:new trie data...          3:v.28 n.3
4: journal of comp...         4:journal of comp...        4:p.379-394
5: v.28 n.3                   5:v.28 n.3                   5:june 1984
6: p.379-394                  6:p.379-394                  6:u00a0[doi>10.1016...
7: june 1984 u00a0[doi>10.... 7:june 1984
                              8:u00a0[doi>10.101...
```

Fig. 37. An example of processing a reference description at each stage. The first fragment on the left shows the text split into tokens, the second in the sequence is the list of tokens after the *chopItems* procedure, and the third shows the result after the parsing procedure is completed.

The main idea of this stage is to dispose of all signs which can interrupt further parsing or comparing. The second operation is to replace all unnecessary quotation marks, which often interfere with results. This approach is consistent with a classical approach in data processing systems, e.g. ETL (Extract Tranform Load) in data warehouse systems [2].

The second stage is responsible for the fragmentation of the text into parts, according to certain accepted rules. The first action is to split the text in places with a comma. After that a list of tokens appears, but it contains many problems, e.g. single initials exist as tokens (e.g. "L.") or two initials in form of one token (e.g. "L.S."), etc. Therefore, in order to handle these problems, a procedure named *chopItems* was applied to each element of this list. This procedure gives the possibility to run searching rules for patterns related to problems in question. These rules, based on regular expressions (regexp), allow one to identify places with a given problem [38] and simultaneously activate a subprocedure of handling a given problem. For example, if you find a token in which there is a single initial, the procedure of combining two tokens in one is implemented. The procedure *chopItems* facilitates the control of this subprocess by means of an external file with processing commands. There are three types of instructions: *CHOP*, *RECURSIVE* and *CHOP-LONG*. All of them are designated to parse a token into smaller parts, ready for classification. In addition to commands, regular expressions are defined (regexp) [3] as well as several additional conditions. Command *CHOP;[a-z]\.[a-z]\.[a-z\ \-]2,\s\band\b\s[a-z]\.[a-z\ \-]2,; and ;5"* is an example, meaning: find a pattern like $x.x.$ $x..x$ and $x.$ $x..x$ and split with *and* phrase skipping 5 characters from the point of splitting. The control file contains any number of instructions. If the first instruction is satisfied, the rest is skipped. Once the *chopItems* operation is completed, unwanted forms of tokens, especially near authors, may exist. Therefore, another procedure, named *joinInitials*, was designed. Its task is to suppress single initials. It is based on regexp rules as well. Figure 37 shows a scheme of the whole procedure. This stage is among the most important ones. The classification process is closely related to preparation of data (parsing).

The third stage is based on a network of compound object comparators (according to the definition in Sect. 4). A structured network [99] was chosen. This kind of network is adapted to processing structured objects. The point is that the network has as many layers as the object has levels. The scheme of the network is shown in Fig. 38. The main point in this solution is to use comparators with the role of classifiers. Each comparator used in the input layer classifies a part of the text to a particular pattern. The input layer consists of the following comparators: Book (B), Country (C), Doi (D), Journal (J), Pages (P), Proceedings (R), Series (S), Title (T), Volume (V), Year (Y), Authors (A). The one dedicated to authors is a composite one. It means that it is a separate network. It is a homogeneous network. Its task is to minimize the amount of a reference set of authors in a way that quantity significantly decreases in subsequent layers, while concurrently maintaining high level of similarity between input and reference objects.

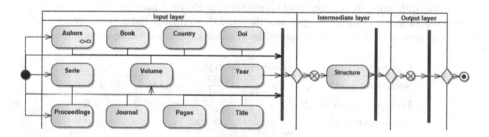

Fig. 38. Diagram of a comparators network which serves to classify the structure of references of scientific publications. The network contains a comparator called *Authors*, which consists of a homogeneous network based on three standard comparators: *Sorted Initials*, *Longest Length* and *Name and Surname*. The other comparators are classified comparators, and their names indicate the features with which they are related.

The respective comparators compute resemblance [112] between an input object and reference objects (dedicated to a particular comparator separately). Two additional phases take place at the end of each layer: aggregation and translation. They form a certain coupling between layers, which keeps the network consistent. Once classification in the first layer is determined, a representation of the text expressed by a string of classified symbols, e.g. ATJPY, is created, which means the text with the following structure: Authors, Title, Journal, Pages, Publication Year. Transformation is done during the aggregation and translation procedure. The intermediate layer processes such structures and compares them with a set of references. Next, aggregation and translation are performed again [101].

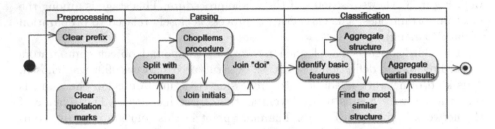

Fig. 39. Activity diagram (UML) of a semantic classification model.

The whole procedure is described in detail in Fig. 39. The results of a particular comparator are saved in the signal granule. At the end, the granule contains a wealth of detailed domain knowledge, which can be used for the purposes of final aggregation of results.

Table 20. List of comparators and base similarity functions. Abbreviations used: diff - Absolute difference in the length of characters, posSupp - positive support, negSupp - negative support, penFactor - penalty factor, DL - editing distance (Levenshtein's), regular expression - similarity measured by regular expression using a set of patterns.

Comparator	Base similarity
SortedInitialsMCOC	$1 - \frac{DL(a,b)}{max(n(a),n(b))} : n(a) \neq 0, n(b) \neq 0$
LongestLengthMCOC	$1 - \frac{diff(n(a),n(b))}{max(n(a),n(b))} : n(a) \neq 0, n(b) \neq 0$
AuthorsMCOC	$\frac{1+posSupp-negSupp}{2+penFactor} : penFactor \geq 0$
BookSCOC	$1 - \frac{DL(a,b)}{max(n(a),n(b))} : n(a) \neq 0, n(b) \neq 0$
CountrySCOC	$1 - \frac{DL(a,b)}{max(n(a),n(b))} : n(a) \neq 0, n(b) \neq 0$
DoiSCOC	Regular expression
JournalSCOC	$1 - \frac{DL(a,b)}{max(n(a),n(b))} : n(a) \neq 0, n(b) \neq 0$
PageSCOC	Regular expression
ProceedingsSCOC	Regular expression
SerieSCOC	Regular expression
TitleSCOC	$1 - \frac{DL(a,b)}{max(n(a),n(b))} : n(a) \neq 0, n(b) \neq 0$
VolumeSCOC	Regular expression
YearsSCOC	$\frac{1+posSupp-negSupp}{2+penFactor} : penFactor \geq 0$
BlobStructureSCOC	$\frac{1+posSupp-negSupp}{2+penFactor} : penFactor \geq 0$

All comparators are as good as a similarity measure applied (Sect. 3.2.2). From this perspective, the network is simply a certain regulated data flow. The real power lies within the single comparator as well as the aggregation and translation procedures in the layer. The difficulty lies in the fact that these things are specific to a problem solved. The framework automates the control of data flow, but the part relating strictly to similarity is always up to the network's designer. In this case, there are several types of measures used (Sect. 3.2.2). They are listed in Table 20.

Two of them require further discussion: year and author similarity functions. Both are based on the same token similarity algorithm. The difference lies in details. The author similarity implements ingredients of this algorithm in the following way:

- *Positive support* is the number of tokens with non-zero similarity to tokens of a pattern. The Levenshtein distance is used for calculating similarity between tokens.
- *Penalty factor* is the number of tokens from the pattern string not satisfied in terms of similarity.
- *Negative support* is the number of tokens from the candidate string not satisfied in terms of similarity.

Example 51. *Let us consider input pair of objects in the form: (u, y_1), where: $x =$ "L Sosnowski", $y_1 =$ "J Sosnowski". Calculation of similarity consists in calculating components, and then introducing results to the formula given in Table 20. Therefore, in order to calculate individual components, we break x and y_1 objects into tokens relative to space and get: $x = \{L, Sosnowski\}$ and $y_1 = \{J, Sosnowski\}$. Then, we compile individual pairs in the form: (L, J), (L, Sosnowski), (Sosnowski, J), (Sosnowski, Sosnowski) and calculate their similarity. For the reference token "J" the maximum similarity is $max\{0,0\}$, because both pairs have zero similarity. For token "Sosnowski", we get $max\{0,1\}$, i.e. finally 1. Next, we compute: POSITIVE SUPPORT $= \frac{0+1}{2}$, NEGATIVE SUPPORT $= \frac{1}{2}$, PENALTY FACTOR $= 1$. Thus, after introducing to the formula, we get the value of similarity $\frac{1}{3}$.*

Similarity of the publication year is based on the same scheme, but an additional tag is supported. It is responsible for checking year, if it is in a valid range. This tag is a part of positive support calculations. All similarity measures return the value from the range $[0, 1]$, which is important from the point of view of the main formula.

The reference set is another important aspect of decision-making classification. The described framework compares the input granule with elements of these sets during network processing. If the set does not contain the proper element, the result is weak. It does not have to be the same object as the one examined, but it should possess appropriate features. The semantic classification of texts of bibliographic data uses several types of reference sets. One is based on regexp rules, another one on instances of objects of a given class. An overview of these reference sets is shown in Table 23.

5.4.3   Description of Results

Two experiments were carried out, which used a data set comprising 400 texts. They are randomly selected from the whole available set of cases. Reference set uses data previously processed in SONCA, so there is a set of authors, titles, journals and so on. It should be noted that these sets have special knowledge of the type of object. They are already classified for a particular group of data. This knowledge is used in semantic text classification. The set is divided into two subsets: a training set and a test set. The first one consists of 132 objects and the second one of 268. The training set is used for developing and improving the reference set of patterns and tuning the parameters of the network. The procedure works on the basis of improving the results for particular comparators. In this case, there is a dedicated unit test, which calculates only one comparator value at a time and checks, if the answer is correct. If not, then the reference set is enriched with a new object, which covers the case in question, or the activation parameter of the current comparator is too rigorous.

The test set is used to measure the effectiveness of the solution. The experiment consists of two stages. The first is to parse and classify particular parts of the texts. It uses reference sets to make a classification decision and to construct patterns for each processed reference text. The second is to match the concatenated, previously classified text parts with patterns gathered in the reference set. An ordered reference subset is the result.

Both experiments use a prepared network of comparator solution, which is similarity-based reasoning. In experiments in question, the most frequent value of similarity was between *0.8* and *0.85*. The described results were achieved by means of default parameters of the network, i.e. the activation function had value p - *0.5*, aggregation method was set at arithmetic mean with $\frac{1}{n}$ weights value. Table 21 presents statistical characteristics of similarity results.

Table 21. Selected characteristics describing similarity of the closest objects.

Characteristic	Value
Max	1.00
Min	0.50
Arithmetic mean	0.85
Median	0.87
Standard deviation	0.12
Variance	0.02

The effectiveness and performance of this solution was verified by means of measures dedicated to classifiers, such as: precision, recall and F1-score. The prepared solution achieved the 0.86 global value of F1-score for the classification phase and 0.78 for the other experiment. In both cases the results are on a good level, which allows to use this solution in practice with sufficient confidence.

Table 22 presents detailed results and shows the best and the worst results for the first experiment (classification) and the second one (matching the classified pattern with reference patterns).

It is clear that results for classification are better than for the matching experiment with reference objects. Certain patterns are not matched at all in the second stage. On the one hand, this means that the reference set does not contain all the patterns used and on the other hand, it makes it clear that input data in some cases may be corrupted or wrong.

Table 22. Top left corner: best results for experiment 1, right upper corner: weakest results for experiment 1, bottom left corner: best results for experiment 2, bottom right corner: the weakest results for experiment 2. Designations used: P_1 - precision for experiment 1, R_1 - recall for experiment 1, $F1_1$ - F1-score for experiment 1, P_2 - precision for experiment 2, R_2 - recall for experiment 2, $F1_2$ - F1-score for experiment 2.

Pattern	P_1	R_1	$F1_1$	P_2	R_2	$F1_2$	Pattern	P_1	R_1	$F1_1$	P_2	R_2	$F1_2$
ATR	1.00	1.00	1.00	0.75	1.00	0.86	ATC	0.33	0.67	0.44	0.00	0.00	0.00
RY	1.00	1.00	1.00	1.00	1.00	1.00	ATYATYP	0.60	0.43	0.50	1.00	0.43	0.60
ATJVY	1.00	1.00	1.00	1.00	0.80	0.89	AYTRY	0.43	0.60	0.50	1.00	0.60	0.75
AT	1.00	0.98	0.99	0.61	0.92	0.73	ATPYC	0.54	0.80	0.64	0.60	0.60	0.60
ATRYP	1.00	0.93	0.96	1.00	0.73	0.83	ATJVYP	1.00	0.50	0.65	0.50	0.25	0.33
ATVPYD	0.91	0.98	0.95	0.92	0.84	0.86	ATVY	0.60	0.75	0.67	1.00	0.75	0.86
ATJVPYD	1.00	0.90	0.94	0.98	0.71	0.81	ATVYPR	0.56	0.83	0.67	1.00	0.50	0.67
ATJVPY	1.00	0.86	0.92	1.00	0.58	0.72	ATJPY	0.94	0.58	0.70	0.89	0.60	0.71
AJVPYD	0.86	1.00	0.92	0.86	1.00	0.92	ATRPYC	0.71	0.77	0.73	1.00	0.77	0.86
ATVPY	1.00	0.85	0.91	1.00	0.60	0.75	ATAT	0.57	1.00	0.73	0.00	0.00	0.00
Pattern	P_1	R_1	$F1_1$	P_2	R_2	$F1_2$	Pattern	P_1	R_1	$F1_1$	P_2	R_2	$F1_2$
RY	1.00	1.00	1.00	1.00	1.00	1.00	AYTJP	1.00	0.70	0.82	0.00	0.00	0.00
ATRY	0.89	0.88	0.86	1.00	0.88	0.93	ATC	0.33	0.67	0.44	0.00	0.00	0.00
ATRPY	0.93	0.90	0.90	0.99	0.88	0.92	ATAT	0.57	1.00	0.73	0.00	0.00	0.00
AJVPYD	0.86	1.00	0.92	0.86	1.00	0.92	ATJYR	1.00	0.60	0.75	0.00	0.00	0.00
ATRJPY	0.83	0.83	0.83	1.00	0.83	0.91	AYT	0.94	0.87	0.87	0.13	0.13	0.13
ATRPYCD	0.84	0.85	0.84	0.97	0.85	0.91	ATJVYP	1.00	0.50	0.65	0.50	0.25	0.33
ATPYD	0.83	1.00	0.91	0.83	1.00	0.91	AYTP	0.73	0.85	0.76	0.37	0.30	0.33
ATY	0.88	0.91	0.87	0.90	0.90	0.90	ATYR	1.00	0.75	0.86	0.50	0.50	0.50
ATJVY	1.00	1.00	1.00	1.00	0.80	0.89	ATYP	0.83	0.69	0.75	0.61	0.47	0.53
ATRP	1.00	0.75	0.86	0.88	0.88	0.88	ATPYC	0.54	0.80	0.64	0.60	0.60	0.60

5.4.4 Summary

The results presented show that effectiveness of the method from experiment 1 of this example was greater than effectiveness of the method from experiment 2. The analysis of results on a case-by-case basis showed that the reference set used in the second layer of the network for experiment 2 was too poor. Among the processed objects there were several references, for which there was no appropriate representative in this set. Other cases, however, showed errors in the reference record, which at the first stage of processing were impossible to detect, while the intermediate layer of the network correctly filtered them, unfortunately reducing the rates of effectiveness. In general, however, the method resulted in very good classification, which allowed to automate this complicated problem of text data processing.

Table 23. Example of samples of reference sets in the comparators network used.

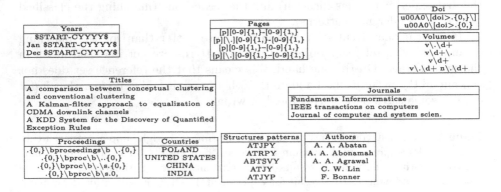

Years
\$START-CYYYY\$
Jan \$START-CYYYY\$
Dec \$START-CYYYY\$

Pages
[p][0-9]{1,}-[0-9]{1,}
[p][\.][0-9]{1,}-[0-9]{1,}
[p][0-9]{1,}-[0-9]{1,}
[p][\.][0-9]{1,}-[0-9]{1,}

Doi
u00A0\[doi>.{0,}\]
u00A0\[doi>.{0,}

Volumes
v\.\d+
v\d+\.
v\d+
v\.\d+ n\.\d+

Titles
A comparison between conceptual clustering and conventional clustering
A Kalman-filter approach to equalization of CDMA downlink channels
A KDD System for the Discovery of Quantified Exception Rules

Journals
Fundamenta Informormaticae
IEEE transactions on computers
Journal of computer and system scien.

Proceedings	Countries	Structures patterns	Authors
.{0,}\bproceedings\b \.{0,}	POLAND	ATJPY	A. A. Abatan
.{0,}\bproc\b\..{0,}	UNITED STATES	ATRPY	A. A. Abonamah
.{0,}\bproc\b\.\s.{0,}	CHINA	ABTSVY	A. A. Agrawal
.{0,}\bproc\b\s.0,	INDIA	ATJY	C. W. Lin
		ATJYP	F. Bonner

5.5 Risk Recognition for Fire and Rescue Actions

5.5.1 Description of the Problem

A fire and rescue (F&R) action is considered one of the greatest challenges in modelling for decision support systems. There were only a few attempts so far to at least partially automate the decision-making process in this area [31, 42]. One such attempt is the R&D project called ICRA[9]. The main goal of ICRA is to build a modern AI-based, risk-informed decision support system for the Incident Commander (IC), which improves safety of firefighters and extends situational awareness of the IC during F&R actions.

The assumptions and a detailed description of the operation of the ICRA system were presented in paper [94]. For the purposes of this dissertation, a general overview of the system and the context related to the conducted experiment is presented.

The Incident Commander (IC) interacts with the system, acquires information needed and asks for details in order to make better decisions. The highest level concepts, which are used to describe the situation at the fire ground, originate from the risk management approach [33]. However, for the needs of the ICRA system, the so-called *threat matrix* groups the most important threats and entities that may be vulnerable to a given threat. This matrix is the proprietary development of the method used in the German fire department, carried out by the ICRA team. The difference lies in the introduction of risk grading, which better reflects the current state of the action. This extended version is called a *risk matrix*. The version used in the ICRA system is shown in Table 6. The filled matrix indicates all the risks under consideration during the F&R action (Table 24).

Figure 40 presents a general shape of a subsystem responsible for the recognition of risk during F&R action. An important element is the repository of F&R actions, which stores historical actions in a well-described and affordable form

---

[9] http://www.icra-project.org.

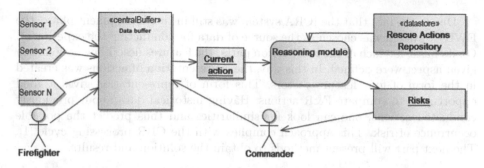

Fig. 40. The general scheme of parts of the ICRA system, associated with the experiment. The list of sensors represents data at the atomic level. Data buffer representing data evidence and aggregated, current action described by means of data from the data buffer, reasoning module responsible for recognizing similar actions and, consequently, searching for risks associated with them in order to warn the commander.

Table 24. Risk matrix defining the occurrence of risk divided into threats and subjects susceptible to threats. Notation: A1 - fear, A2 - toxic smoke, A3 - radiation, A4 - fire spreading, C - chemical substances, E1 - collapse, E2 - electricity, E3 - disease or injury, E4 - explosion.

Risk/object	A1	A2	A3	A4	C	E1	E2	E3	E4
People (ME)									
Animals (T)									
Environment (U)	–								
Property (S)	–	–				–	–	–	
Rescuers (MA)								–	
Equipment (G)	–	–							–

for processing at a later date. Actions are stored in a dedicated register, ready to be searched for recommendations related to the occurrence of a given risk. The form of their record is similar to real characteristics of an action, e.g. it is stored and divided into four stages: notification, disposal, recognition and actions.

The experiment is based on the assumption that two similar F&R actions are associated with similar types of risk. The main goal is to determine the set of risks for the ongoing F&R action. The experiment technically involves examining similarity between a given action and a repository of historical actions in order to select the most similar one. If similarity is good enough, the types of risk assigned to the action identified become the risk candidates for a given F&R action. Consequently, one should build a representation for the action in the form of feature vectors, so that it is possible to make a comparison between two actions. The end result of the experiment is the measured effectiveness of the method and comparison to other selected methods that can be used to solve this problem.

Due to the fact that the ICRA system was still in the development phase, the EWID[10] system was chosen as the source of data for conducting the experiment. On its basis, for each of the F&R action parts, the features describing actions in a given aspect were defined. In this way, the representation of actions was created in the form of four feature vectors. This form of representation gives a good opportunity to compare F&R actions. Having historical data about previously conducted actions, one can look for similarities and thus predict the possible occurrence of risk. This approach complies with the CBR processing cycle [1]. The next part will present methods to obtain the solution and results.

5.5.2 Proposed Methods and Solutions

In this experiment, action is defined by means of compound objects [108]. This form makes it possible to model the structure and dependencies between particular parts as well as to define attributes describing particular parts.

A procedure for defining attributes of objects is quite complex. Balance should be maintained between a detailed description and usefulness of representation for processing in a computer system. Too detailed representation will be impractical in terms of matching with other objects and similarity assessment. On the other hand, a too general representation will result in finding many falsely similar objects in spite of the lack of information. Therefore, a kind of granulation of available historical data to create clusters of processes is used. The list of attributes (features) used, which are processed at particular stages of comparing the sub-processes, can be found in Fig. 41 in form of comparators. This is the result of research on the representation of a fire rescue action and the available data in the EWID system.

A comparator network was used in the experiment [99]. It models similarity of F&R actions. In this case, a network with two layers is used. The first layer is composed of four comparators responsible for examining similarity of particular sub-processes: notifications, disposals, recognition, actions. Each comparator examines the selected features and returns results which are independent of one another. In this case, composite comparators are used, which means that they are independent subnetworks of comparators. This, in turn, means that they have their own layers and comparators. Figure 41 shows details of the network and subnetworks constructed. It presents all comparators proposed for examining particular features. Default values of parameter $p = 0.5$ were used in the solution, but in the future its value can be optimized. Different types of similarity measures implemented as the μ functions are used.

The first type is based on a similarity matrix. It is expert-driven knowledge, limited to the cases in question. For example, for the purposes of comparing building heights, one can use the following similarity matrix:

---

[10] Reporting and evidence system used by the State Fire Service.

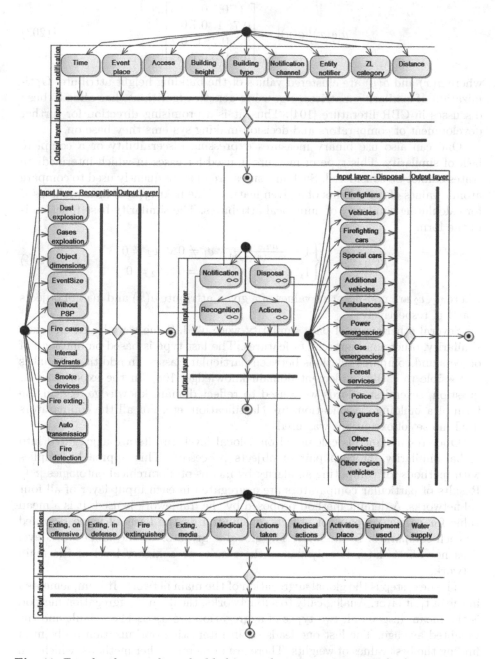

Fig. 41. Four local networks embedded in one heterogeneous network of comparators for finding similar F&R actions. The long black lines are aggregators, oval boxes - comparators of particular features, diamonds - translators of results. Abbreviations used: PSP - State fire brigade, ZL category - threat to humans.

$$\mu[a(x)][a(y)] = \begin{bmatrix} 1 & 0.7 & 0.5 & 0.3 \\ 0.7 & 1 & 0.7 & 0.5 \\ 0.5 & 0.7 & 1 & 0.7 \\ 0.3 & 0.5 & 0.7 & 1 \end{bmatrix}, \tag{203}$$

where $a(x)$ and $a(y)$ are clustered values of the building height attribute. Optimization of similarity matrices, e.g. using genetic algorithms, has already been discusses in CBR literature [104]. Thus, it is a promising direction for further development of comparators and decision-making systems they base on.

One can also use binary measures expressing discernibility or a complete lack of similarity. This type of measure is used for cases in which intermediate states cannot be considered. Such measures are most frequently used to compare atomic values of ingredients of a given feature. The next type of measure is used for calculating similarity of numerical attributes. The similarity base function is in the form:

$$\mu(v_1, v_2) = \begin{cases} 1 - \frac{|v_1 - v_2|}{\max(v_1, v_2)}, & v_1 \neq 0 \vee v_2 \neq 0 \\ 1, & v_1 = 0 \wedge v_2 = 0 \end{cases}, \tag{204}$$

where v_1, v_2 are non-negative values of a given attribute $a(x)$ and $a(y)$ of objects x and y, respectively.

Several other types of base similarity were used. Some were based on token similarity, others on hierarchy of features. The last type is based on ontology of objects and compares relations between particular classes. In addition, it makes it possible to easily implement domain knowledge [108]. In the experiment in question, several standards were used to reflect domain knowledge, e.g. in the form of a building classification, fire classification, etc. For all the comparators used, no set of exceptions was used.

Once resemblance is calculated on a local level, results are aggregated into global similarity for each pair of objects processed. This approach complies with methods for computing similarity by means of hierarchical ontologies [80]. Results of particular comparators are aggregated in each input layer of all four subnetworks. A default method proposed by the framework is used. It is a mean value for particular pairs of objects, calculated on the basis of value returned by each comparator in the layer. The global aggregator on the subnetwork level does not affect data. It only passes the results to a higher level of the main network.

The next step is the global aggregation of the main network. It is implemented in the output layer. Analogically to subnetworks, the default aggregation method is the mean. In general, two types of aggregation are used. The default and the weighted average. The last one leads to an interesting optimization problem of finding the best values of weights. There are numerous other methods, which can be used to aggregate partial results described in Sect. 4.8. Weights are calculated by means of an evolutionary algorithm. Proximity vector v presented in formula (155) is the result of aggregation.

Once the most similar pairs of objects for a given action are identified, risk labels are retrieved from the most similar reference object, e.g. the 1-NN approach [56] is used. The said risks are considered probable with an assigned chance of materialization.

5.5.3 Description of Results

Two experiments were carried out. One was performed on a small data set, which was obtained from EWID. It consisted of 406 reports from F&R actions. This set was used in order to compare results with previous experiments [43]. The second experiment was performed on a bigger set of EWID data. It consisted of 3736 reports. The leave-one-out method was used for both experiments. In total, the data sets contained 164430 and 13953960 pairs of objects, respectively.

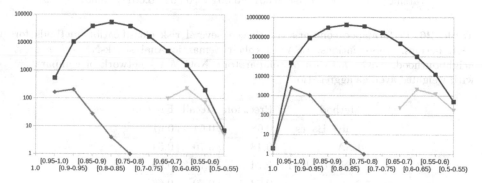

Fig. 42. Similarity distribution for: all, the best and the worst object pairs, designated by means of the *leave-one-out* method. The chart on the left corresponds to a data set with 406 F&R actions and the right one corresponds to data that describe F&R interventions. Both charts use logarithmic scale. The Y axis shows the number of pairs and the X axis shows the clustered similarity value.

This is not a full Cartesian product, because some pairs were eliminated by activation functions. The main statistical characteristics for both data sets are shown in Table 25. Distribution of similarity values for results is presented analogically and is illustrated in Fig. 42.

The results described were achieved with default parameters of the network, e.g. the activation function value p - 0.5 and global aggregation method - *mean*, with $\frac{1}{n}$ weights for NoC $\frac{1}{n}$. On the other hand, the NoC WA results were obtained using the weighted average global aggregator and the evolutionary algorithm for learning weights. The winning weight values are: $\frac{2}{89}, \frac{60}{89}, \frac{1}{89}, \frac{26}{89}$ for notification, disposals, recognition and actions comparators, respectively. Weights were learned by means of a training set consisting of 136 out of 406 F&R actions (33%). The procedure was repeated ten times. The final weights are the ones with maximum values. The evaluation function took into account the number of recognized risks as well as the overall prediction quality.

All parameters of the model can be tuned further. The p value can limit the quantity of processed objects, but on the other hand, weights can control the

Table 25. Selected central tendency indicators for the set of results, * - results for 406 objects, ** - results for 3736 objects. All - all F&R pairs, best - pairs with maximum similarity value taken for each F&R action, worst - pairs with minimum similarity value taken for each F&R action.

Indicator	*-all	*-best	*-worst	**-all	**-best	**-worst
Max	0.985	0.985	0.690	1.000	1.000	0.676
Min	0.530	0.791	0.530	0.477	0.797	0.477
Arithmetic mean	0.815	0.940	0.627	0.811	0.956	0.606
Median	0.820	0.946	0.630	0.824	0.961	0.608
Standard deviation	0.061	0.028	0.031	0.062	0.023	0.030
Variance	0.004	0.001	0.001	0.004	0.001	0.001

Table 26. Performance comparison between several risk classification methods for F&R actions. Abbreviations: ESA - Explicit Semantic Analysis, k-NN - k nearest neighbourhood, NoC - network of comparators, NoC WA - network of comparators with weighted average aggregation.

Method	Precision	Recall	F1-score
Naive Bayes	0.68	0.64	0.61
ESA	0.48	0.70	0.54
k-NN canberra	0.74	0.74	0.69
NoC	0.73	0.70	0.66
NoC WA	**0.79**	**0.75**	**0.71**

importance of feature selection. The final results are verified by a risks similarity layer. The final result was validated in the framework of similarity of the types of risk. Risks classified by particular methods described in paper [43] and in this experiment were compared.

The final analysis was divided into two steps. Firstly, performance was calculated by means of Precision, Recall and F1-score separately for each F&R action and results were averaged. Table 26 presents the values obtained and compares them with those obtained in earlier research performed on the same set of data. Secondly, the analysis of performance for particular risks in the *risk matrix* used in ICRA was carried out. Table 27 presents a comparison of results obtained for various methods.

5.5.4 Interpretation of Results

According to Tables 26 and 27, the best results for the problem in question were achieved by means of the NoC WA method. In the first experiment methods were evaluated by means of three known measures from the information retrieval domain [72]: Precision, Recall and F1-score. For all these measures the approach adopted showed the best performance. The average Precision was 0.79, while the

Table 27. Comparison (using F1-score) of the classification methods relative to risks of F&R. Risks are defined as a Cartesian product of threats and objects from the threats matrix [2]. Abbreviations of algorithm names: ESA - Explicit Semantic Analysis, kNN - k nearest neighbourhood, NoC $\frac{1}{n}$ - network of comparators with arithmetic mean aggregator, NoC WA - network of comparators with weighted average aggregator, * - results for 406 F&R actions, ** - results for 3736 F&R actions.

Risk	Naive Bayes*	ESA*	kNN Canberra*	NoC $\frac{1}{n}$*	NoC WA*	NoC $\frac{1}{n}$**
A1_MA	0.38	0.45	0.34	0.36	**0.39**	0.47
A1_ME	0.86	0.82	**0.91**	**0.91**	0.90	0.91
A1_T	–	0.07	0.09	0.12	**0.16**	0.14
A2_MA	0.81	0.84	**0.89**	0.85	0.88	0.70
A2_ME	0.83	0.84	**0.90**	0.89	0.89	0.84
A2_S	**0.29**	0.22	0.09	0.17	0.1	0.20
A2_T	0.05	0.14	0.09	0.13	**0.17**	0.17
A2_U	0.39	0.30	0.44	0.34	**0.45**	0.38
A4_G	–	0.08	0.21	0.11	**0.25**	0.14
A4_MA	0.30	0.22	0.35	0.24	**0.47**	0.34
A4_ME	0.27	0.17	**0.41**	0.16	0.36	0.33
A4_S	–	–	**0.40**	0.23	0.30	0.34
A4_T	–	0.13	–	–	**0.67**	0.06
E1_MA	–	0.11	**0.48**	0.22	0.42	0.40
E1_ME	–	–	**0.22**	–	–	0.21
E2_MA	0.11	**0.31**	0.17	0.07	0.24	0.30
E2_ME	–	**0.24**	0.20	0.09	0.14	0.24
E2_S	–	**0.15**	0.13	–	0.12	0.03
E3_G	–	0.12	**0.40**	–	–	0.08
E3_MA	–	**0.50**	0.16	0.13	0.17	0.28
E3_ME	–	–	0.12	**0.28**	0.13	0.42
E4_MA	–	–	–	–	**0.14**	0.13
E4_ME	–	–	–	–	–	0.14
E4_S	–	–	–	–	–	0.12

score of the second best classification algorithm was about 0.05 worse. Analogical results were obtained for the other two measures. These results mean that the NoC WA method correctly classified the biggest number of risks and that it had the smallest number of false classifications (type-2 errors according to the description in Sect. 2.4.3).

Further analysis revealed that even from the point of view of particular risks, the approach adopted produces the highest results. The analysis of detailed results in Table 27 reveals that the comparison of each pair of presented methods ranks the NoC WA algorithm first. Two evaluation methods are used: the number of classified risks and the number of risks with a greater F1-score. In the first case, the method applied is ranked first, ex aequo with k-NN, by the Canberra similarity measure. Both methods classified 20 risks and in two cases these methods were the only ones. One of them is E1_ME (k-NN Canberra) and the other is E4_MA (NoC WA). The second evaluation method compares classifiers in pairs. The assessment for each pair and each risk comes down to evaluating which score is better. In this paper emphasis is put on the comparison of the NoC method and other methods. In this comparison, the NoC WA wins as well. The second best is k-NN Canberra. In this pair NoC WA wins in 12 risk categories and k-NN Canberra wins in 8.

6 Summary

This dissertation presents an approach to solving the problem of identification and recognition of compound objects complementary to the already existing approaches. Methodology presented is based on similarity of objects and reasoning based on knowledge of their similarity. This paper includes an introduction to the subject of comparing objects, which clarifies the most important problems and challenges. It also contains a chapter presenting the necessary mathematical apparatus to describe mathematical solutions presented in this paper. The subsequent chapters present a mathematical description of a proprietary solution to measure similarity and build complex structures for the processing of compound objects. The paper also contains three examples of application of the described method, with a description and interpretation of the results obtained. These examples come from real commercial as well as scientific and research projects.

The main results of the dissertation should be considered in the context of objectives set in the thesis (Sect. 1.2). First of all, it is necessary to present a mathematical description of a single comparator of compound objects. This description provides basis for understanding the operation of a comparator as well as for expansion, in order to create complex processing structures. This description is presented in Sect. 3. In addition, the paper covers functional descriptions of comparators, which provided basis for the description of a comparator divided into components.

Another goal achieved was to describe and characterize a network of comparators, which is used for modelling similarities for the purposes of recognition of compound objects. It resulted in a mathematical description of a network, divided into a heterogeneous and a homogeneous network. The description is presented in Sect. 4. In addition to the basic mathematical apparatus presenting a functional description, the paper also presents fuzzy model interpretations. The issues of synthesis of data processed by a network of comparators are described as well. They constitute the basic mechanism of the model's reasoning.

A generalized methodology for designing a network of compound object comparators is formulated and a model of operation of these networks is described in Sect. 5. This chapter is the implementation of the objective of creating a model for the recognition and identification of compound objects based on comparators. The same chapter presents three selected examples of application, which prove the fulfilment of the purpose of performing experiments to verify effectiveness of using comparators in real commercial projects as well as scientific and research projects. Individual applications operate on different types of compound objects. The first one processes images, the second text objects, and the last one describes fire and rescue actions in the context of risk management. Results obtained from all the three experiments were very good. They are described and discussed in Sects. 5.3.3, 5.4.3 and 5.5.3, respectively. They have also been published during numerous international conferences and in scientific magazines.

By implementing all the auxiliary goals, the thesis about the possibility of using comparators and network structures constructed from them to identify objects using similarity was proven. This part is also discussed in detail in the chapter on comparator networks.

The paper contains a number of solutions that are original achievements of the author. From components of a comparator of compound objects in the form of a combination of a fuzzy relation with the mechanism of exceptions rules, through function of thresholding, filtration of solutions, to sharpening the results obtained. Such configuration of various mechanisms is the result of author's research, ensuring effectiveness of the solution.

It is similar with the construction of complex network structures based on comparators. The paper discusses two types of networks, which are also the result of author's research. The description of these structures made it possible to solve problems of recognizing and identifying compound objects by means of comparators. This dissertation also contains a mathematical description of the required properties of comparators used for creating practical solutions in question.

Another element of the author's achievements is the creation of a universal methodology for designing solutions for identification problems, based on similarity between compound objects. The proposed approach is presented in the form of a standardized procedure for designing the network model for different identification problems, also different in terms of types of compound objects.

Adaptations of the network model learning algorithms known from other solutions of artificial intelligence have also been proposed, both in the field of network architecture and internal parameters, which influence the results obtained. Although the learning methods themselves are known, their application to the network of comparators has been described for the first time.

This dissertation contains a complete theoretical material supported by practical experiments. However, in the long-term perspective, a graphical research platform can be developed to ensure freedom of access to the described architecture and conducting experiments with new problems in a greatly simplified way. Environment of this kind will make it easy to design the network model and perform calculations. The respective known examples of measures should be implemented in the platform in question. They should also be ready to use through configuration by means of a graphical interface.

Another problem to be further developed is the preparation or adaptation of standardized test data sets, to make it much easier to compare the results with other methods.

References

1. Aamodt, A., Plaza, E.: Case-based reasoning: foundational issues, methodological variations, and system approaches. Artif. Intell. Commun. **7**(1), 39–59 (1994)
2. Agosta, L.: The Essential Guide to Data Warehousing. Essential Guide Series, Prentice Hall PTR (2000). https://books.google.pl/books?id=p492QgAACAAJ
3. Aho, A.V.: Algorithms for finding patterns in strings. In: van Leeuwen, J. (ed.) Handbook of Theoretical Computer Science, vol. A, pp. 255–300. MIT Press, Cambridge (1990)
4. Allemang, D., Hendler, J.: Semantic Web for the Working Ontologist: Effective Modeling in RDFS and OWL. Morgan Kaufmann Publishers Inc., San Francisco (2008)
5. Arabas, J.: Wykłady z algorytmów ewolucyjnych. Wydawnictwo WNT, Warszawa (2004)
6. Ayodele, T.: Introduction to Machine Learning. INTECH Open Access Publisher (2010). http://books.google.pl/books?id=LqS_oAEACAAJ
7. Barbie, M., Puppe, C., Tasnadi, A.: Non-manipulable domains for the borda count. No. 13 in Bonn econ discussion papers (2003)
8. Bembenik, R., Skonieczny, Ł., Rybiński, H., Niezgódka, M. (eds.): Intelligent Tools for Building a Scientific Information Platform. Springer, Heidelberg (2012). https://doi.org/10.1007/978-3-642-35647-6
9. Bergmann, M.: An Introduction to Many-Valued and Fuzzy Logic: Semantics, Algebras, and Derivation Systems. Cambridge University Press (2008). http://www.amazon.com/Introduction-Many-Valued-Fuzzy-Logic-Derivation/dp/0521707579%3FSubscriptionId%3D0JYN1NVW651KCA56C102%26tag%3Dtechkie-20%26linkCode%3Dxm2%26camp%3D2025%26creative%3D165953%26creativeASIN%3D0521707579
10. Berry, M.J., Linoff, G.: Data Mining Techniques: For Marketing, Sales, and Customer Support. Wiley, New York (1997)
11. Berson, A., Smith, S.J.: Data Warehousing, Data Mining, and Olap, 1st edn. McGraw-Hill Inc., New York (1997)
12. Bishop, C.: Neural Networks for Pattern Recognition. Neural Networks for Pattern Recognition. Oxford University Press, Incorporated (1995). http://books.google.es/books?id=-aAwQO_-rXwC
13. Böhm, C., Berchtold, S., Keim, D.A.: Searching in high-dimensional spaces: index structures for improving the performance of multimedia databases. ACM Comput. Surv. **33**(3), 322–373 (2001)
14. Brams, S.J., Fishburn, P.C.: Going from theory to practice: the mixed success of approval voting. Soc. Choice Welfare **25**(2–3), 457–474 (2005)
15. Brun, M., et al.: Model-based evaluation of clustering validation measures. Pattern Recogn. **40**(3), 807–824 (2007). http://www.sciencedirect.com/science/article/pii/S0031320306003104
16. Büttcher, S., Clarke, C.L.A., Cormack, G.V.: Information Retrieval: Implementing and Evaluating Search Engines. MIT Press, Cambridge (2010). http://www.worldcat.org/title/information-retrieval-implementing-and-evaluating-search-engines/oclc/473652398?lang=de

17. Cantú-Paz, E., Cheung, S.C.S., Kamath, C.: Retrieval of similar objects in simulation data using machine learning techniques. In: Image Processing: Algorithms and Systems, pp. 251–258 (2004)
18. Cornelis, C., Jensen, R., Martín, G.H., Slęzak, D.: Attribute selection with fuzzy decision reducts. Inf. Sci. **180**(2), 209–224 (2010). https://doi.org/10.1016/j.ins.2009.09.008
19. Cross, V., Yu, X., Hu, X.: Unifying ontological similarity measures: a theoretical and empirical investigation. Int. J. Approx. Reason. **54**(7), 861–875 (2013)
20. Dasgupta, D., Michalewicz, Z.: Evolutionary Algorithms in Engineering Applications. Springer, Heidelberg (1997). https://doi.org/10.1007/978-3-662-03423-1. https://books.google.pl/books?id=6C09oNmYiAgC
21. Deb, S.: Multimedia Systems and Content-based Image Retrieval. Idea Group Publishing (2004). http://books.google.pl/books?id=GcO4HGbMi7UC
22. Elkind, E., Lang, J., Saffidine, A.: Choosing collectively optimal sets of alternatives based on the condorcet criterion. In: Walsh, T. (ed.) IJCAI, pp. 186–191. IJCAI/AAAI (2011). http://dblp.uni-trier.de/db/conf/ijcai/ijcai2011.html#ElkindLS11
23. Faliszewski, P., Hemaspaandra, E., Hemaspaandra, L.A.: Using complexity to protect elections. Commun. ACM **53**(11), 74–82 (2010)
24. Fodora, J.C., Ovchinnikov, S.: On aggregation of T-transitive fuzzy binary relations. Fuzzy Sets Syst. **72**(2), 135–145 (1995). http://www.sciencedirect.com/science/article/pii/0165011494003469
25. Fokina, E.B., Friedman, S.-D.: Equivalence relations on classes of computable structures. In: Ambos-Spies, K., Löwe, B., Merkle, W. (eds.) CiE 2009. LNCS, vol. 5635, pp. 198–207. Springer, Heidelberg (2009). https://doi.org/10.1007/978-3-642-03073-4_21
26. Goldberg, D.: Genetic Algorithms in Search, Optimization, and Machine Learning. Artificial Intelligence, Addison-Wesley (1989). http://books.google.pl/books?id=3_RQAAAAMAAJ
27. Gomolinska, A., Wolski, M.: Rough inclusion functions and similarity indices. In: CS&P, pp. 145–156 (2013)
28. Gruber, M.: Mastering SQL, 1st edn. SYBEX Inc., Alameda (2000)
29. Gupta, K., Gupta, R.: Fuzzy equivalence relation redefined. Fuzzy Sets Syst. **79**(2), 227–233 (1996). http://www.sciencedirect.com/science/article/pii/0165011495001557
30. Gwiazda, T.: Algorytmy genetyczne: kompendium. Operator krzyżowania dla problemów numerycznych. No. t. 1, Wydawnictwo Naukowe PWN (2007). https://books.google.pl/books?id=16-JGgAACAAJ
31. Han, L., et al.: Firegrid: an e-infrastructure for next-generation emergency response support. J. Parallel Distrib. Comput. **70**(11), 1128–1141 (2010)
32. Hegenbarth, F.: Examples of free involutions on manifolds. Math. Ann. **224**(2), 117–128 (1976). https://doi.org/10.1007/BF01436193
33. ISO 31000 - Risk management (2009)
34. Iwata, T., Saito, K., Yamada, T.: Modeling user behavior in recommender systems based on maximum entropy. In: WWW, pp. 1281–1282 (2007)
35. Janusz, A., Ślęzak, D., Nguyen, H.S.: Unsupervised similarity learning from textual data. Fundam. Inform. **119**(3–4), 319–336 (2012)
36. Kacprzyk, J.: Multistage Fuzzy Control: A Model-based Approach to Fuzzy Control and Decision Making. Wiley, Hoboken (2012)
37. Klement, E.P., Pap, E., Mesiar, R.: Triangular Norms. Kluwer Academic Publishers, Dordrecht (2000). http://opac.inria.fr/record=b1104736

38. Kolpakov, R., Raffinot, M.: Faster text fingerprinting. In: Amir, A., Turpin, A., Moffat, A. (eds.) SPIRE 2008. LNCS, vol. 5280, pp. 15–26. Springer, Heidelberg (2008). https://doi.org/10.1007/978-3-540-89097-3_4

39. Komorowski, J., Pawlak, Z., Polkowski, L., Skowron, A.: Rough sets: a tutorial (1998)

40. Koronacki, J., Mielniczuk, J.: Statistics: for students of technical and natural sciences (in polish). Wydawnictwa Naukowo-Techniczne (2001). http://books.google.pl/books?id=TI4NAQAACAAJ

41. Kosiński, R.: Sztuczne sieci neuronowe: dynamika nieliniowa i chaos. Wydawnictwa Naukowo-Techniczne (2004). https://books.google.pl/books?id=BgmKtwAACAAJ

42. Krasuski, A., Jankowski, A., Skowron, A., Ślęzak, D.: From sensory data to decision making: a perspective on supporting a fire commander. In: Web Intelligence/IAT Workshops, pp. 229–236 (2013)

43. Krasuski, A., Janusz, A.: Semantic tagging of heterogeneous data: labeling fire & rescue incidents with threats. In: FedCSIS, pp. 77–82 (2013)

44. Kulikowski, J.L.: Toward computer-aided interpretation of situations. In: Burduk, R., Jackowski, K., Kurzynski, M., Wozniak, M., Zolnierek, A. (eds.) Proceedings of the 8th International Conference on Computer Recognition Systems CORES 2013. Advances in Intelligent Systems and Computing, vol. 226. Springer, Heidelberg (2013). https://doi.org/10.1007/978-3-319-00969-8_1

45. Levitin, G., Lisnianski, A.: Reliability optimization for weighted voting system. Rel. Eng. Sys. Saf. **71**(2), 131–138 (2001)

46. Lin, X., Yacoub, S., Burns, J., Simske, S.: Performance analysis of pattern classifier combination by plurality voting. Pattern Recogn. Lett. **24**(12), 1959–1969 (2003)

47. Luckham, D.: The Power of Events: an Introduction to Complex Event Processing in Distributed Enterprise Systems. The Power of Events: An Introduction to Complex Event Processing in Distributed Enterprise Systems, ADDISON WESLEY Publishing Company Incorporated (2002). http://books.google.es/books?id=AN1QAAAAMAAJ

48. MacParthalain, N., Jensen, R.: Simultaneous feature and instance selection using fuzzy-rough bireducts. In: FUZZ-IEEE 2013, IEEE International Conference on Fuzzy Systems, Hyderabad, India, 7–10 July 2013, Proceedings, pp. 1–8 (2013). http://dx.doi.org/10.1109/FUZZ-IEEE.2013.6622500

49. Maedche, A., Staab, S.: Comparing ontologies – similarity measures and a comparison study. Technical report, Institute AIFB, University of Karlsruhe, March 2001

50. Mallik, A., Chaudhury, S., Ghosh, H.: Nrityakosha: preserving the intangible heritage of indian classical dance. JOCCH **4**(3), 11 (2011)

51. Malmstadt, H., Enke, C., Crouch, S.: Electronic Analog Measurements and Transducers: Instrumentation for Scientists Series 1. Analog Measurements and Transducers. Benjamin (1973). http://books.google.pl/books?id=U9XkSAAACAAJ

52. Marin, N., Medina, J.M., Pons, O., Sanchez, D., Vila, M.A.: Complex object comparison in a fuzzy context. Inf. Softw. Technol. **45**, 431–444 (2003)

53. Mas, M., Monserrat, M., Torrens, J.: Modus ponens and modus tollens in discrete implications. Int. J. Approx. Reason. **49**(2), 422–435 (2008). https://www.sciencedirect.com/science/article/pii/S0888613X08000637

54. McKelvey, R.D., Patty, J.W.: A theory of voting in large elections. Game Econ. Behav. **57**(1), 155–180 (2006). https://www.sciencedirect.com/science/article/pii/S0899825606000698

55. Michalewicz, Z.: Genetic Algorithms + Data Structures = Evolution Programs, 3rd edn. Springer-Verlag, London (1996). https://doi.org/10.1007/978-3-662-03315-9

56. Mitchell, T.M.: Machine Learning. McGraw Hill Series in Computer Science. McGraw-Hill, New York (1997)

57. Molodtsov, D.: Soft set theory - first results. Comput. Math. Appl. **37**(4–5), 19–31 (1999). http://www.sciencedirect.com/science/article/pii/S0898122199000565

58. Navarro, G.: A guided tour to approximate string matching. ACM Comput. Surv. **33**(1), 31–88 (2001). https://doi.org/10.1145/375360.375365

59. Nesenbergs, M., Mowery, V.O.: Logic synthesis of some high-speed digital comparators. Bell Syst. Tech. J. **38**, 19–44 (1959)

60. Nguyen, S.H., Bazan, J., Skowron, A., Nguyen, H.S.: Layered learning for concept synthesis. LNCS Trans. Rough Sets **1**(3100), 187–208 (2004)

61. Pal, S., Shiu, S.: Foundations of Soft Case-Based Reasoning. Wiley Series on Intelligent Systems, Wiley (2004). http://books.google.es/books?id=LqZkJ_snUiYC

62. Pawlak, Z.: On rough sets. Bull. EATCS **24**, 94–108 (1984)

63. Pawlak, Z.: Rough set theory. KI **15**(3), 38–39 (2001)

64. Pawlak, Z., Skowron, A.: Rough sets: some extensions. Inf. Sci. **177**(1), 28–40 (2007)

65. Pedrycz, W., Skowron, A., Kreinovich, V.: Handbook of Granular Computing. Wiley (2008). http://books.google.fr/books?id=CpMrHqMPe2UC

66. Pekalska, E., Duin, R.P.W.: The Dissimilarity Representation for Pattern Recognition. Foundations and Applications (Machine Perception and Artificial Intelligence). World Scientific Publishing Co., River Edge (2005)

67. Peters, J.F.: Near sets: an introduction. Math. Comput. Sci. **7**(1), 3–9 (2013)

68. Polkowski, L.: Approximate Reasoning by Parts: An Introduction to Rough Mereology. Intelligent Systems Reference Library. Springer, Heidelberg (2011). https://doi.org/10.1007/978-3-642-22279-5

69. Polkowski, L., Artiemjew, P.: Granular Computing in Decision Approximation. ISRL, vol. 77. Springer, Cham (2015). https://doi.org/10.1007/978-3-319-12880-1

70. Quackenbush, R.W.: On the composition of idempotent functions. Algebra Univers. **1**(1), 7–12 (1971). http://dx.doi.org/10.1007/BF02944949

71. Rasmusen, E.: Games and Information: An Introduction to Game Theory. Blackwell (2001). https://books.google.pl/books?id=7ylayBG9sa4C

72. Rinaldi, A.M.: An ontology-driven approach for semantic information retrieval on the web. ACM Trans. Internet Technol. **9**, 10:1–10:24 (2009). http://doi.acm.org/10.1145/1552291.1552293

73. Riza, L.S., et al.: Implementing algorithms of rough set theory and fuzzy rough set theory in the R package "roughsets". Inf. Sci. **287**, 68–89 (2014). http://dx.doi.org/10.1016/j.ins.2014.07.029

74. Rumbaugh, J., Jacobson, I., Booch, G.: Unified Modeling Language Reference Manual, 2nd edn. Pearson Higher Education, London (2004)

75. Rumelhart, D.E., Hinton, G.E., Williams, R.J.: Neurocomputing: Foundations of Research. Learning Representations by Back-propagating Errors, pp. 696–699. MIT Press, Cambridge (1988). http://dl.acm.org/citation.cfm?id=65669.104451

76. Russ, J.: The Image Processing Handbook, 6th edn. Taylor & Francis, Abingdon-on-Thames (2011). http://books.google.pl/books?id=gxXXRJWfEsoC

77. Rutkowski, L.: Computational Intelligence: Methods and Techniques. Springer, Heidelberg (2008). https://doi.org/10.1007/978-3-540-76288-1. http://books.google.es/books?id=iRTGlFXt1lwC

78. Saari, D.G.: The Optimal Ranking Method is the Borda Count. Discussion Papers 638, Northwestern University, Center for Mathematical Studies in Economics and Management Science, January 1985. https://ideas.repec.org/p/nwu/cmsems/638.html
79. Saari, D.G., Merlin, V.R.: The Copeland Method. I: Relat. Dictionary **8**, 51–76 (1996)
80. Schickel-Zuber, V., Faltings, B.: OSS: a semantic similarity function based on hierarchical ontologies. In: Proceedings of the 20th International Joint Conference on Artificial Intelligence, IJCAI 2007, pp. 551–556. Morgan Kaufmann Publishers Inc., San Francisco (2007). http://dl.acm.org/citation.cfm?id=1625275.1625363
81. Serpico, S., Bruzzone, L., Roli, F.: An experimental comparison of neural and statistical non-parametric algorithms for supervised classification of remote-sensing images. Pattern Recogn. Lett. **17**(13), 1331–1341 (1996). http://www.sciencedirect.com/science/article/pii/S0167865596000906. Special Issue on Non-conventional Pattern Analysis in Remote Sensing
82. Shannon, C.E.: A mathematical theory of communication. The Bell Syst. Tech. J. **27**, 379–423, 623–656 (July, October 1948). http://cm.bell-labs.com/cm/ms/what/shannonday/shannon1948.pdf
83. Skowron, A., Polkowski, L.: Rough mereological foundations for design, analysis, synthesis, and control in distributed systems. In: Proceedings The Second Joint Annual Conference on Information Sciences, Wrightsville Beach, NC, pp. 129–156 (1998)
84. Ślęzak, D., Sosnowski, Ł.: SQL-based compound object comparators: a case study of images stored in ICE. In: Kim, T., Kim, H.-K., Khan, M.K., Kiumi, A., Fang, W., Ślęzak, D. (eds.) ASEA 2010. CCIS, vol. 117, pp. 303–316. Springer, Heidelberg (2010). https://doi.org/10.1007/978-3-642-17578-7_30
85. Ślęzak, D.: Approximate reducts in decision tables. In: 6th International Conference on Information Processing and Management of Uncertainty in Knowledge-Based Systems, pp. 1159–1164. Universidad de Granada (1996)
86. Ślęzak, D., Janusz, A.: Ensembles of bireducts: towards robust classification and simple representation. In: Kim, T., et al. (eds.) FGIT 2011. LNCS, vol. 7105, pp. 64–77. Springer, Heidelberg (2011). https://doi.org/10.1007/978-3-642-27142-7_9
87. Ślęzak, D., Szczuka, M.: Rough neural networks for complex concepts. In: An, A., Stefanowski, J., Ramanna, S., Butz, C.J., Pedrycz, W., Wang, G. (eds.) RSFDGrC 2007. LNCS (LNAI), vol. 4482, pp. 574–582. Springer, Heidelberg (2007). https://doi.org/10.1007/978-3-540-72530-5_69
88. Slowinski, R.: A generalization of the indiscernibility relation for rough set analysis of quantitative information. Riv. Matematica Economiche e Sociali **15**(1), 65–78 (1992)
89. Smith, W.D.: Range voting (2000)
90. Sosnowski, Ł.: Identification with compound object comparators technical aspects. In: Hołubiec, J. (ed.) Techniki informacyjne teoria i zastosowania, vol. 1, pp. 168–179. IBS PAN (2011)
91. Sosnowski, Ł.: Characters recognition based on network of comparators. In: Myśliński, A. (ed.) Techniki informacyjne teoria i zastosowania, vol. 4, pp. 123–134. IBS PAN (2012)
92. Sosnowski, Ł.: Applications of comparators in data processing systems. Technical Transactions Automatic Control, pp. 81–98 (2013)
93. Sosnowski, Ł.: Framework of compound object comparators. Intell. Decis. Technol. **9**(4), 343–363 (2015)

94. Sosnowski, Ł., Pietruszka, A., Krasuski, A., Janusz, A.: A resemblance based approach for recognition of risks at a fire ground. In: Ślęzak, D., Schaefer, G., Vuong, S.T., Kim, Y.-S. (eds.) AMT 2014. LNCS, vol. 8610, pp. 559–570. Springer, Cham (2014). https://doi.org/10.1007/978-3-319-09912-5_47
95. Sosnowski, Ł., Pietruszka, A., Łazowy, S.: Election algorithms applied to the global aggregation in networks of comparators. In: M. Ganzha, L., Maciaszek, M.P., (ed.), Proceedings of the 2014 Federated Conference on Computer Science and Information Systems. Annals of Computer Science and Information Systems, vol. 2, pp. 135–144. IEEE (2014). http://dx.doi.org/10.15439/2014F494
96. Sosnowski, Ł., Ślęzak, D.: Comparators for compound object identification. In: Kuznetsov, S.O., Ślęzak, D., Hepting, D.H., Mirkin, B.G. (eds.) RSFDGrC 2011. LNCS (LNAI), vol. 6743, pp. 342–349. Springer, Heidelberg (2011). https://doi.org/10.1007/978-3-642-21881-1_53
97. Sosnowski, Ł., Ślęzak, D.: Learning in comparator networks. In: Kacprzyk, J., Szmidt, E., Zadrożny, S., Atanassov, K.T., Krawczak, M. (eds.) IWIFS-GN/EUSFLAT -2017. AISC, vol. 643, pp. 316–327. Springer, Cham (2018). https://doi.org/10.1007/978-3-319-66827-7_29
98. Sosnowski, Ł., Ślęzak, D.: RDBMS framework for contour identification. In: Szczuka, M., Czaja, L., Skowron, A., Kacprzak, M. (eds.) CS&P, pp. 487–498. Białystok University of Technology, Pułtusk (2011). electronic edition
99. Sosnowski, Ł., Ślęzak, D.: How to design a network of comparators. In: Imamura, K., Usui, S., Shirao, T., Kasamatsu, T., Schwabe, L., Zhong, N. (eds.) BHI 2013. LNCS (LNAI), vol. 8211, pp. 389–398. Springer, Cham (2013). https://doi.org/10.1007/978-3-319-02753-1_39
100. Sosnowski, Ł., Ślęzak, D.: Networks of compound object comparators. In: FUZZ-IEEE, pp. 1–8 (2013)
101. Sosnowski, Ł., Ślęzak, D.: Fuzzy set interpretation of comparator networks. In: Kryszkiewicz, M., Bandyopadhyay, S., Rybinski, H., Pal, S.K. (eds.) PReMI 2015. LNCS, vol. 9124, pp. 345–353. Springer, Cham (2015). https://doi.org/10.1007/978-3-319-19941-2_33
102. Sosnowski, L., Szczuka, M.S.: Recognition of compound objects based on network of comparators. In: Proceedings of FedCSIS 2016, Position Papers, pp. 33–40 (2016)
103. Staab, S., Maedche, A.: Knowledge portals: ontologies at work. AI Mag. **22**(2), 63–75 (2001)
104. Stahl, A., Gabel, T.: Using evolution programs to learn local similarity measures. In: Ashley, K.D., Bridge, D.G. (eds.) ICCBR 2003. LNCS (LNAI), vol. 2689, pp. 537–551. Springer, Heidelberg (2003). https://doi.org/10.1007/3-540-45006-8_41
105. Sundaram, N., et al.: Streaming similarity search over one billion tweets using parallel locality-sensitive hashing. PVLDB **6**(14), 1930–1941 (2013)
106. Szczuka, M., Ślęzak, D.: Feedforward neural networks for compound signals. Theoret. Comput. Sci. **412**(42), 5960–5973 (2011)
107. Szczuka, M.: The use of rough set methods in knowledge discovery in databases. In: Kuznetsov, S.O., Ślęzak, D., Hepting, D.H., Mirkin, B.G. (eds.) RSFDGrC 2011. LNCS (LNAI), vol. 6743, pp. 28–30. Springer, Heidelberg (2011). https://doi.org/10.1007/978-3-642-21881-1_6
108. Szczuka, M.S., Sosnowski, Ł., Krasuski, A., Kreński, K.: Using domain knowledge in initial stages of KDD: optimization of compound object processing. Fundam. Inform. **129**(4), 341–364 (2014)
109. Tho, D.: Perceptron Problem in Neural Network. GRIN Verlag (2010). https://books.google.pl/books?id=eLWmQfpgansC

110. Tietze, U., Schenk, C., Gamm, E.: Electronic Circuits. Springer, Heidelberg (2008). https://doi.org/10.1007/978-3-540-78655-9. http://books.google.pl/books?id=NB5GAQAAIAAJ
111. Turksen, I.: Interval valued fuzzy sets based on normal forms. Fuzzy Sets Syst. **20**(2), 191–210 (1986). http://www.sciencedirect.com/science/article/pii/0165011486900771
112. Tversky, A., Shafir, E.: Preference, Belief, and Similarity: Selected Writings. MIT Press, Cambridge (2004)
113. Wilkinson, B.: The Essence of Digital Design. Essence of Engineering. Prentice Hall, Upper Saddle River (1998). http://books.google.es/books?id=-BNTAAAAMAAJ
114. Yager, R.R., Filev, D.: Summarizing data using a similarity based mountain method. Inf. Sci. **178**(3), 816–826 (2008)
115. Yang, X., Lin, T.Y., Yang, J., Li, Y., Yu, D.: Combination of interval-valued fuzzy set and soft set. Comput. Math. Appl. **58**(3), 521–527 (2009). http://www.sciencedirect.com/science/article/pii/S0898122109003228
116. Zadeh, L.A.: The concept of a linguistic variable and its application to approximate reasoning. J. Inf. Sci. **8**(3), 199–249 (1975)
117. Zadeh, L.A.: Fuzzy sets. Inf. Control **8**(3), 338–353 (1965)
118. Zadeh, L.A.: Toward a theory of fuzzy information granulation and its centrality in human reasoning and fuzzy logic. Fuzzy Sets Syst. **90**(2), 111–127 (1997)
119. Zadeh, L.A.: Computing with Words - Principal Concepts and Ideas. Studies in Fuzziness and Soft Computing, vol. 277. Springer, Heidelberg (2012). https://doi.org/10.1007/978-3-642-27473-2
120. Zadeh, P.D.H., Reformat, M.: Feature-based similarity assessment in ontology using fuzzy set theory. In: FUZZ-IEEE, pp. 1–7 (2012)
121. Zhao, Y., Luo, F., Wong, S.K.M., Yao, Y.: A general definition of an attribute reduct. In: Yao, J.T., Lingras, P., Wu, W.-Z., Szczuka, M., Cercone, N.J., Ślęzak, D. (eds.) RSKT 2007. LNCS (LNAI), vol. 4481, pp. 101–108. Springer, Heidelberg (2007). https://doi.org/10.1007/978-3-540-72458-2_12
122. Zhou, J., Pedrycz, W., Miao, D.: Shadowed sets in the characterization of rough-fuzzy clustering. Pattern Recogn. **44**(8), 1738–1749 (2011). http://dx.doi.org/10.1016/j.patcog.2011.01.014

Rseslib 3: Library of Rough Set and Machine Learning Methods with Extensible Architecture

Arkadiusz Wojna[1]($\boxtimes$) and Rafał Latkowski[2]

[1] Security On-Demand, 12121 Scripps Summit Dr 320, San Diego, CA 92131, USA
[2] Loyalty Partner, Złota 59, 00-120 Warsaw, Poland
{wojna,rlatkows}@mimuw.edu.pl

Abstract. The paper presents a new generation of Rseslib library - a collection of rough set and machine learning algorithms and data structures in Java. It provides algorithms for discretization, discernibility matrix, reducts, decision rules and for other concepts of rough set theory and other data mining methods. The third version was implemented from scratch and in contrast to its predecessor it is available as a separate open-source library with API and with modular architecture aimed at high reusability and substitutability of its components. The new version can be used within Weka and with a dedicated graphical interface. Computations in Rseslib 3 can be also distributed over a network of computers.

Keywords: Rough set · Discernibility matrix · Reduct ·
k nearest neighbors · Machine learning · Java · Weka ·
Distributed computing · Open source

1 Introduction

Rough set theory [20] was introduced by Pawlak as a methodology for data analysis based on approximation of concepts in information systems. Discernibility is a key concept in this methodology, which is the ability to distinguish objects, based on their attribute values. Along with theoretical research rough sets were developed in practical directions as well. To facilitate applications software tools implementing rough set concepts and methods have been developed. This paper describes one of such tools.

Rseslib 3 is a library of rough set and machine learning algorithms and data structures implemented in Java [35,36]. It is the successor of Rseslib 2 used in Rough Set Exploration System (RSES) [2]. The first version of the library started in 1993 and was implemented in C++. It was used as the core of Rosetta system [19]. Rseslib 2 was the first version of the library implemented in Java and it stands for the core of RSES. The third version of the library was entirely redesigned and all the methods available in this version were implemented from

J. F. Peters and A. Skowron (Eds.): TRS XXI, LNCS 10810, pp. 301–323, 2019.
https://doi.org/10.1007/978-3-662-58768-3_7

scratch. It provides algorithms for discretization, discernibility matrix, reducts, decision rules and rule-based classifiers as well as very fast implementation of the k nearest neighbors method with high accuracy distance measure and many well-known classical classification methods. The following features are distinguishing the version 3 from its predecessor:

- available as a library with an API
- open source distributed under GNU GPL license
- modular component-based architecture
- easy-to-reuse data representations and methods
- easy-to-substitute components
- available in Weka.

As an open source library of rough set methods in Java Rseslib 3 fills in an uncovered gap in the spectrum of rough set software tools. The algorithms in Rseslib 3 can be used both by users who need to apply ready-to-use rough set methods in their data analysis tasks as well as by researchers interested in extension of the existing rough set methods who can use the source code of the library as the basis for their extended implementations. The library can be used also within the following external tools: Weka [8], the dedicated graphical interface Qmak and Simple Grid Manager distributing computations over a network of computers.

The library is not limited to rough sets, it contains and is open to concepts and algorithms from other areas of machine learning and data mining. That is related to another goal of the project which is to provide a universal library of highly reusable and substitutable components at a very elementary level unmet in open source data mining Java libraries available today.

The paper is organised as follows. Other software implementing rough set related methods is discussed in Sect. 2. The types of data handled by the library and the data related notation used in the paper are presented in Sect. 3. Section 4 describes all discretization methods available in Rseslib. Section 5 discusses the types of discernibility matrix and indiscernibility relations provided by the library. Section 6 defines the types of reducts available in Rseslib and describes all the implemented algorithms computing reducts. Section 7 presents the algorithms computing rules from reducts. Section 8 describes the classification models implemented in Rseslib including the reduct-based method. Section 9 enumerates other available algorithms. Section 10 introduces to modularity of the library and discusses reusability and substitutability of its components. Section 11 presents the tools that can be used with the library: Weka, the dedicated graphical interface Qmak and a tool running Rseslib-based experiments on many computers or cores. Section 12 provides examples of Rseslib usage in independent research and software projects. Section 13 concludes the paper and outlines the future plans for the project.

2   Related Work

Looking for analogous open source Java projects one can find Modlem[1] and Richard Jensen's programs[2].

Modlem is a sequential covering algorithm inducing decision rules that contains some aspects of rough set theory. Numerical values are handled without discretization. Modlem as a classification method is available as Weka package.

Richard Jensen implemented a number of rough-fuzzy feature selection methods in Java. That includes a variety of search techniques, e.g. hill-climbing, ant colony optimization, genetic algorithm, as well as metrics and measures. Jensen provides also his own version of Weka with some methods included.

There are two useful libraries developed in other programming languages.

RoughSets package [23] implemented in the R programming language provides rough set and fuzzy rough models and methods. It implements the concepts of indiscernibility relations, lower and upper approximations, positive region and discernibility matrix. Using these concepts it provides the algorithms for discretization, feature selection, instance selection, rule induction, prediction and classification. RoughSets package was extended with RapidRoughSets [11] — an extension facilitating the use of the package in RapidMiner, a popular java platform for data mining, machine learning and predictive analytics.

NRough library [32] implemented in C# provides algorithms computing decision reducts, bireducts, decision reduct ensembles and decision rules. The algorithms can be used as feature selection and classification methods.

There are a number of tools providing rough set methods within graphical interface.

Rosetta [19] is the graphical tool based on the first version of Rseslib library. It provides functions for tabular data analysis supporting the overall data mining and knowledge discovery process. It provides methods computing exact and approximate reducts and generating if-then rules from computed reducts.

Rough Set Data Explorer (ROSE) [22] is the graphical tool for rough set based analysis of data. It provides methods for data processing including discretization, core and reduct computation, decision rule induction from rough approximations, and rule-based classification. As a distinctive feature ROSE includes variable precision rough set model.

Rough Set Exploration System (RSES) [2] is the graphical tool based on the second version of Rseslib library. It provides wide range of methods for data discretization, reduct computation, rule induction and rule-based classification.

3   Data

The concept of the library is based on classical representation of data in machine learning. It is assumed that a finite set of objects U, a finite set of conditional attributes $A = \{a_1, \ldots, a_n\}$ and a decision attribute dec are given. Each object

---

[1] https://sourceforge.net/projects/modlem.
[2] http://users.aber.ac.uk/rkj/?page_id=79.

$x \in U$ is represented by a vector of values $(x_1, \ldots, x_n)$. The value x_i is the value of the attribute a_i on the object x belonging to the domain of values V_i corresponding to the attribute a_i: $x_i \in V_i$. The type of a conditional attribute a_i can be either numerical, if its values are comparable and can be represented by numbers $V_i \subseteq \mathbb{R}$ (e.g.: age, temperature, height), or nominal, if its values are incomparable, i.e., if there is no linear order on V_i (e.g.: color, sex, shape).

The library contains many algorithms implementing various methods of supervised learning. These methods assume that each object $x \in U$ is assigned with a value of the decision attribute $dec(x)$ called a decision class and they learn from the objects in U a function approximating the real function dec on all objects outside U. At present the algorithms in the library assume that the domain of values of the decision attribute dec is discrete and finite: $V_{dec} = \{d_1, \ldots, d_m\}$.

The library reads data from files. Three data formats are accepted by the library:

– **ARFF**
The format of the popular open source machine learning software WEKA [8] widely adopted in the machine learning community.
– **CSV (Comma Separated Version)**
A popular format that can be exchanged between databases, spreadsheet programs like Microsoft Excel or Libre Office and software recognizing this format like Rseslib. To read this format Rseslib needs the description of columns called Rseslib header [36]. The header can be provided inside the file with data or in a separated file. The option of the header in a separate file enables to use the file with data by other programs without any extra conversion and eliminates the inconvenience of editing large files in case of very large data sets. Unlike in ARFF listing the values of the decision attribute is optional. The decisions can be collected directly from data if not given in the header.
– **RSES2**
The format of RSES system.

4   Discretizations

Some algorithms require data in form of nominal attributes, e.g. some rule based algorithms like the rough set based classifier. Discretization (known also as quantization or binning) is data transformation converting data from numeric attributes into nominal attributes.

The library provides a number of discretization methods. Each method splits domain of a numerical attribute into a number of disjoint intervals. New nominal attribute is formed by encoding a numerical value into an identifier of an interval.

The discretization methods available in Rseslib are described below.

4.1   Equal Width

The range of values of a numerical attribute in a data set is divided into k intervals of equal length. The number of intervals k is the parameter of the method.

4.2   Equal Frequency

The range of values of a numerical attribute in a data set is divided into k intervals containing the same number of objects from a data set. The number of objects in particular intervals may differ by one if the size of the data set does not divide by k. The number of intervals k is the parameter of the method.

4.3   One Rule

Holte's 1R algorithm [9] tries to cut the range of values of a numerical attribute into intervals containing training objects with the same decision but it avoids very small intervals. The minimal number n of training objects that must fall into each interval is the parameter of 1R algorithm. The algorithm executes the following steps:

1. Sort the objects by the values of a numerical attribute to be discretized
2. Scan the objects in the ascending order adding them to an interval until one of the decision classes, denote it by d, has n representatives in the interval
3. While the decision of the object next in the ascending order is d add the object to the interval
4. Start the next interval as empty and go to 2.

4.4   Static Entropy Minimization

Static entropy minimization [5] is a top-down local method discretizing a single numerical attribute. It starts with the whole range of values of the attribute in a data set and divides it into smaller intervals. At each step the algorithm remembers which objects from the data set fall into each interval. In a single step the algorithm searches all possible cuts in all intervals and selects the new cut c maximizing information gain, i.e. minimizing entropy:

$$E(a_i, c, S) = \frac{|S_1|}{|S|} Ent(S_1) + \frac{|S_2|}{|S|} Ent(S_2) \qquad (1)$$

where

$$Ent(S) = -\sum_{j=1}^{m} \frac{|\{x \in S : dec(x) = d_j\}|}{|S|} \log \left(\frac{|\{x \in S : dec(x) = d_j\}|}{|S|} \right)$$

a_i is the attribute to be discretized, S is the set of the objects falling into the interval on a_i containing a candidate cut c, $S_1 = \{x \in S : x_i \leq c\}$, $S_2 = \{x \in S : x_i > c\}$.

The method applies the minimum description length principle to decide when to stop the algorithm.

4.5 Dynamic Entropy Minimization

Dynamic entropy minimization method [5] is similar to static entropy minimization but it discretizes all numerical attributes at once. It starts with the whole
set of objects and splits it into two subsets with the optimal cut selected from all
numerical attributes. Then the algorithm splits each subset recursively scanning
all possible cuts over all numerical attributes at each split. To select the best cut
the algorithm minimizes the same formula 1 as the static method.

On average the dynamic method is faster than the static method and produces fewer cuts.

4.6 ChiMerge

ChiMerge [13] ia a bottom-up discretization method using χ^2 statistics to test
whether neighbouring intervals have significantly different decision distributions.
If the distributions are similar the algorithm merges the intervals into one interval. The method discretizes each numerical attribute independently.

The method has two parameters. The first parameter n is the minimal number of final intervals. The second parameter is the confidence level $(0.0 - 1.0)$
used to recognize two neighbouring intervals as different and not to merge them.

First, the algorithm calculates the threshold θ from χ^2 distribution with
$m - 1$ degrees of freedom and a given confidence level and starts with a separate
interval for each value of a numerical attribute occuring in a data set U. At each
step it merges the pair of neighbouring intervals with the minimal χ^2 value as
long as this minimal value is less then θ and the number of intervals does not
drop below n. χ^2 value is defined as:

$$\chi^2(S_1, S_2) = \sum_{j=1}^{m} \frac{\left(\left|S_1^j\right| - ES_1^j\right)^2}{ES_1^j} + \sum_{j=1}^{m} \frac{\left(\left|S_2^j\right| - ES_2^j\right)^2}{ES_2^j}$$

where S_1, S_2 are the sets of objects from U falling into two neighbouring intervals,
$S_k^j = \{x \in S_k : dec(x) = d_j\}$ and ES_k^j is the expected number of objects in S_k
with the decision d_j:

$$ES_k^j = |S_k| \frac{|\{x \in S_1 \cup S_2 : dec(x) = d_j\}|}{|S_1 \cup S_2|}$$

4.7 Global Maximal Discernibility Heuristic

Global maximal discernibility heuristic method [17] is a top-down dynamic
method discretizing all numerical attributes at once. At each step it evaluates
cuts globally with respect to the whole training set. It starts with the set S^* of
all pairs of objects with different decisions defined as:

$$S^* = \{\{x, y\} \subseteq U : dec(x) \neq dec(y)\}$$

At each step the algorithm finds the cut c that discerns the greatest number of pairs in the current set S^*, adds the cut c to the result set and removes all pairs discerned by the cut c from the set S^*. The optimal cut is searched among all possible cuts on all numerical attributes. The algorithm stops when the set S^* is empty.

4.8 Local Maximal Discernibility Heuristic

Local maximal discernibility heuristic method [17] selects the cuts optimizing the number of pairs of discerned objects like the global method but the procedure selecting the best cut is applied recursively to the subsets of objects obtained by splitting the data set by the previously selected cuts.

It starts with the best cut for the whole training set U splitting it into subsets U_1 and U_2. Next the discretization algorithm selects the best cut splitting U_1 and recursively the best cuts with respect to the subsets of U_1. Next it searches independently for the best cuts for U_2. At each step the best cut is searched over all attributes.

5 Discernibility Matrix

Computation of reducts is based on the concept of discernibility matrix [27]. The library provides 4 types of discernibility matrix including types handling inconsistencies in data [21,26]. Each type is $|U| \times |U|$ matrix defined for all pairs of objects $x, y \in U$. The fields of discernibility matrix $M(x, y)$ are defined as the subsets of the set of conditional attributes: $M(x, y) \subseteq A$. If a data set contains numerical attributes discernibility matrix can be computed using either the original or the discretized numerical attributes.

The first type of discernibility matrix M^{all} depends on the values of the conditional attributes only, it does not take the decision attribute into account:

$$M^{all}(x, y) = \{a_i \in A : x_i \neq y_i\}$$

In many applications, e.g. in object classification, we want to discern objects only if they have different decisions. The second type of discernibility matrix M^{dec} discerns objects from different decision classes:

$$M^{dec}(x, y) = \begin{cases} \{a_i \in A : x_i \neq y_i\} & \text{if } dec(x) \neq dec(y) \\ \emptyset & \text{if } dec(x) = dec(y) \end{cases}$$

If data are inconsistent, i.e. if there are one or more pairs of objects with different decisions and with equal values on all conditional attributes:

$$\exists x, y \in U : \forall a_i \in A : x_i = y_i \wedge dec(x) \neq dec(y)$$

then $M^{dec}(x, y) = \emptyset$ like for pairs of objects with the same decision. To overcome this inconsistency the concept of generalized decision was introduced [21,26]:

$$\partial(x) = \{d \in V_{dec} : \exists y \in U : \forall a_i \in A : x_i = y_i \wedge dec(y) = d\}$$

If U contains inconsistent objects x, y they have the same generalized decision. The next type of discernibility matrix M^{gen} is based on generalized decision:

$$M^{gen}(x,y) = \begin{cases} \{a_i \in A : x_i \neq y_i\} & \text{if } \partial(x) \neq \partial(y) \\ \emptyset & \text{if } \partial(x) = \partial(y) \end{cases}$$

This type of discernibility matrix removes inconsistencies but discerns pairs of objects with the same original decision, e.g. an inconsistent object from a consistent object. The fourth type of discernibility matrix M^{both} discerns a pair of objects only if they have both the original and the generalized decision different:

$$M^{both}(x,y) = \begin{cases} \{a_i \in A : x_i \neq y_i\} & \text{if } \partial(x) \neq \partial(y) \wedge dec(x) \neq dec(y) \\ \emptyset & \text{if } \partial(x) = \partial(y) \vee dec(x) = dec(y) \end{cases}$$

Data can contain missing values. All types of discernibility matrix available in the library have 3 modes to handle missing values [14,15,29]:

- different value — an attribute a_i discerns x, y if the value of one of them on a_i is defined and the value of the second one is missing (missing value is treated as yet another value): $a_i \notin M(x,y) \Leftrightarrow x_i = y_i \vee (x_i = * \wedge y_i = *)$
- symmetric similarity — an attribute a_i does not discern x, y if the value of any of them on a_i is missing: $a_i \notin M(x,y) \Leftrightarrow x_i = y_i \vee x_i = * \vee y_i = *$
- nonsymmetric similarity — asymmetric discernibility relation between x and y: $a_i \notin M(x,y) \Leftrightarrow (x_i = y_i \wedge y_i \neq *) \vee x_i = *$

The first mode treating missing value as yet another value keeps indiscernibility relation transitive but the next two modes make it intransitive. Such a relation is not an equivalence relation and does not define correctly indiscernibility classes in the set U. To eliminate that problem the library provides an option to transitively close an intransitive indiscernibility relation.

6 Reducts

Reduct [27] is a key concept in rough set theory. It can be used to remove some data without loss of information or to generate decision rules.

Definition 1. *The subset of attributes $R \subseteq A$ is a (global) reduct in relation to a discernibility matrix M if each pair of objects discernible by M is discerned by at least one attribute from R and no proper subset of R holds that property:*

$$\forall x, y \in U : M(x,y) \neq \emptyset \Rightarrow R \cap M(x,y) \neq \emptyset$$

$$\forall R' \subsetneq R \, \exists x, y \in U : M(x,y) \neq \emptyset \wedge R' \cap M(x,y) = \emptyset$$

If M is a decision-dependent discernibility matrix the reducts related to M are the reducts related to the decision attribute dec.

Reducts defined in Definition 1 called also global reducts are sometimes too large and generate too specific rules. To overcome this problem the notion of local reducts was introduced [40].

Definition 2. *The subset of attributes $R \subseteq A$ is a local reduct in relation to a discernibility matrix M and an object $x \in U$ if each object $y \in U$ discerned from x by M is discerned from x by at least one attribute from R and no proper subset of R holds that property:*

$$\forall y \in U : M(x, y) \neq \emptyset \Rightarrow R \cap M(x, y) \neq \emptyset$$

$$\forall R' \subsetneq R \,\exists y \in U : M(x, y) \neq \emptyset \wedge R' \cap M(x, y) = \emptyset$$

It may happen that local reducts are still too large. In the extreme situation there is only one global or local reduct equal to the whole set of attributes A. In such situations partial reducts [16, 18] can be helpful.

Let P be the set of all pairs of objects $x, y \in U$ discerned by a discernibility matrix M: $P = \{\{x, y\} \subseteq U : M(x, y) \neq \emptyset\}$ and let $\alpha \in (0; 1)$.

Definition 3. *The subset of attributes $R \subseteq A$ is a global α-reduct in relation to a discernibility matrix M if it discerns at least $(1 - \alpha) |P|$ pairs of objects discernible by M and no proper subset of R holds that property:*

$$|\{\{x, y\} \subseteq U : R \cap M(x, y) \neq \emptyset\}| \geq (1 - \alpha) |P|$$

$$\forall R' \subsetneq R : |\{\{x, y\} \subseteq U : R' \cap M(x, y) \neq \emptyset\}| < (1 - \alpha) |P|$$

Let $P(x)$ be the set of all objects $y \in U$ discerned from $x \in U$ by a discernibility matrix M: $P(x) = \{y \in U : M(x, y) \neq \emptyset\}$ and let $\alpha \in (0; 1)$.

Definition 4. *The subset of attributes $R \subseteq A$ is a local α-reduct in relation to a discernibility matrix M and an object $x \in U$ if it discerns at least $(1 - \alpha) |P(x)|$ objects discernible from x by M and no proper subset of R holds that property:*

$$|\{y \in U : R \cap M(x, y) \neq \emptyset\}| \geq (1 - \alpha) |P(x)|$$

$$\forall R' \subsetneq R : |\{y \in U : R' \cap M(x, y) \neq \emptyset\}| < (1 - \alpha) |P(x)|$$

The following algorithms computing reducts are available in Rseslib:

- **All Global Reducts**

  The algorithm computes all global reducts from a data set. The algorithm is based on the fact that a set of attributes is a reduct if and only if it is a prime implicant of a boolean CNF formula generated from the discernibility matrix [25]. First the algorithm calculates the discernibility matrix and then it transforms the discernibility matrix into a boolean CNF formula. Finally it applies an efficient algorithm finding all prime implicants of the formula using well-known in the field of boolean reasoning advanced techniques accelerating computations [4]. All found prime implicants are global reducts.

- **All Local Reducts**

  The algorithm computes all local reducts for each object in a data set. Like the algorithm computing global reducts it uses boolean reasoning. The first step is the same as for global reducts: the discernibility matrix specified by parameters is calculated. Next for each object x in the data set the row of the discernibility matrix corresponding to the object x is transformed into a CNF formula and all local reducts for the object x are computed with the algorithm finding prime implicants.

– **One Johnson Reduct**
  The method computes one reduct with greedy Johnson algorithm [12]. The algorithm starts with the empty set of attributes called the candidate set and adds iteratively one attribute maximizing the number of discerned pairs of objects according to the semantics of a selected discernibility matrix. It stops when all objects are discerned and checks if any of the attributes in the candidate set can be removed. The final candidate set is a reduct.

– **All Johnson Reducts**
  A version of the greedy Johnson algorithm in which the algorithm branches and traverses all possibilities rather than selecting one of them arbitrarily when more than one attribute cover the maximal number of uncovered fields of the discernibility matrix. The result is the set of the reducts found in all branches of the algorithm.

– **Global Partial Reducts**
  The algorithm finding global α-reducts described in [16]. The value α is the parameter of the algorithm.

– **Local Partial Reducts**
  The algorithm finding local α-reducts described in [16]. The value α is the parameter of the algorithm.

Table 1. Time (in seconds) of computing decision-related reducts by Rseslib algorithms on exemplary data sets.

Dataset	Attrs	Objects	All global	All local	Global partial	Local partial
Segment	19	1540	0.6	0.9	0.2	0.2
Chess	36	2131	4.1	66.1	0.2	0.4
Mushroom	22	5416	2.9	4.9	0.8	1.5
pendigits	16	7494	10.4	23.2	2.2	4.3
Nursery	8	8640	6.5	6.7	1.5	2.8
Letter	16	15000	44.6	179.7	9.7	20.5
Adult	13	30162	62.1	70.1	18.0	33.0
Shuttle	9	43500	91.8	92.5	22.7	48.4
Covtype	12	387342	8591.9	8859.0	903.7	7173.7

Algorithms computing reducts are the most time-consuming among rough set algorithms, the time cost of other steps in the overall knowledge discovery process is often negligible when compared to reduct computations. Hence it is important to provide an efficient implementation of the algorithms computing reducts. Table 1 presents the time of computing decision-related reducts by the algorithms available in Rseslib on data sets from UCI machine learning repository[3]. Numerical attributes were discretized with the local maximal discernibility method. The experiments were run on Intel Core i7-4790 3.60 GHz processor.

---

[3] https://archive.ics.uci.edu/ml.

7   Rules Generated from Reducts

Reducts described in the previous section can be used in Rseslib to generate decision rules. As reducts can be generated from a discernibility matrix using generalized decision Rseslib uses generalized decision rules:

Definition 5. *A decision rule indicates the probabilities of the decision classes at given values of some conditional attributes:*

$$a_{i_1} = v_1 \wedge \ldots \wedge a_{i_p} = v_p \Rightarrow (p_1, \ldots, p_m)$$

where p_j is defined as:

$$p_j = \frac{\left|\{x \in U : x_{i_1} = v_1 \wedge \ldots \wedge x_{i_p} = v_p \wedge dec(x) = d_j\}\right|}{\left|\{x \in U : x_{i_1} = v_1 \wedge \ldots \wedge x_{i_p} = v_p\}\right|} \tag{2}$$

A data object x is said to match a rule if the premise of the rule is satisfied by the attribute values of x: $x_{i_1} = v_1, \ldots, x_{i_p} = v_p$.

Rseslib provides new functionality regarding the semantics of missing descriptor values and missing attribute values. If rules were induced with use of a discernibility matrix then this matrix specifies similarity measure between objects (c.f. Sect. 5). This similarity relation is used for rule matching in such a way that a rule matches an object if the description of the rule is similar to the object with respect to the used similarity relation. In case of the different-value similarity relation implemented in the classic discernibility matrix the behaviour of rule matching is exactly as described above and compatibile with all other implementations not using special missing attribute value handling. If other similarity relations and other discernibility matrices are used then different semantics of missing attribute values can be used. In such circumstances rule matching is defined according to a specified similarity relation (c.f. [15]).

Each decision rule r: $a_{i_1} = v_1 \wedge \ldots \wedge a_{i_p} = v_p \Rightarrow (p_1, \ldots, p_m)$ in Rseslib is assigned with its support in the data set U used to generate rules:

$$support(r) = \left|\{x \in U : x_{i_1} = v_1 \wedge \ldots x_{i_p} = v_p\}\right|$$

Rseslib provides two algorithms generating decision rules from reducts:

– **Rules from global reducts** (Johnson reducts are global reducts). Given a set of global reducts GR the algorithm finds all templates in the data set:

$$Templates(GR) = \left\{\bigwedge_{a_i \in R} a_i = x_i : R \in GR, x \in U\right\}$$

For each template the algorithm generates one rule with the decision probabilities p_j as defined in Formula 2:

$$Rules(GR) = \{t \Rightarrow (p_1, \ldots, p_m) : t \in Templates(GR)\}$$

- **Rules from local reducts.** For each object $x \in U$ the algorithm applies the selected algorithm $LR : U \mapsto \mathcal{P}(A)$ computing local reducts $LR(x)$ for x and generates the set of templates as the union of the sets of templates from all objects in U:

$$Templates(LR) = \left\{ \bigwedge_{a_i \in R} a_i = x_i : R \in LR(x), x \in U \right\}$$

The set of decision rules is obtained from the set of templates in the same way as in case of global reducts:

$$Rules(LR) = \{ t \Rightarrow (p_1, \ldots, p_m) : t \in Templates(LR) \}$$

8   Classification

8.1   Rough Set Classifier

Rough set classifier provided in Rseslib uses the algorithms computing discernibility matrix, reducts and rules generated from reducts described in the previous sections. It enables to apply any of the discretization methods described in Sect. 4 to transform numerical attributes into nominal attributes. A user of the classifier selects a discretization method, a type of discernibility matrix and an algorithm generating reducts. The classifier computes a set of decision rules and the support of each rule in the training set.

Let *Rules* denote the computed set of decision rules. The rules are used in classification to determine a decision value when provided with an object x to be classified. First, the classifier calculates the vote of each decision class $d_j \in V_{dec}$ for the object x:

$$vote_j(x) = \sum_{\{t \Rightarrow (p_1, \ldots, p_m) \in Rules: \, x \text{ matches } t\}} p_j \cdot support(t \Rightarrow (p_1, \ldots, p_m))$$

Then the classifier assigns to x the decision with the greatest vote:

$$dec_{roughset}(x) = \max_{d_j \in V_{dec}} vote_j(x)$$

8.2   K Nearest Neighbors/RIONA

Rseslib provides an originally extended version of the k nearest neighbors (k-nn) classifier [34]. It can work with data containing both numerical and nominal attributes and implements very fast neighbor search [33] that make the classifier work in reasonable time for large data sets.

In the learning phase the algorithm induces a distance measure from a training set and constructs an indexing tree used for fast neighbor search. Optionally, the algorithm can learn the optimal number k of nearest neighbors from the training set. The distance measure is the weighted sum of distances between

values of two objects on all conditional attributes. The classifier provides two
metrics for nominal attributes: Hamming metric and Value Difference Metric
(VDM), and three metrics for numerical attributes: the city-block Manhattan
metric, Interpolated Value Difference Metric (IVDM) and Density-Based Value
Difference Metric (DBVDM). IVDM and DBVDM metrics are adaptations of
VDM metric to numerical attributes. For computation of the weights in the distance measure three methods are available: distance-based method, accuracy-
based method and a method using perceptron.

While classifying an object the classifier finds k nearest neighbors in the
training set according to the induced distance measure and it applies one of three
methods of voting for the decision by the found neighbors: equally weighted, with
inverse distance weights or with inverse square distance weights.

The classifier has also the mode to work as RIONA algorithm [6]. This mode
implements a classifier combining the k-nn method with rule induction where
the nearest neighbors not validated by additional rules are excluded from voting.

Table 2. Comparison of the nearest neighbor ($k = 1$) search methods from Rseslib
and Weka for the *covtype* dataset (387342 training instances, 193670 test instances).
The BallTree and CoverTree search methods available for Weka IBk were also used in
the test but they failed reporting errors.

Classifier	Search method	Training time (sec)	Classification time (sec)	Accuracy
Weka IBk	Linear search	0.63	2674.86	93.9%
Weka IBk	KDTree	120.32	20.54	93.9%
Rseslib KNN	Rseslib KNN	27.64	3.06	96.5%

K nearest neighbors method in Rseslib implements very fast nearest neighbors search algorithm based on center-based indexing of training instances and
using double criterion to prune searching. Table 2 presents time comparison
between 1-nn search methods from Rseslib and Weka on an exemplary large
data set. The training time of the Rseslib method is over 4 times shorter than
the training time of the Weka method and the classification time is almost 7
times shorter. It is worth mentioning that at the same time the distance measure induced by the Rseslib method gives a significantly higher classification
accuracy than the distance measure of the Weka method.

8.3 K Nearest Neighbors with Local Metric Induction

K nearest neighbors with local metric induction is the k nearest neighbors method
extended with an extra step - the classifier computes a local metric for each classified object [28]. While classifying an object, first the classifier finds a large set
of the nearest neighbors (according to a global metric). Then it generates a new,
local metric from this large set of neighbors. At last, the k nearest neighbors are
selected from this larger set of neighbors according to the locally induced metric
and used to vote for the decision.

In comparison to the standard k-nn algorithm this method improves classification accuracy particularly for the case of data with nominal attributes. It is reasonable to use this method rather for large data sets (2000 training objects or more).

8.4   Classical Classifiers

Rseslib delivers also implementations of classifiers well-known in the machine learning community (see [36] for more details):

- **C4.5** - decision tree developed by Quinlan
- **AQ15** - rule-based classifier with a covering algorithm
- **Neural network** - classical backpropagation algorithm
- **Naive Bayes** - simple Bayesian network
- **SVM** - support vector machine
- **PCA** - classifier using principal component analysis
- **Local PCA** - classifier using local principal component analysis
- **Bagging** - metaclassifier combining a number of "weak" classifiers
- **AdaBoost** - another popular metaclassifier.

9   Other Algorithms

Beside rough set and classification methods Rseslib provides many other machine learning and data mining algorithms. Each algorithm is available as separate class or method and easy to use as an independent component. That includes:

- **Data transformation:** missing value completion (non-invasive data imputation by Gediga and Duentsch), attribute selection, numerical attribute scaling, new attributes (radial, linear and arithmetic transformations)
- **Data filtering:** missing values filter, Wilson's editing, Minimal Consistent Subset (MSC) by Dasarathy, universal boolean function based filter
- **Data sampling:** with repetitions, without repetitions, with given class distribution
- **Data clustering:** k approximate centers algorithm
- **Data sorting:** attribute value related, distance related
- **Rule induction:** from global reducts, from local reducts, AQ15 algorithm
- **Metric induction:** Hamming and Value Difference Metric (VDM) for nominal attributes, city-block Manhattan, Interpolated Value Difference Metric (IVDM) and Density-Based Value Difference Metric (DBVDM) for numerical attributes, attribute weighting (distance-based, accuracy-based, perceptron)
- **Principal Component Analysis (PCA):** OjaRLS algorithm
- **Boolean reasoning:** two different algorithms generating prime implicant from a CNF boolean formula
- **Genetic algorithm scheme:** a user provides cross-over operation, mutation operation and fitness function only
- **Classifier evaluation:** single train-and-classify test, cross-validation, multiple test with random train-and-classify split, multiple cross-validation (all types of tests can be executed on many classifiers).

10   Extensible Modular Component-Based Architecture

Providing a collection of rough set and machine learning algorithms is not the only goal of Rseslib. It is designed also to assure maximum reusability and substitutability of the existing components in new components of the library. Hence a strong emphasis is put on its modularity. The code is separated into loosely related elements as small as possible so that each element can be used independently of other elements. For each group of the elements of the same type a standardizing interface is defined so that each element used in an algorithm can be easily substituted by any other element of the same type. Code separation and standardization is applied both to the algorithms and to the objects.

The previous sections presented the range of algorithms available in Rseslib. Below there is a list of the objects in the library implementing various data-related mathematical concepts that can be used as isolated components:

- **Basic:** attribute, data header, data object, boolean data object, numbered data object, data table, nominal attribute histogram, numeric attribute histogram, decision distribution
- **Boolean functions/operators:** attribute value equality, numerical attribute interval, nominal attribute value subset, binary discrimination, metric cube, negation, conjunction, disjunction
- **Real functions/operators:** scaling, perceptron, radius function, multiplication, addition
- **Integer functions:** discrimination (discretization, 3-value cut)
- **Decision distribution functions:** nominal value to decision distribution, numeric value to vicinity-based decision distribution, numeric value to interpolated decision distribution
- **Vector space:** vector, linear subspace, principal components subspace, vector function
- **Linear order**
- **Indiscernibility relations**
- **Distance measures:** Hamming, Value Difference Metric, city-block Manhattan, Interpolated Value Difference Metric, Density-Based Value Difference Metric, metric-based indexing tree
- **Rules:** boolean function based rule, equality descriptors rule, partial matching rule
- **Probability:** gaussian kernel function, hypercube kernel function, m-estimate.

The structure of rough set algorithms in Rseslib is one of the examples of the component-based architecture (see Fig. 1). Each of the six modules: *Discretization, Logic, Discernibility, Reducts, Rules* and *Rough Set Classifier* provides well-abstracted algorithms with clearly defined interfaces that allow algorithms from other modules to use them as their components. For example, the algorithms computing reducts from the *Reducts* module use a discernibility matrix from the *Discernibility* module and one of the methods computing prime implicants of a CNF boolean formula from the *Logic* module. It is easy to extend

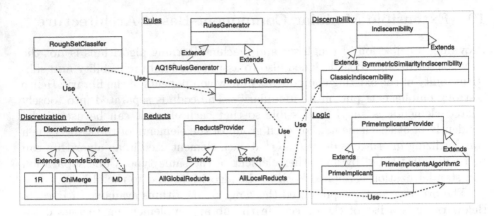

Fig. 1. Examples of relations between Rseslib modules containing rough set algorithms on simplified UML diagram

each module with implementation of a new method and to add the new method as an alternative in all components using the module.

The component-based architecture of Rseslib makes it possible to implement unconventional combinations of data mining methods. For example, perceptron learning is used as one of the attribute weighting methods in the algorithm computing a distance measure between data objects. Estimation of value probability at given decision is another example of such combination: it uses k nearest neighbors voting as one of the methods defining conditional value probability.

11   Tools

11.1   Rseslib Classifiers in Weka

Weka [8] is a non-commercial suite of open source machine learning and data mining software written in Java. It is one of the most popular platforms used by data scientists and researchers for data analysis and predictive modeling with downloads counted in millions per year. It provides four graphical interfaces and one command line interface for its users. Weka has the system of external packages updated independently of the core of Weka that allows people all over the world to contribute to Weka and maintain easily their Weka extensions. Such extensions can be easily dowloaded and installed in each Weka installation.

Rseslib is such an official Weka package available from Weka repository. Rseslib version 3.1.2 (the latest at the moment of preparing this paper) provides three Rseslib classifiers with full configuration in Weka:

- Rough set classifier
- K nearest neighbours / RIONA
- K nearest neighbours with local metric induction.

These three classifiers can be used, tested and compared with other classifiers within all Weka interfaces.

11.2   Graphical Interface Qmak

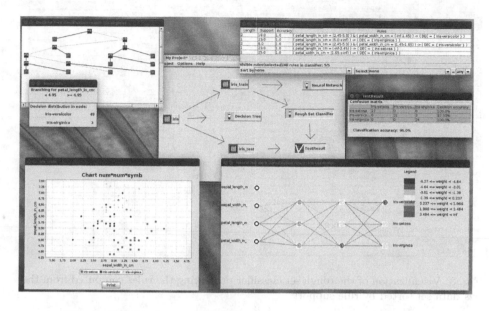

Fig. 2. Graphical user interface Qmak dedicated to Rseslib library

Qmak is a graphical user interface dedicated to Rseslib library (see Fig. 2). It is a tool for data analysis, data classification, classifier evaluation and interaction with classifiers. Qmak provides the following features:

- visualization of data, classifiers and single object classification
- interactive classifier modification by a user
- classification of test data with presentation of misclassified objects
- experiments on many classifiers: single train-and-classify test, cross-validation, multiple test with random train-and-classify split, multiple cross-validation.

Qmak 1.0.0 (the latest at the moment of preparing this paper) with Rseslib version 3.1.2 provides visualization of 5 classifiers: rough set classifier, k nearest neighbors, C4.5 decision tree, neural network and principal component analysis classifier. Visualization of a rough set classifier presents the decision rules of the classifier (see Fig. 3). The rules can be filtered and sorted by attribute occurrence, attribute values, length, support and accuracy. Visualization of classification by rough set classifier shows the decision rules matching a classified object enabling the same types of filtering and sorting criteria as visualization of the classifier.

Users can implement new classifiers and their visualization and add them easily to Qmak. It does not require any change in Qmak itself. A new classifier can be added using GUI or in the configuration file.

Qmak is available from Rseslib homepage. Help on Qmak can be found in the main menu of the application.

Length	Support	Accuracy	Rules
2	1.0	1.0	(sepallength = [5.95-6.05)) & (petalwidth = [1.65-+inf)) -> (DEC = { Iris-virginica })
2	1.0	1.0	(sepalwidth = (-inf-2.45)) & (petallength = [4.95-5.05)) -> (DEC = { Iris-virginica })
2	1.0	1.0	(sepalwidth = [2.9-+inf)) & (petallength = [4.95-5.05)) -> (DEC = { Iris-versicolor })
2	1.0	1.0	(sepallength = [6.4-+inf)) & (petallength = [4.95-5.05)) -> (DEC = { Iris-versicolor })
2	1.0	1.0	(petallength = [4.95-5.05)) & (petalwidth = (-inf-1.65)) -> (DEC = { Iris-versicolor })
2	1.0	1.0	(sepallength = [6.05-6.4)) & (petallength = [4.95-5.05)) -> (DEC = { Iris-virginica })
2	1.0	1.0	(sepallength = [5.95-6.05)) & (petallength = [5.05-+inf)) -> (DEC = { Iris-versicolor })
2	1.0	1.0	(sepallength = [5.95-6.05)) & (petallength = [4.95-5.05)) -> (DEC = { Iris-virginica })
2	1.0	1.0	(sepallength = [5.95-6.05)) & (sepalwidth = [2.45-2.9)) -> (DEC = { Iris-versicolor })
2	1.0	1.0	(sepallength = (-inf-5.95)) & (petallength = [4.95-5.05)) -> (DEC = { Iris-virginica })
3	2.0	1.0	(sepallength = [6.4-+inf)) & (sepalwidth = [2.45-2.9)) & (petallength = (-inf-1.65)) -> (DEC = { Iris-versicolor })
2	2.0	1.0	(sepalwidth = [2.45-2.9)) & (petallength = [4.95-5.05)) -> (DEC = { Iris-versicolor })
3	2.0	1.0	(sepallength = [5.95-6.05)) & (sepalwidth = [2.9-+inf)) & (petallength = (-inf-1.65)) -> (DEC = { Iris-versicolor })
2	2.0	1.0	(sepallength = [6.05-6.4)) & (sepalwidth = (-inf-2.45)) -> (DEC = { Iris-versicolor })
3	4.0	1.0	(sepallength = [6.05-6.4)) & (sepalwidth = [2.9-+inf)) & (petallength = (-inf-1.65)) -> (DEC = { Iris-versicolor })
2	4.0	1.0	(sepallength = (-inf-5.95)) & (petallength = [5.05-+inf)) -> (DEC = { Iris-virginica })
2	6.0	1.0	(sepallength = [6.05-6.4)) & (petallength = [5.05-+inf)) -> (DEC = { Iris-virginica })
2	8.0	1.0	(sepallength = [6.05-6.4)) & (petalwidth = [1.65-+inf)) -> (DEC = { Iris-virginica })
3	8.0	1.0	(sepallength = (-inf-5.95)) & (sepalwidth = [2.9-+inf)) & (petallength = [2.45-4.95)) -> (DEC = { Iris-versicolor })
2	9.0	1.0	(sepalwidth = (-inf-2.45)) & (petallength = [2.45-4.95)) -> (DEC = { Iris-versicolor })
2	10.0	1.0	(sepallength = [6.4-+inf)) & (petallength = [2.45-4.95)) -> (DEC = { Iris-versicolor })
3	12.0	1.0	(sepallength = (-inf-5.95)) & (sepalwidth = [2.45-2.9)) & (petalwidth = (-inf-1.65)) -> (DEC = { Iris-versicolor })
2	16.0	1.0	(sepalwidth = [2.45-2.9)) & (petalwidth = [1.65-+inf)) -> (DEC = { Iris-virginica })
2	29.0	1.0	(sepalwidth = [2.9-+inf)) & (petallength = [5.05-+inf)) -> (DEC = { Iris-virginica })
2	31.0	1.0	(sepallength = [6.4-+inf)) & (petallength = [5.05-+inf)) -> (DEC = { Iris-virginica })
2	38.0	1.0	(petallength = [5.05-+inf)) & (petalwidth = [1.65-+inf)) -> (DEC = { Iris-virginica })
2	47.0	1.0	(petallength = [2.45-4.95)) & (petalwidth = (-inf-1.65)) -> (DEC = { Iris-versicolor })
1	50.0	1.0	(petallength = (-inf-2.45)) -> (DEC = { Iris-setosa })

Visible rules(selected)/All rules in classifier: 28/28

| Sort by | rule support | ▼ | Select: | none | ▼ | = | any | ▼ |
| then sort by: | none | ▼ | | | | | | |

Fig. 3. Visualization of the rough set classifier presenting the rules computed from the *iris* data set sorted by rule support

11.3   Computing in Cluster

Simple Grid Manager is a tool for running massive Rseslib-based experiments on all available computers. It allows to create ad-hoc cluster with no prior configuration or additional cluster resource manager. SGM is the successor of DIXER — the previous version of software dedicated to Rseslib 2 [3]. Using SGM a user can create an ad-hoc cluster of computers by running the server module on one machine and the client module on all machines assigned to run the experiments (the elements of the cluster, see Fig. 4).

The server reads experiment lists from script files, distributes tasks between all available client machines, collects results of executed tasks and stores them in a result file. The main features of the tool are:

– Executes train-and-test experiments with any set of classifiers from Rseslib library (or user written classifiers compatible with Rseslib standards)
– Allows ad-hoc cluster creation without any configuration and maintenance
– Automatically resumes failed jobs and skips completed jobs in case of restart
– Uses robust communication and allows relying by nodes launched on NAT/Firewall knots
– Enables utilizing multi-core architectures by executing many client instances on one machine.

In order to use Simple Grid Manager a user needs only to create a list of experiments in a text file. Each experiment is specified with the name of a

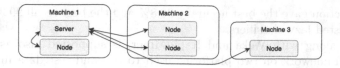

Fig. 4. Simple Grid Manager allows running experiments on many computers by creating ad-hoc cluster with no prerequisites on cluster configuration.

classifier class, training data file name, test data file name and a list of options for the classifier. SGM executes the experiments on the cluster in such way that each experiment is executed on one node. The classification results are stored in the result file. SGM can execute experiments with any class (i.e. also user-written) that extends the interface *Classifier*.

Simple Grid Manager is available from Rseslib homepage. The guide on how to run the distributed experiments can be found in [36].

12   Rseslib Usage Examples

Rseslib has been successfully used in various independent research projects and software tools.

Hu [10] used the rough set classifier from Rseslib to build a classifier that uses a decision table obtained by pairwise comparisons between training objects based on the preference relation from the PROMETHEE method.

Adamczyk [1] used the rough set classifier to evaluate the candidate attribute sets generated in subsequent populations in the parallel algorithm for feature selection based on asynchronous particle swarm optimization.

K nearest neighbors method from Rseslib was successfully applied to environmental sound recognition [7] that can be used, for example, in an intruder detection system. The Rseslib classifier gave the best accuracy winning with 8 other Weka classifiers including Support Vector Machine and Random Forest. The classifiers were tested with 8 different sets of features, the classification accuracy of the method from Rseslib was close to perfect (in the range 99.6%–99.79%) regardless of the number of features used in the tests. Moreover, the best accuracy of K-NN was not paid by the computational time - the classifier ranked also as one of the fastest among the tested methods.

The effectiveness of the Rseslib classifier in sound recognition was confirmed by another study conducted for the problem of context awareness of a service robot [24]. K-NN method was selected among 8 classification algorithms as one of the two methods giving satisfactory classification accuracy. Further tests were carried out to find the optimal parameter values of the K-NN method for the problem of context awareness of a service robot [30].

K nearest neighbors method with local metric induction from Rseslib was applied to the problem of liver cancer diagnosis and compared with the IBkLG method from Weka and the K-NN method from Rseslib [31]. K-NN with local metric induction gave the best accuracy 98.8% and the best recall 99.3% among the three tested lazy classifiers.

Rseslib, along with Weka and RapidMiner, is supported also as the programming framework on two platforms used to run and test data mining and machine learning algorithms. The first one, Debellor, is an open source platform with stream-oriented architecture for scalable data mining [37,38]. The second one, TunedIT, is the platform for automated evaluation, benchmarking and comparison of machine learning algorithms [39]. In particular, the existing Rseslib algorithms can be run and tested both on Debellor and on TunedIT.

Rseslib was used also as the programming framework for discretization and computation of reducts in mahout-extensions[4], a library extending Mahout with attribute selection methods. Mahout is an extensible programming environment and framework for building scalable algorithms in machine learning. The mahout-extensions library uses also some algorithms from Rseslib, e.g. the algorithm computing a discernibility matrix and the ChiMerge discretization method.

13   Conclusions and Future Work

The paper presents the contents of Rseslib 3 library that is designed to be used both by users who need to apply ready-to-use rough set or other data mining methods in their data analysis tasks as well as by researchers interested in extension of the existing methods. More information on Rseslib 3 and its tools can be found on the home page[5] and in the user guide [36].

The development of Rseslib 3 is continued. The repository of the library[6] is maintained by GitHub and is open to new contributions from all researchers and developers willing to extend the library. There is ongoing work on a classifier specialized in imbalanced data. The algorithms computing reducts are planned to be added to Weka package as attribute selection methods. Discretizations are also to be added to Weka package as separate algorithms. We are going to add Rseslib to Maven repository and to investigate the possibility of connecting Rseslib to RapidMiner.

Acknowledgment. We would like to thank Professor Andrzej Skowron for his support and mentorship over the project and for his advice on the development and Professor Dominik Ślęzak for his remarks to this paper. It must be emphasized that the library is the result of joint effort of many people and we express our gratitude to all the contributors: Jan Bazan, Rafał Falkowski, Grzegorz Góra, Wiktor Gromniak, Marcin Jałmużna, Łukasz Kosson, Łukasz Kowalski, Michał Kurzydłowski, Łukasz Ligowski, Michał Mikołajczyk, Krzysztof Niemkiewicz, Dariusz Ogórek, Marcin Piliszczuk, Maciej Próchniak, Jakub Sakowicz, Sebastian Stawicki, Cezary Tkaczyk, Witold Wojtyra, Damian Wójcik and Beata Zielosko.

---

[4] https://github.com/wgromniak/mahout-extensions.
[5] http://rseslib.mimuw.edu.pl.
[6] https://github.com/awojna/Rseslib.

References

1. Adamczyk, M.: Parallel feature selection algorithm based on rough sets and particle swarm optimization. In: Proceedings of the 2014 Federated Conference on Computer Science and Information System. In: ACSIS, vol. 2, pp. 43–50 (2014)
2. Bazan, J.G., Szczuka, M.: The rough set exploration system. In: Peters, J.F., Skowron, A. (eds.) Transactions on Rough Sets III. LNCS, vol. 3400, pp. 37–56. Springer, Heidelberg (2005). https://doi.org/10.1007/11427834_2
3. Bazan, J.G., Latkowski, R., Szczuka, M.: DIXER – distributed executor for rough set exploration system. In: Ślęzak, D., Yao, J.T., Peters, J.F., Ziarko, W., Hu, X. (eds.) RSFDGrC 2005. LNCS (LNAI), vol. 3642, pp. 39–47. Springer, Heidelberg (2005). https://doi.org/10.1007/11548706_5
4. Brown, F.M.: Boolean Reasoning: The Logic of Boolean Equations. Kluwer Academic Publishers, Dordrecht (1990)
5. Fayyad, U., Irani, K.: Multi-interval discretization of continuous-valued attributes for classification learning. In: Proceedings of the 13th International Joint Conference on Artificial Intelligence, pp. 1022–1027. Morgan Kaufmann (1993)
6. Góra, G., Wojna, A.: RIONA: a new classification system combining rule induction and instance-based learning. Fundamenta Informaticae 51(4), 369–390 (2002)
7. Grama, L., Rusu, C.: Choosing an accurate number of mel frequency cepstral coefficients for audio classification purpose. In: Proceedings of the 10th International Symposium on Image and Signal Processing and Analysis, pp. 225–230. IEEE (2017)
8. Hall, M., Frank, E., Holmes, G., Pfahringer, B., Reutemann, P., Witen, I.: The weka data mining software: an update. SIGKDD Explor. 11(1), 10–18 (2009)
9. Holte, R.C.: Very simple classification rules perform well on most commonly used datasets. Mach. Learn. 11(1), 63–90 (1993)
10. Hu, Y.C.: Rough sets for pattern classification using pairwise-comparison-based tables. Appl. Math. Model. 37(12–13), 7330–7337 (2013)
11. Janusz, A., Stawicki, S., Szczuka, M., Ślęzak, D.: Rough set tools for practical data exploration. In: Ciucci, D., Wang, G., Mitra, S., Wu, W.-Z. (eds.) RSKT 2015. LNCS (LNAI), vol. 9436, pp. 77–86. Springer, Cham (2015). https://doi.org/10.1007/978-3-319-25754-9_7
12. Johnson, D.S.: Approximation algorithms for combinatorial problems. J. Comput. Syst. Sci. 9(3), 256–278 (1974)
13. Kerber, R.: Chimerge: discretization of numeric attributes. In: Proceedings of the 10th National Conference on Artificial Intelligence, pp. 123–128. AAAI Press (1992)
14. Kryszkiewicz, M.: Properties of incomplete information systems in the framework of rough sets. In: Polkowski, L., Skowron, A. (eds.) Rough Sets in Knowledge Discovery 1: Methodology and Applications, pp. 422–450. Physica-Verlag, Heidelberg (1998)
15. Latkowski, R.: Flexible indiscernibility relations for missing attribute values. Fundamenta Informaticae 67(1–3), 131–147 (2005)
16. Moshkov, M., Piliszczuk, M., Zielosko, B.: Partial Covers, Reducts and Decision Rules in Rough Sets: Theory and Applications. SCI, vol. 145. Springer, Heidelberg (2008)
17. Nguyen, H.S.: Discretization of real value attributes: a boolean reasoning approach. Ph.D. thesis, Warsaw University (1997)

18. Nguyen, H.S., Ślęzak, D.: Approximate reducts and association rules - correspondence and complexity results. In: Zhong, N., Skowron, A., Ohsuga, S. (eds.) Proceedings of the International Workshop on Rough Sets, Fuzzy Sets, Data Mining, and Granular-Soft Computing. LNCS, pp. 137–145. Springer, Heidelberg (1999)
19. Øhrn, A., Komorowski, J., Skowron, A., Synak, P.: The design and implementation of a knowledge discovery toolkit based on rough sets - the rosetta system. In: Polkowski, L., Skowron, A. (eds.) Rough Sets in Knowledge Discovery 2: Applications, Case Studies and Software Systems, pp. 376–399. Physica-Verlag, Heidelberg (1998)
20. Pawlak, Z.: Rough Sets - Theoretical Aspects of Reasoning about Data. Kluwer Academic Publishers, Dordrecht (1991)
21. Pawlak, Z., Skowron, A.: Rudiments of rough sets. Inf. Sci. **177**(1), 3–27 (2007)
22. Prędki, B., Wilk, S.: Rough set based data exploration using ROSE system. In: Raś, Z.W., Skowron, A. (eds.) ISMIS 1999. LNCS, vol. 1609, pp. 172–180. Springer, Heidelberg (1999). https://doi.org/10.1007/BFb0095102
23. Riza, L.S., et al.: Implementing algorithms of rough set theory and fuzzy rough set theory in the R package "roughsets". Inf. Sci. **287**, 68–89 (2014)
24. Rusu, C., Grama, L.: Recent developments in acoustical signal classification for monitoring. In: Proceedings of the 5th International Symposium on Electrical and Electronics Engineering. IEEE (2017)
25. Skowron, A.: Boolean reasoning for decision rules generation. In: Komorowski, J., Raś, Z.W. (eds.) ISMIS 1993. LNCS, vol. 689, pp. 295–305. Springer, Heidelberg (1993). https://doi.org/10.1007/3-540-56804-2_28
26. Skowron, A., Grzymała-Busse, J.W.: From rough set theory to evidence theory. In: Yager, R.R., Kacprzyk, J., Fedrizzi, M. (eds.) Advances in the Dempster-Shafer Theory of Evidence, pp. 193–236. Wiley, New York (1994)
27. Skowron, A., Rauszer, C.: The discernibility matrices and functions in information systems. In: Slowinski, R. (ed.) Intelligent Decision Support, Handbook of Applications and Advances of the Rough Sets Theory, pp. 331–362. Kluwer Academic Publishers, Dordrecht (1992)
28. Skowron, A., Wojna, A.: K nearest neighbor classification with local induction of the simple value difference metric. In: Tsumoto, S., Słowiński, R., Komorowski, J., Grzymała-Busse, J.W. (eds.) RSCTC 2004. LNCS (LNAI), vol. 3066, pp. 229–234. Springer, Heidelberg (2004). https://doi.org/10.1007/978-3-540-25929-9_27
29. Stefanowski, J., Tsoukiàs, A.: On the extension of rough sets under incomplete information. In: Zhong, N., Skowron, A., Ohsuga, S. (eds.) RSFDGrC 1999. LNCS (LNAI), vol. 1711, pp. 73–81. Springer, Heidelberg (1999). https://doi.org/10.1007/978-3-540-48061-7_11
30. Telembici, T., Grama, L.: Detecting indoor sound events. Acta Technica Napocensis - Electron. Telecommun. **59**(2), 13–17 (2018)
31. Tiwari, M., Chakrabarti, P., Chakrabarti, T.: Performance analysis and error evaluation towards the liver cancer diagnosis using lazy classifiers for ILPD. In: Zelinka, I., Senkerik, R., Panda, G., Lekshmi Kanthan, P.S. (eds.) ICSCS 2018. CCIS, vol. 837, pp. 161–168. Springer, Singapore (2018). https://doi.org/10.1007/978-981-13-1936-5_19
32. Widz, S.: Introducing Nrough framework. In: Polkowski, L., et al. (eds.) IJCRS 2017. LNCS (LNAI), vol. 10313, pp. 669–689. Springer, Cham (2017). https://doi.org/10.1007/978-3-319-60837-2_53
33. Wojna, A.: Center-based indexing for nearest neighbors search. In: Proceedings of the 3rd IEEE International Conference on Data Mining, pp. 681–684. IEEE Computer Society Press (2003)

34. Wojna, A.: Analogy-based reasoning in classifier construction. In: Peters, J.F., Skowron, A. (eds.) Transactions on Rough Sets IV. LNCS, vol. 3700, pp. 277–374. Springer, Heidelberg (2005). https://doi.org/10.1007/11574798_11
35. Wojna, A., Latkowski, R.: Rseslib 3: open source library of rough set and machine learning methods. In: Nguyen, H.S., Ha, Q.-T., Li, T., Przybyła-Kasperek, M. (eds.) IJCRS 2018. LNCS (LNAI), vol. 11103, pp. 162–176. Springer, Cham (2018). https://doi.org/10.1007/978-3-319-99368-3_13
36. Wojna, A., Latkowski, R., Kowalski, L.: RSESLIB: User Guide. http://rseslib.mimuw.edu.pl/rseslib.pdf
37. Wojnarski, M.: Debellor: a data mining platform with stream architecture. In: Peters, J.F., Skowron, A., Rybiński, H. (eds.) Transactions on Rough Sets IX. LNCS, vol. 5390, pp. 405–427. Springer, Heidelberg (2008). https://doi.org/10.1007/978-3-540-89876-4_22
38. Wojnarski, M.: Debellor: Open source modular platform for scalable data mining. In: Proceedings of the 17th International Conference on Intelligent Information Systems (2009)
39. Wojnarski, M., Stawicki, S., Wojnarowski, P.: TunedIT.org: system for automated evaluation of algorithms in repeatable experiments. In: Szczuka, M., Kryszkiewicz, M., Ramanna, S., Jensen, R., Hu, Q. (eds.) RSCTC 2010. LNCS (LNAI), vol. 6086, pp. 20–29. Springer, Heidelberg (2010). https://doi.org/10.1007/978-3-642-13529-3_4
40. Wróblewski, J.: Covering with reducts - a fast algorithm for rule generation. In: Polkowski, L., Skowron, A. (eds.) RSCTC 1998. LNCS (LNAI), vol. 1424, pp. 402–407. Springer, Heidelberg (1998). https://doi.org/10.1007/3-540-69115-4_55

Author Index

Printed in the United States
By Bookmasters